W0080837

Traditional Chinese Medicine
Scientific Basis for Its Use

RSC Drug Discovery Series

Editor-in-Chief
Professor David Thurston, *Kings College, London, UK*

Series Editors:
Dr David Fox, *Vulpine Science ad Learning, UK*
Professor Salvatore Guccione, *University of Catania, Italy*
Professor Ana Martinez, *Instituto de Quimica Medica-CSIC, Spain*
Professor David Rotella, *Northeastern University, USA*

Advisor to the Board:
Professor Robin Ganellin, *University College London, UK*

Titles in the Series:
1: Metabolism, Pharmacokinetics and Toxicity of Functional Groups
2: Emerging Drugs and Targets for Alzheimer's Disease; Volume 1
3: Emerging Drugs and Targets for Alzheimer's Disease; Volume 2
4: Accounts in Drug Discovery
5: New Frontiers in Chemical Biology
6: Animal Models for Neurodegenerative Disease
7: Neurodegeneration
8: G Protein-Coupled Receptors
9: Pharmaceutical Process Development
10: Extracellular and Intracellular Signaling
11: New Synthetic Technologies in Medicinal Chemistry
12: New Horizons in Predictive Toxicology
13: Drug Design Strategies: Quantitative Approaches
14: Neglected Diseases and Drug Discovery
15: Biomedical Imaging
16: Pharmaceutical Salts and Cocrystals
17: Polyamine Drug Discovery
18: Proteinases as Drug Targets
19: Kinase Drug Discovery
20: Drug Design Strategies: Computational Techniques and Applications
21: Designing Multi-Target Drugs
22: Nanostructured Biomaterials for Overcoming Biological Barriers
23: Physico-Chemical and Computational Approaches to Drug Discovery
24: Biomarkers for Traumatic Brain Injury
25: Drug Discovery from Natural Products
26: Anti-Inflammatory Drug Discovery
27: New Therapeutic Strategies for Type 2 Diabetes: Small Molecules
28: Drug Discovery for Psychiatric Disorders
29: Organic Chemistry of Drug Degradation
30: Computational Approaches to Nuclear Receptors
31: Traditional Chinese Medicine

How to obtain future titles on publication:
A standing order plan is available for this series. A standing order will bring
delivery of each new volume immediately on publication.

For further information please contact:
Book Sales Department, Royal Society of Chemistry, Thomas Graham House,
Science Park, Milton Road, Cambridge, CB4 0WF, UK
Telephone: +44 (0)1223 420066, Fax: +44 (0)1223 420247,
Email: booksales@rsc.org
Visit our website at www.rsc.org/books

Traditional Chinese Medicine
Scientific Basis for Its Use

Edited by

James David Adams Jr
School of Pharmacy, University of Southern California, Los Angeles, CA, USA
Email: jadams@usc.edu

Eric J. Lien
School of Pharmacy, University of Southern California, Los Angeles, CA, USA
Email: elien@usc.edu

RSCPublishing

RSC Drug Discovery Series No. 31

ISBN: 978-1-84973-661-9
ISSN: 2041-3203

A catalogue record for this book is available from the British Library

© The Royal Society of Chemistry 2013

All rights reserved

Apart from fair dealing for the purposes of research for non-commercial purposes or for private study, criticism or review, as permitted under the Copyright, Designs and Patents Act 1988 and the Copyright and Related Rights Regulations 2003, this publication may not be reproduced, stored or transmitted, in any form or by any means, without the prior permission in writing of The Royal Society of Chemistry or the copyright owner, or in the case of reproduction in accordance with the terms of licences issued by the Copyright Licensing Agency in the UK, or in accordance with the terms of the licences issued by the appropriate Reproduction Rights Organization outside the UK. Enquiries concerning reproduction outside the terms stated here should be sent to The Royal Society of Chemistry at the address printed on this page.

The RSC is not responsible for individual opinions expressed in this work.

Published by The Royal Society of Chemistry,
Thomas Graham House, Science Park, Milton Road,
Cambridge CB4 0WF, UK

Registered Charity Number 207890

For further information see our web site at www.rsc.org

Printed in the United Kingdom by Henry Ling Limited, Dorchester, DT1 1HD, UK

Preface

"I believe in the goodness of the American and Chinese people who remain good friends."

Dr Hua Chuen Mei
from "Dr Hua Chuen Mei – May 30, 1925" available at
www.abeduspress.com.

Linda Mei Adams, my wife, is the grand-daughter of Dr Hua Chuen Mei, who was an American lawyer in Shanghai who defended several Chinese students accused of rioting in the International Settlement at the Louza Police Station on 30 May 1925. He won the case in June 1925, which led to the May 30 Movement in China. My wife and her mother, Eva Lum Mei, taught me to speak Cantonese. My wife's grand-mother, Polly Lum, used Traditional Chinese Medicine even though her son, son-in-law and grand-son were all Western-trained MDs. I am grateful to the Mei and Lum families for helping me learn the utility of Traditional Chinese Medicine, even when American Medicine may not recognize the utility. I was able to learn the philosophy for using Traditional Chinese Medicine, which differs greatly from the medicine I learned as a Graduate Student of Pharmacology at the University of California San Francisco. This philosophy depends on living in balance in order to be healthy. I was also taught the importance of spirituality in health, since health requires a healthy body, mind and spirit. My poem below is "be kind to yourself, be kind to others, allow God to love you." My Chinese name, below, is Ha Cheen Sing, Summer Thousand Honest, and was given to me by my mother-in-law.

跟自己做慈善
跟第二的人做慈善
讓上帝愛你
夏千诚

RSC Drug Discovery Series No. 31
Traditional Chinese Medicine: Scientific Basis for Its Use
Edited by James David Adams, Jr. and Eric J. Lien
© The Royal Society of Chemistry 2013
Published by the Royal Society of Chemistry, www.rsc.org

I am grateful to Professor Eric Lien for encouraging me to think about the Science involved in Traditional Chinese Medicine. This has been a great challenge for me and resulted in this book. I am grateful to all the authors of the chapters who have applied their great expertise in this book.

James David Adams Jr, PhD
Associate Professor of Pharmacology and Pharmaceutical Sciences
School of Pharmacy, University of Southern California
Los Angeles, CA, USA

Knowledge and Health

Knowledge and learning have no limits.
Teaching and learning reinforce each other.
Education should not discriminate against anyone.
Doing good deeds brings the most happiness.
New knowledge can promote good health, enhance productivity and contribute to longevity.
Sustain vitality and youthfulness as long as possible.
Good memories will last forever.

Eric J. Lien, PhD
Professor Emeritus
School of Pharmacy, University of Southern California
Los Angeles, CA, USA

Contents

RSC Drug Discovery Series No. 31
Traditional Chinese Medicine: Scientific Basis for Its Use
Edited by James David Adams, Jr. and Eric J. Lien
© The Royal Society of Chemistry 2013
Published by the Royal Society of Chemistry, www.rsc.org

               Eri Oshima

          12.1   Background   238
                 12.1.1   Introduction   238
                 12.1.2   Uses in Industry   239
                 12.1.3   General Characteristics   242
                 12.1.4   Nomenclature   242
          12.2   TCM Theory   242
                 12.2.1   Classics   242
                 12.2.2   Concept of Interdependence and Balance   245
                 12.2.3   Herbal Theory   248
                 12.2.4   Medicinal Uses of Seaweed   248
          12.3   Modern Research   250
                 12.3.1   Bioactive Substances in Seaweed   250
                 12.3.2   Human Clinical Studies with Dietary Seaweed   256
                 12.3.3   TCM Clinical Studies – Herbal Uses of
                          Seaweed   259
          12.4   Safety   261
          12.5   Conclusion – Totality of Evidence   263
          Acknowledgements   263
          References   264

**Chapter 13   The Preventive Effect of Traditional Chinese Medicinal
               Herbs on Type 2 Diabetes Mellitus   268**
               Zhijun Wang, Patrick Chan and Jeffrey Wang

          13.1   Introduction   268
                 13.1.1   Diabetes Statistics   268
                 13.1.2   Pathology and Therapeutics   269
          13.2   Prevention of T2DM   270
          13.3   The Historical Use of TCM Herbs for DM Treatment   273
          13.4   Frequently Used TCM Herbs   273
                 13.4.1   Individual TCM Herbs   274
                 13.4.2   TCM Herbal Formulas   299
                 13.4.3   Combination of TCM with Western Drugs   301
          13.5   Concerns and Future Perspectives   301
          13.6   Conclusion   303
          References   303

CHAPTER 1

The Traditional and Scientific Bases for Traditional Chinese Medicine: Communication Between Traditional Practitioners, Physicians and Scientists

JAMES D. ADAMS JR* AND ERIC J. LIEN

Department of Pharmacology and Pharmaceutical Sciences, School of Pharmacy, University of Southern California, Los Angeles, CA, 90089, USA
*E-mail: jadams@pharmacy.usc.edu

1.1 Introduction

The species *Homo sapiens* has existed for 200 000 years or so. Medicinal plants have been important to humans as indicated by plants in prehistoric burial sites, mummy wrappings and ancient legends. It is likely that during the entire period of human existence, plants have been used as medicines. This means there has been a tremendous natural selection, such that those who responded to plant medicines survived. Our genome has been altered by this natural selection. It has only been during the past 50 years or so that Science has tried to displace traditional medicine. However, even today, the majority of purified drugs used clinically are derived directly or indirectly from plants and other

RSC Drug Discovery Series No. 31
Traditional Chinese Medicine: Scientific Basis for Its Use
Edited by James David Adams, Jr. and Eric J. Lien
© The Royal Society of Chemistry 2013
Published by the Royal Society of Chemistry, www.rsc.org

natural sources. It is critical to realize that the alteration of the human genome by the use of plant medicines makes humans more responsive to plant medicines today. We should continue to use plant medicines.

There is an old Chinese saying, "Thousands of prescriptions are easy to come by. A really good drug is hard to find." This is still true today in spite of modern advances in biomedical technology. It is humbling to realize that while drugs can treat symptoms, only Mother Nature, the body, can cure and heal. Extension of life expectancy is accompanied by many lifestyle-related diseases such as diabetes, obesity, hypertension, cancer, drug addiction and degeneration of various organs and tissues. None of these diseases can be easily overcome by a quick fix with a miracle pill or surgical procedure. Fundamental changes are needed to reverse the downward spiral trend in healthcare effectiveness.

It is a sad fact that modern lifestyles cause the incidence of heart disease, diabetes, arthritis and cancer to increase every year, as shown by Centers for Disease Control statistics. In fact, heart disease, diabetes and arthritis are enabled by non-traditional medicine, since only the symptoms are treated, allowing patients to live with their disease. Efforts to cure these diseases are almost completely ineffective. Traditional Chinese medicine has much to teach in terms of preventing and curing these conditions.

1.2   The Basis of Traditional Chinese Medicine

There is a strong scientific basis for traditional Chinese medicine. However, the Science of traditional Chinese medicine is rarely discussed and needs to be more widely known. Currently, the scientific approach to traditional Chinese plant medicines is to purify them into single components. This is simply modern drug discovery in Chinese plants. However, traditional Chinese medicine usually uses complex mixtures of plant extracts, not single purified drugs. This is based on yin, yang and chi theories that are not widely understood by scientists.

Yin is cold, wet and female. Yang is hot, dry and male. Yin and yang are constant influences in the body and must be kept in balance to prevent disease.[1,2] Disease is treated by re-establishing the balance of yin and yang. Yin and yang are the balance of endogenous agonists and antagonists. This includes the balance between the sympathetic and parasympathetic nervous systems and also other systems (Figure 1.1). For instance, pain can be caused by prostaglandins, whereas pain relief can come from lipoxins.[3,4] Prostaglandins interact with prostaglandin receptors such as DP, EP, FP and IP. Lipoxins interact with lipoxin receptors, such as ALXR. Both prostaglandins and lipoxins are endogenous compounds, are products of arachidonic acid and exist in a balance that controls pain (Figure 1.2). Pain is a necessary part of normal life, but should not be excessive. Pain relief can result from balancing prostaglandins and lipoxins.

Chi is the life force, flows in the body in acupuncture channels and has several components. Chi derives from signaling processes in the body, such as phosphorylation/dephosphorylation, G protein signaling, cAMP production and degradation, calcium release and sequestration and others, and regulates

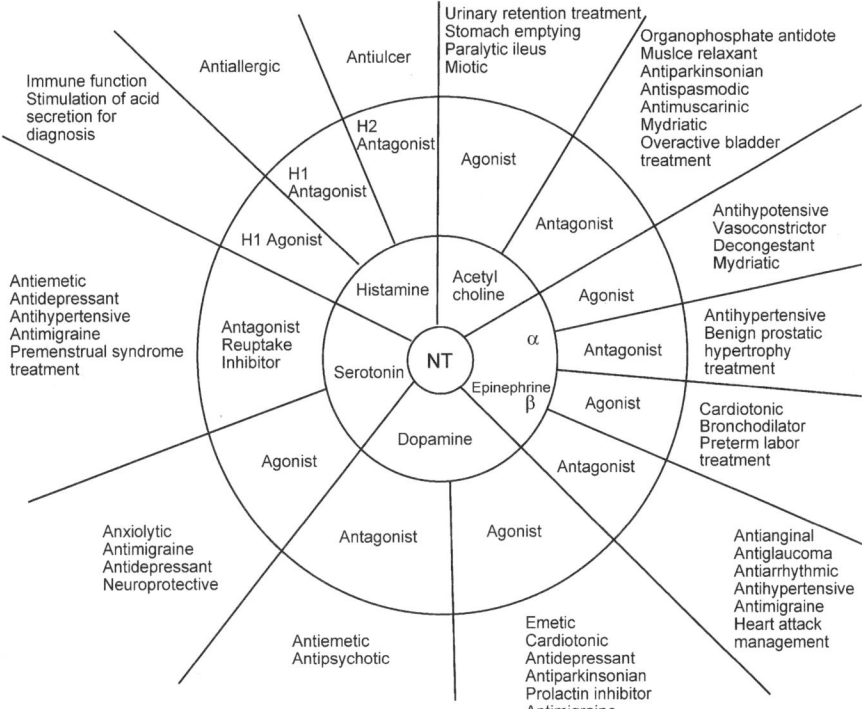

Figure 1.1 Many therapeutic applications have be developed based on neurotransmitter (NT) agonists and antagonists. Each receptor has an agonist, NT, that is made locally, excreted when it is needed and cleared quickly. Each receptor also has an endogenous antagonist that is made locally, excreted when it is needed and cleared quickly. However, the endogenous antagonists for the NT shown are not known. Modified from Kuo *et al.*[14]

bodily functions.[5] For instance, anxiety is a learned response that alters the functions of several receptor signal transduction systems in the brain, serotonergic, GABAergic, noradrenergic, dopaminergic and others.[6] Anxiety can greatly increase the risk of developing some diseases. It is necessary to unlearn anxiety in order to have a productive life. This involves a volitional act of restoring the balance or chi, of signal transduction systems. Chi also has a genetic component that is passed down from ancestors. Therefore, chi is controlled by genes. Chi controls yin and yang. Conversely, yin and yang influence chi. Chi can also be described as the Gibbs free energy ($\triangle G$) available to do useful work according to the laws of thermodynamics.[7]

Yin, yang and chi constantly interact in the body to maintain health. Western medicine understands very well how to use drugs that are agonists and antagonists to modify signaling processes. However, Western medicine does not accept the use of complex plant extracts to perform these functions. Traditional Chinese medicine has learned experimentally over the centuries how to restore the chi, yin and yang of a patient to restore health with plant medicines.

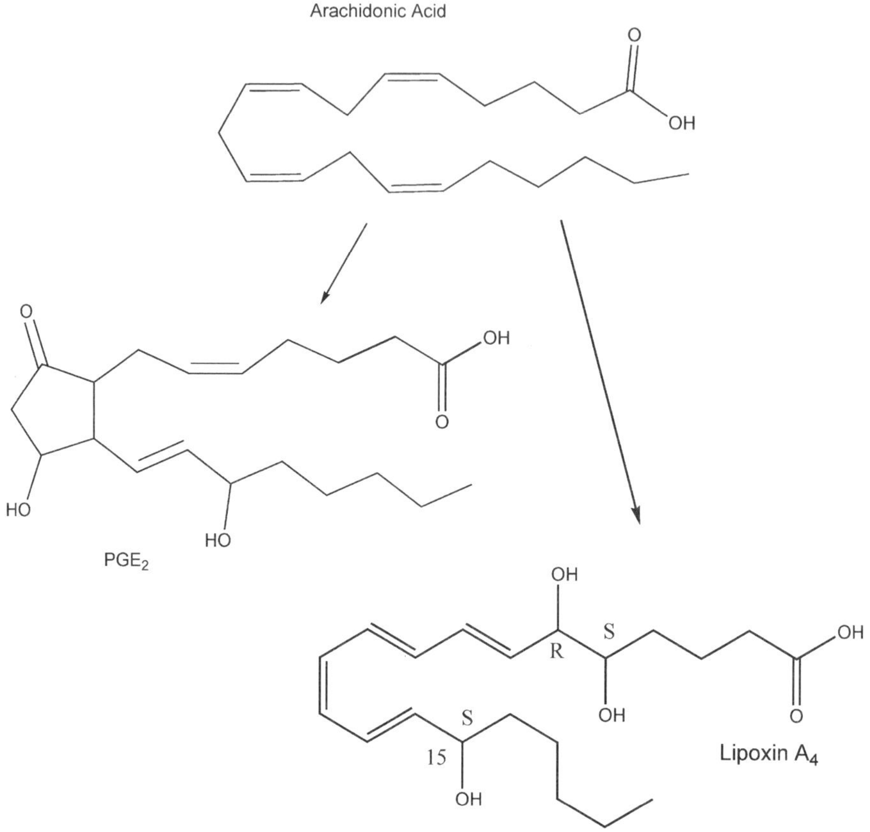

Arachidonic Acid

PGE$_2$

15

Lipoxin A$_4$

Figure 1.2   Arachidonic acid is the source of prostaglandin E$_2$, which causes pain, and lipoxin A$_4$, which relieves pain.

1.3   Disease Causation

In traditional Chinese medicine, disease is caused by an imbalance in the body, usually caused by too much yin or too much yang.[1,2] For instance, a cold may be caused by too much exposure to cold, wet wind. The cold is then counterbalanced by a fever that heats up the body. Yin in this case is an external influence. Yang, the fever, results from endogenous factors in the body. Hypertension may be caused by too much yang that results in more blood flow to cool the body. In this case, yang and yin are endogenous factors.

Science has only learned in the past few years that many diseases are caused by an imbalance in endogenous agonists and antagonists. For instance, arthritis, hypertension, atherosclerosis, diabetes and other conditions are caused by obesity that produces an abundance of toxic adipokines and toxic lipids in the body.[5] These are endogenous agonists and antagonists that work at specific receptors to cause inflammation, increase blood pressure, damage blood vessels and the heart and cause insulin resistance (Figure 1.3). AIDS is

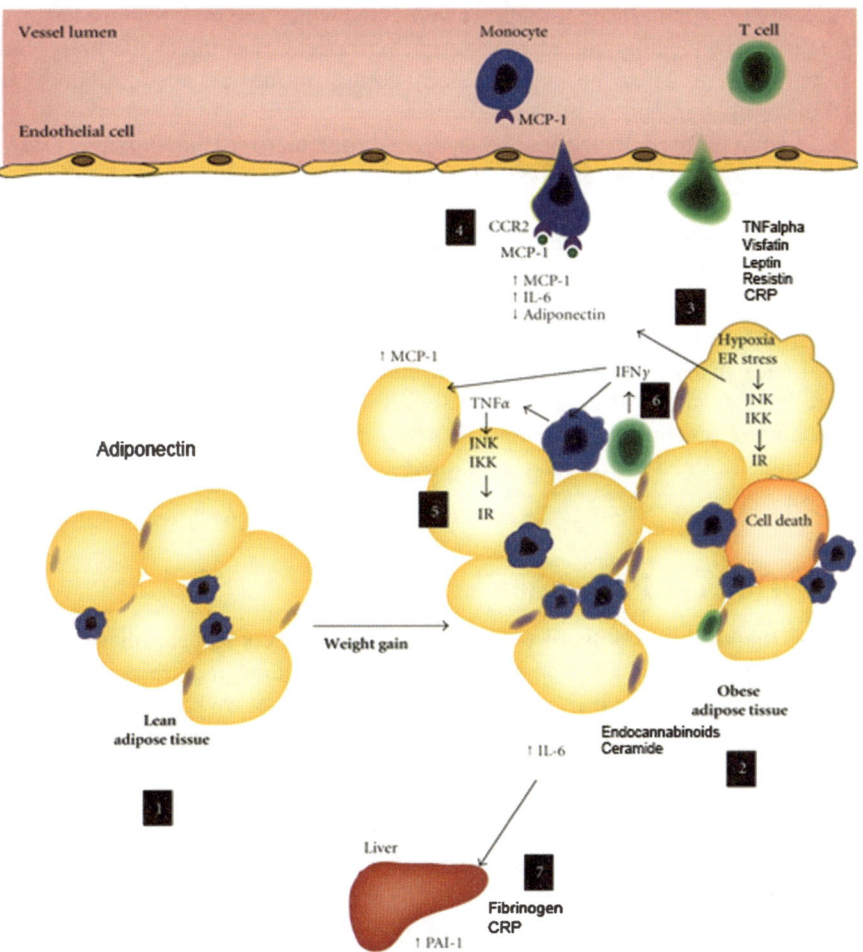

Figure 1.3 Adipokines produced by visceral fat. As visceral fat accumulates, adiponectin, an anti-inflammatory adipokine [1], decreases and many inflammatory adipokines increase [2], including tumor necrosis factor alpha (TNF-α), interleukin-6 (IL-6), visfatin, resistin, C-reactive protein (CRP), leptin and monocyte chemoattractant protein-1 (MCP-1). Toxic lipids are also produced excessively throughout the body as visceral fat accumulates, including ceramide and the endocannabinoids, anandamide and 2-arachidonylglycerol. Expanding adipocytes may become hypoxic and die through endoplasmic reticular stress (ER stress) that involves activation of inflammatory kinases, JNK and IKK [3]. Macrophages [4] and T cells [6] accumulate and secrete cytokines as visceral fat increases and activate inflammatory kinases in fat [5]. Macrophages express the MCP-1 receptor, chemokine C-C motif receptor 2 (CCR2). Adipokines in the plasma can increase liver production of inflammatory mediators and coagulation factors [7], including fibrinogen, CRP and plasminogen activator inhibitor-1 (PAI-1). Modified from Rocha and Folco.[15]

caused by a virus-induced imbalance in T cells, which the body cannot correct. Infections are usually caused by external pathogens that infect the body, but are only able to infect because the body cannot mount an adequate immune response through endogenous leukocytes.

Cancer is caused by endogenous or external factors that damage DNA. Cancer is able to grow only because the body cannot kill the cancer cells with endogenous factors. Recent evidence has found that obesity increases cancer risk and mortality.[8] Visceral fat releases adipokines, including IL-6, TNF and leptin, that increase the growth and malignancy of tumor-initiating stem cells.[9] This leads to tumor growth, malignancy and mortality.

In non-traditional medicine, diseases are detected too late, after the disease has already progressed to an intolerable state. This is true of diabetes, arthritis, heart disease, cancer and most other diseases. The patient may benefit from an early disease prevention program that prevents diseases from progressing to an intolerable state. Patients can be taught to avoid toxic lifestyles that cause diseases.

1.4   Disease Prevention

One of the vital functions of traditional practitioners is to keep their patients healthy and working. They do this with strong programs in disease prevention. Most of us are born healthy and could remain healthy if we knew how. Disease prevention is the key to staying healthy. Traditional practitioners have learned over the centuries that people who keep themselves thin and strong stay healthy, people who stay away from alcohol stay healthy, people who stay away from abuse of some plant-derived medicines and smoking stay healthy and people who minimize stress in their lives stay healthy. Toxic lifestyles lead to disease.[5] All people should be encouraged to perform 60–70 minutes' of aerobic exercise daily for disease prevention and to encourage healing. Exercise increases the production of stem cells in every organ examined so far.

Today, keeping people alive is the priority in many non-traditional medical practices, not disease prevention. Even people with substantial multiple organ pathology from diabetes, heart disease, congestive heart failure, stroke and other conditions are kept alive with modern medicines. Many non-traditional medical practitioners ignore the traditional approach to disease prevention since it is easier simply to treat the patient with a drug that manages symptoms. Hypertension, high blood cholesterol, high blood sugar and other specific symptoms can be managed with drugs. This approach does not cure the underlying disease, but keeps the patient alive. These drugs are used for the rest of the patient's life. The problem then becomes managing the toxicity of the drugs. Some statins used in hypercholesterolemia cause muscle toxicity and diabetes. Pioglitazone, used in diabetes, causes congestive heart failure and kidney cancer. Some calcium channel blockers, used in hypertension, can cause bradycardia. Most drugs used to control blood sugar in diabetes can cause hypoglycemia. Non-steroidal anti-inflammatory drugs (NSAIDs) used in arthritis cause ulcers, kidney damage, heart attacks and strokes. Acetaminophen causes liver damage, kidney damage, heart attacks and strokes.

1.5   The Traditional Approach to the Treatment of Disease

Science still has much to learn from traditional Chinese medicine.[10,11] The traditional approach is pragmatic, experimental and based on what helps the patient get back to a productive life. For instance, a patient crippled by stroke may be treated for up to 6 months with plant preparations and physical therapy in order to decrease paralysis and get them back to work. Scientists frequently do not understand the mechanisms involved in traditional medicine and should continue searching.

Traditional Chinese medicine requires expertise to find the proper mixture of plant medicines to achieve balance in the body. This is a difficult experimental process in each patient. Each disease in each patient requires an individualized approach to achieve balance.

Meditation, tai chi, yoga, prayer and other calming influences are critical to disease treatment and prevention in the traditional approach. Chi increases because of these calming influences and can maintain the balance of yin and yang. These calming influences help treat anxiety that greatly magnifies the risk of developing some diseases and decreases the chance of recovering from many diseases.

1.6   Randomized, Placebo-Controlled Clinical Trials

Non-traditional medicine is currently based on randomized, double-blind, placebo-controlled clinical trials. The theory is that a drug must be more powerful than a placebo at curing a disease. Traditional medicine rejects randomized, placebo-controlled clinical trials since it is not ethical to administer a placebo or other compound that is not effective in a sick patient. This has led to the rejection of traditional Chinese medicine by non-traditional medicine.

It is important to remember that drugs do not cure diseases. The body heals itself. No matter how powerful the drug, the body heals itself. Drugs facilitate the ability of the body to heal itself. The body is equipped with endogenous agonists and antagonists in addition to signal transduction mechanisms that heal the body; in other words yin, yang and chi work in harmony. Even the most powerful antibiotics cannot cure infections without a functional immune system. Antibiotics facilitate the ability of the body to recover from bacterial, viral, protozoal and other infections. The basic fallacy of randomized, placebo-controlled clinical trials is that drugs do not cure diseases. Non-traditional medicine relies on ever more powerful drugs, in other words, more dangerous drugs, to help the patient. This is frequently the wrong approach. Traditional

Chinese medicine relies instead on finding a different combination of plant medicines, diet, physical therapy and psychological healing to facilitate the balance that the body requires to heal itself. Once the patient is healed, the practitioner works with the patient to prevent further disease.

It is also important to realize that placebos are actually drugs that work through endogenous mechanisms. Placebos alter opioid, serotonergic, dopaminergic and other systems by stimulating endogenous agonists and antagonists. This is why placebos can be as powerful as drugs in some clinical trials. Even non-traditional practitioners know that tender loving care is the most powerful medicine. There is currently an interest in China in writing standards for randomized clinical trials of traditional Chinese medicines, similar to the CONSORT standards.[12] Hopefully, these standards will be based on the Chinese principles of medicine. Clinical trials should compare drugs with no treatment or with standard treatment, not to placebos.[13]

1.7 Therapy Must Allow the Body to Heal Itself

Modern therapy is dominated by the carpenter approach: if the hammer does not work, get a bigger hammer. Patient therapy is rapidly escalated to the most powerful drugs available, with inadequate attempts to heal the patient. Congestive heart failure patients are treated with digoxin, which is very effective. However, digoxin is also very toxic, causing arrhythmias that damage the heart and prevent the heart from healing itself. Bipolar patients are treated with lithium, which is very effective but very toxic. Lithium causes seizures that damage the brain and prevent the brain from healing itself. Arthritis patients are treated with steroids that are very effective, but cause visceral fat accumulation. Visceral fat secretes inflammatory adipokines that make arthritis worse.[5] Pain patients are treated with hydrocodone, oxycodone, fentanyl and other opioids that are effective temporarily, until tolerance makes them ineffective. Tolerance to administered opioids also makes the body tolerant to endogenous opioids, such as endorphins, enkephalins and dynorphins. This tolerance can be long term, lasting weeks or months. Cessation of opioid therapy leaves patients in much worse pain than before, due to long-term tolerance to endogenous opioids and opioid-induced hyperalgesia. Children with attention deficits are treated with amphetamine and methylphenidate: they become addicted to these drugs and have little chance of recovering from their attention deficits. Insulin is very effective in the treatment of type II diabetes. However, administration of insulin exacerbates insulin receptor insensitivity and does not allow the body to heal itself. Hypertension and hypercholesterolemia can be treated with effective drugs. These conditions are caused by adipokines secreted by visceral fat.[5] Drug therapy does not adequately inhibit the effects of the adipokines. Only losing weight and exercising can inhibit adipokine secretion and rebalance the body.

Cancer is another issue in healing. Combinations of very toxic drugs are used to kill cancer cells, in addition to surgery and radiation therapy. Healing cancer is difficult, since the drugs must kill the cancer but leave the body healthy enough to heal. Bone marrow toxicity is frequently an issue in cancer therapy, such that the immune system may become compromised and unable to mount a defense against the cancer. A healthy immune system is required for healing. Cancer is frequently caused or promoted by toxic lifestyles, including obesity, smoking, drinking alcohol and exposure to toxic agents.[8]

The body must be allowed to heal itself. Toxic lifestyles must be changed to healthy lifestyles that allow the body to reach the proper balance needed in healing. Prevention must be the first approach to disease. This involves losing weight, practicing daily aerobic exercise, stopping smoking and drinking no more than one alcoholic beverage per day. Mild therapy can be the second approach to disease. Mild therapies may involve acupuncture, dream therapy, physical therapy, tai chi, nutritional therapy and other therapies that rebalance the body. Placebos may be used since they work only by allowing the body to heal itself. Nutritional therapy and functional foods are important in allowing the body to heal itself. Patients may need to learn to eat more fruit and vegetables to control blood sugar and cholesterol. Many people eat too much meat and fat, but not enough whole grains, fruits and vegetables. Plant medicines should be the third therapy when the first two therapies are not enough. Drug therapy should be the fourth therapy that is used only temporarily in patients with severe diseases. Drugs should only be used long enough to allow the patient to recover and begin to use milder therapies that allow the body to heal itself.

An example of this is that a clinical trial of 340 cancer patients treated with 2 g of American ginseng daily for 8 weeks showed improvement in fatigue and wellbeing compared with placebo. This study was presented at the 2012 Meeting of the American Society of Clinical Oncology. The standard therapy for fatigue and anemia in cancer patients is epoetin alfa. This powerful medicine also causes strokes and heart attacks in cancer patients by promoting embolism formation. It appears better to use American ginseng, which is milder therapy and allows the body to heal itself. Ginseng works by decreasing blood levels of cytokines and adipokines that promote cancer growth. In addition, cortisol levels decrease. Cortisol can increase visceral fat that increases adipokine levels.

1.8 Conclusion

This book is designed to provide a basis for communication between traditional and non-traditional practitioners. The scientific basis of plant-derived medicines is well understood and is applied to traditional medicine in this book. Disease prevention must be learned, in part from traditional practitioners, in order to decrease diabetes, heart disease, arthritis and cancer throughout the world. Disease prevention is critical to the current economic crisis being created by overuse of the healthcare system.

References

1. E. J. Lien, *Int. J. Orient. Med.*, 1995, **20**, 126.
2. E. J. Lien, L. L. Lien and J. Wang, *Curr. Drug Discov. Technol.*, 2010, **7**, 13.
3. C. I. Svensson, M. Zattoni and C. N. Serhan, *J. Exp. Med.*, 2007, **204**(2), 245.
4. S. Narumiya and T. Furuyashiki, *FASEB J.*, 2011, **25**(3), 813.
5. J. D. Adams and K. Parker, *Extracellular and Intracellular Signaling*, Royal Society of Chemistry, Cambridge, 2011.
6. J. M. Amiel, S. J. Mathew, A. Garakani, A. Neumeister and D. S. Charney, in *Textbook of Psychopharmacology*, 4th edn, ed. A. F. Schatzberg and C. B. Nemeroff, American Psychiatric Publishing, Arlington, VA, 2009, p. 965.
7. E. J. Lien and S. Ren, *Chin. Pharm. J.*, 1998, **50**, 1.
8. A. G. Renehan, M. Tyson, M. Egger, R. F. Heller and M. Zwahlen, *Lancet*, 2008, **371**, 569.
9. D. E. Feldman, C. Chen, V. Punj, H. Tsukamoto and K. Machida, *Proc. Natl. Acad. Sci. U. S. A.*, 2012, **109**, 829.
10. K. C. Huang, *The Pharmacology of Chinese Herbs*, 2nd edn, CRC Press, Boca Raton, FL, 1999.
11. Z. Zhou, G. Xie, X. Yan and G. W. A. Milne, *Traditional Chinese Medicines, Molecular Structures, Natural Sources and Applications*, 2nd edn, Ashgate Publishing, Aldershot, 2003.
12. Z. Bian, B. Liu, D. Moher, T. Wu, Y. Li, H. Shang and C. Cheng, *Front. Med.*, 2011, **5**, 171.
13. J. D. Adams, *World J. Pharmacol.*, 2011, **1**, 4.
14. C. L. Kuo, R. B. Wang, L. J. Shen, L. L. Lien and E. J. Lien, *J. Clin. Pharm. Ther.*, 2004, **29**, 279.
15. V. Z. Rocha and E. J. Folco, *Int. J. Inflamm.*, 2011, Article ID 529061, doi: 10.4061/2011/529061.

CHAPTER 2

Structure–Activity Relationship Analysis of Plant-Derived Compounds

ERIC J. LIEN*, LINDA L. LIEN AND JAMES D. ADAMS

Department of Pharmacology and Pharmaceutical Sciences, School of Pharmacy, University of Southern California, Los Angeles, CA 90089, USA
*E-mail: elien@usc.edu

2.1 Introduction

In traditional Chinese medicine (TCM), usually several different plants are used in the treatment or prevention of illness, according to the condition and need of the individual patient. The outcome is observed subjectively by the physician so that the prescription can be adjusted according to the experience of the prescriber. This type of empirical approach is quite different from the common practice of modern Western medicine where a specific drug developed through randomized double-blind clinical study is used to treat a specific disease. The only time subjective observation is called for is in the reporting of serious adverse reactions, since it is neither ethical nor legal to design clinical studies to investigate unwanted side effects of a drug. Both subjective empirical and objective unbiased approaches have their own advantages and disadvantages.

As biochemical pharmacology and the molecular basis of drug action became better understood,[1] in TCM there is a trend to identify active ingredients and molecular mechanisms of action in each individual plant drug used and the actions of combinations of different drugs.

RSC Drug Discovery Series No. 31
Traditional Chinese Medicine: Scientific Basis for Its Use
Edited by James David Adams, Jr. and Eric J. Lien
© The Royal Society of Chemistry 2013
Published by the Royal Society of Chemistry, www.rsc.org

The successful discovery of artemisinin (chinghausu) from *Artemisia annua* as effective malaria therapy has saved millions of lives across the globe. This resulted in the award of the prestigious Lasker Award to the most deserving pharmacologist Tu Youyou, in 2011.

2.2 Structure–Activity Relationship Analysis

In this chapter, some specific examples of structure–activity relationship (SAR) analysis of plant materials are presented, using functional group/pharmacophore analysis, comparison of physicochemical parameters relevant to intermolecular interactions and quantitative SAR analysis. It is our belief that a combination of deductive logic and inductive logic can be useful in solving the complex issues of using plant-derived compounds in the treatment/prevention of diseases with the hope of achieving graceful aging and longevity.[2]

Recently, Tan *et al.*[3] reviewed *in vitro* and *in vivo* studies of antitumor natural products isolated from Chinese herbs. The activities reported include inhibition of proliferation, induction of apoptosis, suppression of angiogenesis, retardation of metastasis and enhancement of chemotherapy. These natural products belong to flavonoids (gambogic acid, curcumin, wogonin and silibinin), alkaloids (berberine), terpenes (artemisinin and derivatives, β-elemene, oridonin, triptolide and ursolic acid), quinones (shikonin and emodin) and saponins (ginsenoside R_{g3}), all from Chinese medicinal plants.

2.2.1 From Natural Sources to Synthetic Compounds

The antiasthmatic herb ma huang (*Ephedra sinica*, *E. equisetina* or *E. intermedica*) has been used in TCM as an antiasthmatic and anticough agent, to soothe the lung and to stimulate the heart and mind. In the 1920s, K. K. Chen explored the pharmacology of the plant. The structure of its active principle, ephedrine, was established and became the prototype for sympathomimetic agents. Pseudoephedrine (a stereoisomer of ephedrine used in over-the-counter cold preparations) led to the synthesis of more lipophilic CNS stimulating agents such as amphetamine and methamphetamine, unfortunately resulting in drug abuse. All these would not have been expected when Nagai first isolated ephedrine from *E. vulgaris* in 1888.[4] Lipophilicity as indicated by the octanol–water partition coefficients (log*P*) is as follows: ephedrine 1.43, pseudoephedrine 1.40, amphetamine 1.76 and methamphetamine 2.07.

Because of the presence of a basic amine group in alkaloids, making them water soluble when protonated under acidic pH and lipid soluble under basic conditions, they have been thoroughly investigated. Besides the ease of isolation and characterization, the presence of the positively charged group enables them to interact with G-protein coupled receptors.[5] This explains why most alkaloids have significant biological activities.

2.2.2 From Estradiol, Phytoestrogens to Synthetic Diethylstilbestrol

The natural hormone 17β-estradiol has a phenolic OH and an alcoholic OH group which can be structurally overlapped with two phenolic OH groups present in genistein and daidzein, well-known isoflavone phytoestrogens present in soybean.[6,7] Other isoflavonoids can be found in Leguminosae plants. Because of the antioxidant and free radical scavenging properties of the phenolic function, many different biological activities have been reported *in vitro* and *in vivo*.[8] The synthetic hormone diethylstilbestrol is a symmetrical molecule with two identical phenolic OH groups. It was used clinically but turned out to be carcinogenic. This may be because it was never exposed to humans or other animals during evolution and the body is less able to detoxify it as compared with compounds derived from natural sources. It is worth noting that premarin (containing conjugated estrogenic hormones) obtained from urine of pregnant mares is still widely used in estrogen replacement therapy. Figure 2.1 shows the similarity of natural and synthetic estrogen? and phytoestrogen isoflavonoids. In contrast, flavonoids (such as luteolin, quercetin and others) are more widely present in many different families of

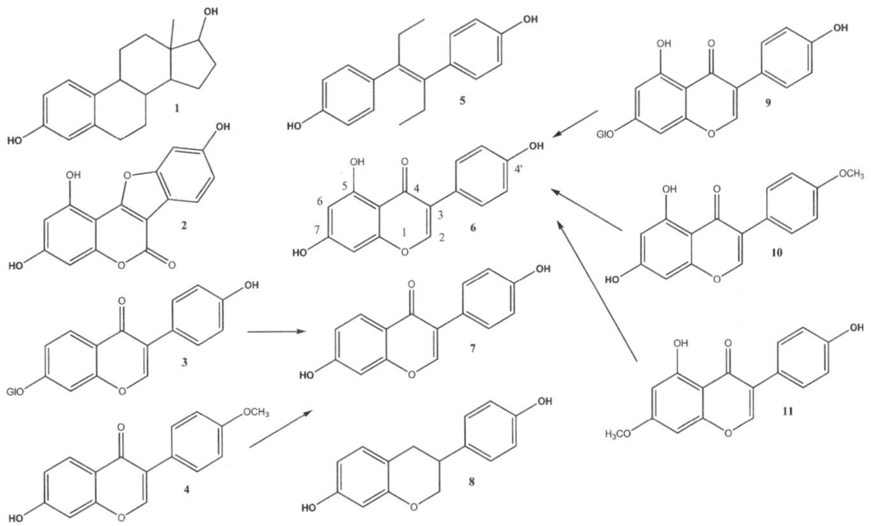

Figure 2.1 Structural similarity of estrogens and phytoestrogen isoflavonoids showing that the 7-OH and 4′-OH groups are in line. **1**, Estradiol (natural); **2**, coumestrol; **3**, daidzin (isoflavone glucoside); **4**, formono-netin (methoxylated isoflavone); **5**, diethylstilbestrol (synthetic); **6**, genistein (isoflavone); **7**, daidzein (isoflavone); **8**, equol (isoflavandiol); **9**, genistin (isoflavone glucoside); **10**, biochanin A (methoxylated isoflavone); **11**, prunetin (methoxylated isoflavone). Arrows show the *in vivo* removal of side chains by oxidative metabolism or hydrolysis of the glycoside group. Adapted from Lien *et al.*[6]

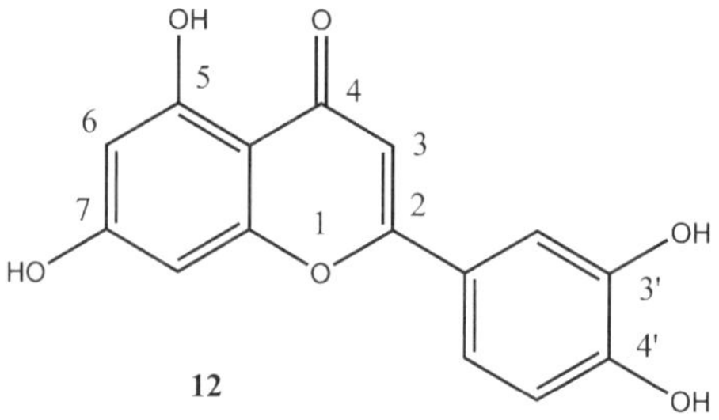

Figure 2.2   Structures of antiestrogenic flavonoids, showing that the 7-OH and 4'-OH groups are not in line. **12**, Luteolin; **13**, quercetin.

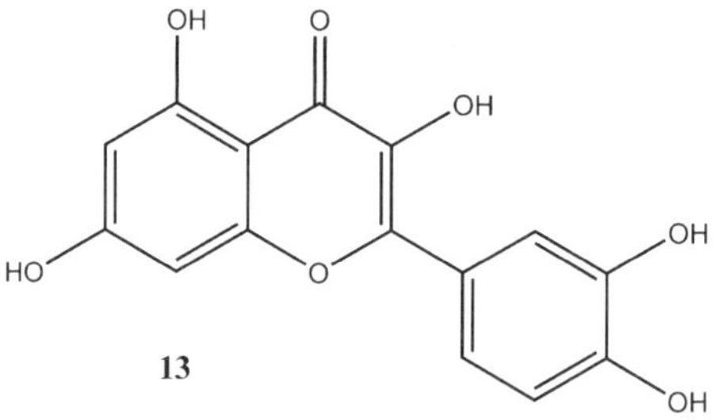

plants. The benzene ring in flavonoids is attached to the 2-position of the benzopyran ring (Figure 2.2) instead of the 3-position as in isoflavonoids (Figure 2.1). Consequently, many flavonoids have antiestrogenic/androgenic activities.[9,10] The best known androgenic plants from the *Epimedium* genus contain many flavonoid glycosides.[10] The wide range of biological activities, including cancer-preventing activities, of flavonoids have been reviewed by Das *et al.*[8]

2.3   Medicinal Diet (Yao San)

In TCM, a properly balanced diet is considered a very important component of health maintenance. Instead of using purified bleached starch, unprocessed mixed grains are preferred. It has been shown recently that in addition to fibers, grains such as oats contain phenolic antioxidants as well as β-glucan. Using hyperlipidemic ICR mice, Li *et al.*[11] showed that both phenol-rich and β-glucan extracts decreased the concentrations of total serum cholesterol, low-density lipoprotein cholesterol and hepatic total cholesterol. Only the phenol extract decreased hepatic triglycerides, inhibited hepatic 3-hydroxy-3-methyl-glutaryl CoA reductase activity and improved the hepatic antioxidant defense system. This finding is in agreement with the reported benefits of lignans (polyphenols) present in many different whole grains and some Chinese herbs, such as Pueraria and Sophora.[12]

2.4 Immunostimulating (Fu-zhen) Herbs

In TCM, it is a common strategy to use fu-zhen and chei-shei drugs concomitantly. Fu-zhen means to strengthen the patient's resistance to disease and to restore the normal function of the body. Chei-shei means to dispel pathological factors or to get rid of abnormal elements. Fu-zhen drugs may enhance cancer treatment either by increasing the curative effects via the immune system or reducing the side effects of chemotherapy and radiotherapy (Figure 2.3).[13]

Positive results on immune restoration and/or augmentation of local graft versus host reaction by TCM herbs have been reported by collaborative efforts between Chinese and American scientists at the M.D. Anderson Tumor Institute.[14] The herbs studied are *Astragalus membranceus* (huang-chi) and *Ligustrum lucidum* (nu-chen-tzu). *Astragalus* was first recorded in "Shen Nung's Herbal" (Shen Nung Pen Tsao Ching) as a superior herb, and *Ligustrum* was listed as a high-grade drug.[15]

Included in fu-zhen herbs are several fungi, such as *Lentinus edodes* (hsiang-ku, shiitake) and fungi of the genera *Ganodermas*, *Tremella*, *Poria* and *Cordyceps*. Their immunostimulant properties have been reported.[16] Active ingredients in these fungi include polysaccharides with complex primary to quaternary structure and molecular weights up to 10^6 kDa or higher. Polysaccharides from over 30 higher plants have also been reported to have immunostimulating activities.[17] Most of the polysaccharides from higher plants are heteroglycans with very complex structures. The anti-complementary activities of these polysaccharides (via the classical and/or alternative pathway) are found to be related to their acidity, molecular weight and hydroxyl groups present. The antitumor activities of some of these

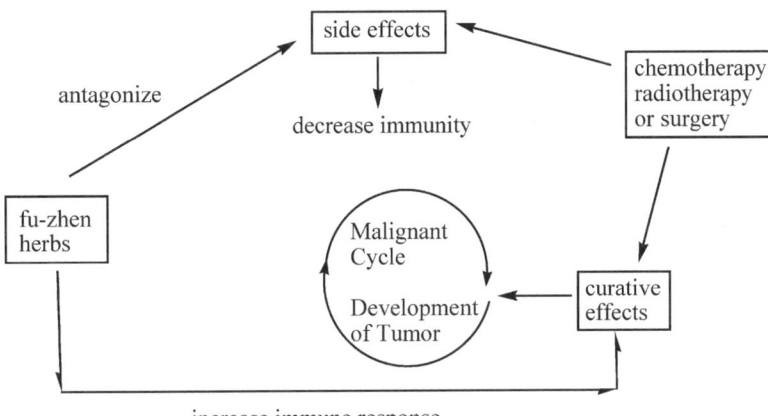

Figure 2.3   The possible role of fu-zhen herbs in cancer treatment. Adapted from Li and Lien.[13]

polysaccharides are closely related to the inhibition of drug-metabolizing enzymes in the liver, and also their immune-enhancing activities.[17]

References

1. K. C. Huang, *The Pharmacology of Chinese Herbs*, 2nd edn, CRC Press, Boca Raton, FL, 1999, p. 328.
2. E. J. Lien, L. L. Lien and J. Wang, *Curr. Drug Discov. Technol.*, 2010, **7**, 13.
3. W. Tan, J. Lu, M. Huang, Y. Li, M. Chen, G. Wu, J. Gong, Z. Zhong, Z. Xu, Y. Dang, J. Guo, X. Chen and Y. Wang, *Chin. Med.*, 2011, **6**, 27.
4. M. Nagai, *Chem. Zentralbl.*, 1888, **59**, 130.
5. C. L. Kuo, R. B. Wang, L. J. Shen, L. L. Lien and E. J. Lien, *J. Clin. Pharm. Ther.* 2004, **29**, 279.
6. E. J. Lien, A. Das and L. L. Lien, *Prog. Drug Res.*, 1996, **46**, 263.
7. L. L. Lien and E. J. Lien, *J. Clin. Pharm. Ther.*, 1996, **21**, 101.
8. A. Das, J. H. Wang and E. J. Lien, *Prog. Drug Res.*, 1994, **42**, 133.
9. H. Wu, R. Wang, C. Kuo, L. L. Lien and E. J. Lien, *Int. J. Orient. Med.*, 2003, **28**, 13.
10. H. Wu, E. J. Lien and L. L. Lien, *Prog. Drug Res.*, 2003, **60**, 1.
11. Y. Li, Y.C. Luo, G. Chen, Z. L. Xiao, Q. Wang, L. Zhao and B. P. Ji, *J. Food Drug Anal.*, 2011, **19**, 49.
12. S. S. C. Huang, K. I. Komatsu, S. S. J. Ren, L. L. Lien and E. J. Lien, *Int. J. Orient. Med.*, 2001, **26**, 66.
13. W. Y. Li and E. J. Lien, *Orient. Healing Arts Inst. Bull.*, 1986, **11**, 1.
14. Y. Sun, E. M. Hersh, M. Talpaz, S. L. Lee, W. Wong, T. L. Loo and G. M. Mavligit, *Cancer*, 1983, **52**, 70.
15. H. Y. Hsu, *Oriental Materia Medica: a Concise Guide*, Oriental Healing Arts Institute, Long Beach, CA, 1983.
16. E. J. Lien, *Prog. Drug Res.*, 1990, **34**, 395.
17. E. J. Lien and H. Gao, *Int. J. Orient. Med.*, 1990, **15**, 123.

CHAPTER 3

Huang Di Nei Jing and the Treatment of Low Back Pain

MINCHEN WANG (VINCE)

Pacific College of Oriental Medicine, 7445 Mission Valley Drive, Suite 105, San Diego, CA 92108, USA
E-mail: pbacupuncture@gmail.com

3.1 Significance of Low Back Pain (LBP)

3.1.1 Prevalence of LBP

LBP is very common, and is one of the most common illnesses for which patients seek medical help. According to a 2002 survey in the United States, about 80% of the population reported experiencing an LBP episode in their lifetime. Each year, one in two people experiences LBP severe enough to make him or her aware of it.

3.1.2 Human Cost

Many patients have LBP but do not seek medical help.[1,2] Some do get better with time. Up to one-third of patients report persistent back pain of at least moderate intensity 1 year after an acute episode, and one in five report substantial limitations on activity.[3,4]

People who experience LBP that does not subside gradually alter their posture and reduce their daily activity to minimize aggravating the problem. Such changes impact normal body function adversely.

RSC Drug Discovery Series No. 31
Traditional Chinese Medicine: Scientific Basis for Its Use
Edited by James David Adams, Jr. and Eric J. Lien
© The Royal Society of Chemistry 2013
Published by the Royal Society of Chemistry, www.rsc.org

Furthermore, the low back and knees are intimately connected. People who suffer from LBP tend to develop knee disorders that further limit their mobility. The lack of mobility accelerates body aging. It is a serious health issue that significantly jeopardizes quality of life.

3.1.3 Medical Cost and Societal Cost

The cost of treating LBP has a substantial impact on the US healthcare system. LBP is the second most frequently cited reason for seeking primary care services[3] and is believed to result in more disability among working-age adults than any other health condition.[1] National estimates of the direct costs of care for LBP range from $25 billion to $33 billion annually[5,6] in the United States.

With the "graying" of the United States, the cost will surely continue to increase. For working adults, the incidence of LBP will rise with the increased job stress caused by prolonged intense computer use.

3.2 Scope of the Problem- What Is Involved in LBP?

3.2.1 Diversity of Causes and Diagnosis

LBP is a complaint of discomfort in the low back below the rib cage extending to the posterior upper pelvis. It can be so severe that patients are incapacitated. It actually reflects a whole host of diverse disorders. In a study to evaluate LBP medical care, 60 different ICD9-CM (*International Classification of Diseases*, 9th edition, Clinical Modification) diagnosis codes were found to be used by physicians to refer to LBP.[7,8] Such diversity reflects the complexity of causes that can lead to LBP, and this diversity makes communication difficult between researchers, leading to contradictory findings regarding the efficacy of various modalities of treatments.

3.2.2 Practical Considerations of LBP

3.2.2.1 Two Categories of LBP

From the clinical perspective, it is helpful to separate LBP cases into two categories. Category one mainly involves soft tissue disorders. Here, we are dealing with muscular-skeletal stress and the resulting postural imbalances. These are the majority of clinical cases of LBP.

In the second category, LBP is the byproduct of other pathological issues. The pain may or may not be very severe. Regardless, priority must be placed on dealing with the main pathological sources. An example of such a situation is LBP that occurs when a patient suffers from fever. The pain in the low back is incidental to the fever, and the treatment should be focused on resolving the fever. According to the 2010 edition of *Current Medical Diagnosis and Treatment*[9] (CMDT), such incidents should include "(1) infection, (2) cancer,

(3) inflammatory back diseases such as ankylosing spondylitis or (4) pain referred from abdominal or pelvic processes, such as peptic ulcer or expanding aortic aneurism."

At the same time, for this second category of LBP, concurrent treatments that aim to reduce the pain can be beneficial to the comfort of the patient. This would accelerate the recovery.

3.2.2.2 Holistic Approach

On the surface, the issue of LBP appears to be due to lesions of soft tissues, such as muscles, tendons and associated fascia confined to the low back area. The focus tends to be directed solely on the low back where the pain is. Such a narrow view makes treatment unsatisfactory. Invariably, numerous body components, including the spine, the pelvis, the legs and postural configuration, are deeply involved. As evident from this chapter, a holistic approach is not only desirable but also indispensable in most cases for a good therapeutic outcome.

3.2.2.3 Age and Duration Variation

Since muscle elasticity and the body's metabolic rate decrease with age, LBP issues will become more frequent and severe with age. However, with age, LBP is often surpassed by joint pain. This shift is caused by the fact that long-term LBP together with age-related muscle stiffness exert constant tension on the joints. Such tension exacerbates the deterioration of cartilage protecting the joints. Although the chief complaint may shift, the soft tissue issues should be properly recognized just the same.

With time, postural adaptation due to muscle tension can become semi-permanent and increase susceptibility to chronic pain. The adaptation processes also promote disorders such as bone spurs and fibrosis. Whenever these disorders are present, the treatment is more difficult as the issue is more than just resolving the soft tissue disorder. Longer term therapy would be needed. It is often assumed that such adaptation is irreversible. This is not true. Many people end up seeking surgical intervention. This may not be necessary, as human bones continue to transform throughout life.

3.3   Introduction to Huang Di Nei Jing

3.3.1   Overview

Huang Di Nei Jing (abbreviated to Nei Jing) is a book written in the format of conversations between Huang Di (Yellow Emperor) and his court doctors, mainly Qi Bo, in ancient China. The title is also been translated as Yellow Emperor Inner Cannon. It is about human physiology and the influence of environmental factors such as cold, heat, wind and dampness on the health of human beings. It contains pathology theories of human disorders and the treatment methods for diseases based on acupuncture and moxibustion. Huang Di was a legendary figure in the 27th century BC. He is considered to be the common ancestor of the majority of ethnic Chinese.

3.3.2 Historical Perspectives[5]

It is unlikely that Nei Jing was written in Huang Di's time. In his time, the written language was very primitive. Even in the Shang dynasty ,which was hundreds of years after Huang Di's era, the written language was still primitive. The first recorded history of the existence of Nei Jing was in the Han Dynasty, about 2000 years ago. Upon examining the Nei Jing carefully, it appears likely to be a collection of medical writings, possibly from multiple contributors. It is certain that thousands of years were required for the development of medical knowledge and skills before the Nei Jing was written.

The current edition of Nei Jing was edited and recompiled by Lin Yi and co-workers in the imperial editorial office of the Song dynasty in 1051 AD. His edition was based on an earlier edition by Wang Bing of the Tang Dynasty (762 AD).

3.3.3 The Significance of the Nei Jing

3.3.3.1 Influence

The concepts expressed in the Nei Jing became the source and the ultimate authority of medical writing until today. It is fair to say that the Nei Jing cast a great influence over all Chinese medical writing throughout Chinese history. It provided inspiration for medical practices for over 2000 years.

Although the Nei Jing is regarded as the source of all Traditional Chinese medical practice, its treatment focus is on acupuncture and moxibustion. In fact, the Nei Jing regards herbal treatment as the last resort in treating body disorders.

3.3.3.2 Concept of the Meridians

Central to the Nei Jing is the concept of meridians. Meridians provide the vital connecting link between various parts of the body, and are regarded as the transportation and communication superhighways of the human body.

Through this superhighway system, signs of disorders inside the body are transmitted and registered at some specific points along the meridians. These points are called acupoints. In return, remedial messages called Qi can be sent through these points via the superhighways to these disordered sites to correct the problems. Figure 3.1 illustrates a few selected meridians that have relevance to low back pain. All meridians are in pairs, bilaterally, except for the Du meridian.

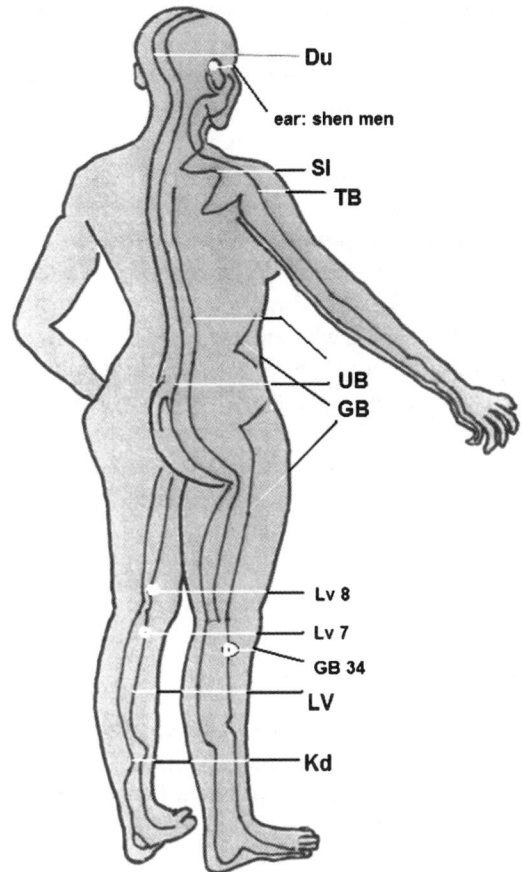

Figure 3.1   Selected meridians related to back pain. All are bilateral meridians except for the Du meridian. Included are four acupoints: GB 34 (Yang Ling Quan), Lv 7 (Xi Guan), Lv 8 (Qu Quan) and ear point Shen Men.

3.3.3.3   Qi

QI has a few connotations. In its original context, Qi was used to refer to an invisible entity that governs proper functioning of any kind of life process. When a body part stops working, most often it is not related to any physical lesion. Such a disorder is called a Qi disorder. It can be a weak Qi that diminishes the ability to function. It may be that this body part stops performing its regular biological function. This is called Qi stagnation or a rebellious Qi. Medically, it is referred to as an idiopathic disorder or disorder of unknown pathological origin. Such disorders are very common. Most LBP belongs in this category. This is why according to the 2010 edition of CMDT,[9] for low back pain "a precise diagnosis cannot be made in the majority of cases."

To resolve Qi-related problems, we need to restore the proper Qi. In acupuncture, needling processes are used to correct this disorderly Qi. The process of restoring Qi is called "Dao Qi" or moving the proper Qi to the site of the disorder. When the proper Qi is restored, it is called "De Qi" or Qi restored. The problem will be corrected as the result.

"De Qi" is often confused with "Qi Gan", the sensation of Qi, which is a soreness, distention or numbness sensation.[10] "Qi Gan" accompanies some needling. Such sensation is produced by the interaction of the needle with nerve endings. It does not guarantee the production of proper Qi, and it can even be an improper Qi that is produced.

There is a tendency for people who are not well versed in science to equate Qi with energy. Scientifically, energy cannot be created. Acupuncture needling does not create energy. It can create the Qi that stimulates the body to produce energy through metabolic processes. The confusion arises from the fact that when Qi disorders are restored, such as in LBP, the patient feels energized. In a situation where a pinched nerve is involved, a warm sensation is often produced as the result of the restoration of blood circulation when Qi is restored. It is easy to associate Qi with energy. In traditional Chinese medicine (TCM), the body energy comes from food and is called Gu Qi, energy from grain.

3.3.3.4   The Fifteen Meridians

There are 12 paired meridians: six pairs of foot-meridians originate from the toes and six pairs of hand-meridians originate from the fingers. These meridians are named after the internal parts, mostly organs, to which they are connected. On the upper limbs, there are heart (Ht), lung (Lu), small intestine (SI), large intestine (LI), triple burner (TB) and pericardium (PC) meridians. Triple burner refers to three segments of the torso; upper, middle and lower. On the lower limbs, there are gall bladder (GB), urinary bladder (UB), stomach (St), spleen (Sp), kidney (Kd) and liver (Lv) meridians.

In addition, three singular meridians exist. Two of them originate from the center of the perineum. One is Du, or the governing meridian, which travels along the spine, up along the center line of the head and ends below the nostril. The other, named Ren or conception, travels up the anterior of the body along the mid line. It ends at the lower lip. The two form a complete circle of the body. As such, both are important to low back health. The last meridian is called Dai or belt meridian and circles almost horizontally through the umbilicus, like a waist belt. This meridian acts like a belt upholding the low abdominal organs and is thus important in regulating all the organs below the belt and the low back. Since the Dai meridian shares acupoints with the GB meridian, this meridian is more of a conception than a utility.

The importance of the meridian system lies in its utility. It serves as a diagnosis tool for body disorders and it provides access ports for treating such disorders. These disorders can be in any place in the body. Body disorders are reflected, through pain sensitivity, to certain spots on the meridian that travels through this part of the body. By manipulating such spots, disorders can be mitigated. The manipulation can be achieved through surface needling, finger pressure or heating devices. This is a very important physiological property that is yet to be recognized in medical science. Such points are acupoints. In Chinese, they are called Xue, meaning caves. The name reflects the properties of acupoints as the gathering sites of Qi. It also reflects the physical feature, since they are mostly located at the crevices between the bones and muscles.

3.3.3.5   Five Shu Acupoints

The most important acupoints are located on the limbs. Each hand and foot meridian possesses five sets of acupoints called the five Shu points. Shu means transport, reflecting the role as major ports of entry for Qi to travel into the body interior. These acupoints are located from the elbows or knees to the extremities.

These acupoints are the most powerful points that should be emphasized in any application of acupuncture for any disorder throughout the body. They are the most effective acupoints for LBP.

3.3.3.6   Acupuncture

Acupuncture utilizes sharp, thin, pointed tools. There were nine styles of needles in ancient China, but modern acupuncturists only use two styles. The most common are thin stainless-steel needles of various lengths and diameters. The other style has a three-edged sharp tip for penetrating the skin and letting out static blood in the veins. In Nei Jing's era, areas that exhibited static blood under the skin needed to be lanced.

Figure 3.2 shows the most commonly used needles in clinics, with diameters from 0.15 up to 0.25 mm. The length used depends on the application; 30 mm is the most versatile. Needles 12 mm long are good for application to the ears, hands and feet where the insertion depth is only about 1.5–5 mm. On the ear or top of the head, the insertion should be slanted, at an angle of no more than 45° with respect to the skin surface.

3.3.3.7   Moxibustion

Moxibustion is a form of heat therapy. Traditionally, it was done through burning a small bead of dried mugwort wool (*Artemisia argyi*) on the skin over ginger cakes or using a burning moxa stick held at a few centimeters above the skin.

In the West, acupuncture implies the process of using fine thin needles to insert into certain locations of the body to achieve a therapeutic result. In the Nei Jing, acupuncture and moxibustion are used interchangeably. Except for a few situations such as over the major arteries or over the skin with a bacterial infection, one can be used to substitute the other. In fact, moxibustion is considered to be more versatile and powerful than needling. Therefore, it is

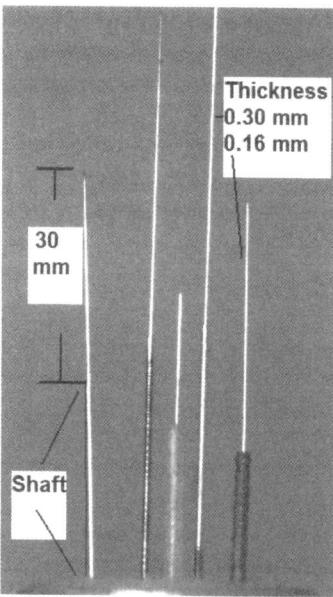

Figure 3.2 Acupuncture needles most commonly used, with lengths ranging from 15 to 75 mm, not including the shaft for handing.

more appropriate to use the term acumoxa instead of acupuncture unless either acupuncture or moxibustion is specifically utilized.

Although moxibustion is important, it is not widely practiced in Western acupuncture clinics for good reasons. One is that traditional moxa sticks produce heavy smoke, and are not suitable for air-conditioned offices. This problem can be avoided with smokeless moxa sticks as shown in Figure 3.3. Unfortunately,

Figure 3.3 Application of smokeless moxibustion. It is desirable to rest the wrist or little finger on the patient to prevent accidental movement from the location. The ignited head is about 10 to 20 mm from the skin surface.

most acupuncturists do not know the proper technique for using them. They use the "sparrow pecking" technique that is designed for smoking moxa sticks, moving the ignited moxa stick up and down, at a distance, over the target site on the patient's body. Rather, the proper way to use a smokeless moxa stick is to hold the ignited moxa stick steady at about 2 cm above the skin of the patient until the patient feels too hot. Depending on the severity of the condition, this could take from tens of seconds to a couple of minutes. It works as effectively as the smoking type of moxa sticks. It is also much safer to use than the smoking moxa.

3.4 Treatment of LBP According to the Nei Jing

3.4.1 Overview of Pain in the Nei Jing

Chapter 41 of Volume 1 of the Nei Jing is entitled "Discussion of Low Back Pain Needling." It discusses exclusively the acumoxa treatment for LBP. The content of this chapter is summarized in Table 3.1.

In addition, several other disorders which provoke LBP were discussed in separate chapters in the Nei Jing, such as in the chapter on "Needling Malaria." In this situation, LBP is incidental to other disorders. In the treatment, LBP becomes the secondary issue; one may or may not have to deal with it.

3.4.2 Pain and Qi Stagnation in the Nei Jing

3.4.2.1 Pain and the Nei Jing

Chapter 34 of Ling Shu, which is Volume 2 of the Nei Jing, is important in understanding the Nei Jing's thoughts about pain. It is entitled "Ju Tong Lun" or "Discussion on Pain." In this chapter, Qi stagnation and coldness are considered the main factors resulting in pain.

3.4.2.2 Qi Stagnation

There are many factors that cause Qi to stagnate, including emotional issues, lack of blood flow and coldness. Lack of blood flow can lead to either coldness or heat through vessel constriction or dilation. In either case, pain is produced. One example of dilation is migraine headache.

3.4.2.3 Coldness

Any tissue or organ that is subject to a cold environment would not function properly. This is a simple thermodynamic issue, applicable to any animal. The coldness issue can be caused by old age, anemia, low blood pressure, pain, blockage of circulation, neurological disorders or simply environmental factors.

Table 3.1 Low back pain treatment according to Huang Di Nei Jing.

Meridian	Problem area	Additional symptoms	Treatment/note
Stomach	Yang Min meridian	Unable to look backward, prone to feel sadness	Bleed St 36 (not in the autumn)
	San Mai near shin bone	Hot sensation, irritable. As if a wood block lays horizontally inside the low back, incontinence	Needle Sp 8
Gall Bladder	Cheng Gu (extra)	Pin and needle on the low back, difficult to bend forward or backward, unable to look back	Bleed tip of the fibula head (not in summer)
	Rou Li Mai	Coughing causes muscle tendon spasm	Needle GB 38 (Yang Fu)
Urinary Bladder	Abdomen, flank tense	Low back pain, unable to bend backward	Needle UB 34 (Xia Liao) on the opposite side of pain
	Heng Luo	Unable to lean forward and bend backward. Leaning backward causes falling sensation. Caused by heavy lifting injury.	Between UB 37 (Yin Men) and UB 39 (Wei Yang) blocked, severe blood stasis
	Jie Mai (part of the UB meridian)	Shoulder pain, blurry vision, incontinence. or Feels like a tight belt is breaking up the low back	Bleed the horizontal vein in the popliteal fossa, or, Bleed the veins with pellet-like appearance
	Pain along both sides of spine	Dizzy, feeling lack of balance. (Stretched neck, spine and sacra, swollen back sensation)	Blood letting at UB 40
	Yang Wei Meridian	Swelling at the painful spot on the back	Needle UB 57, Cheng Shan
	Tai Yang meridian	Neck, upper, lower back, sacrum tight, like carrying a heavy weight.	Bleed UB 40 (not in spring)
Kidney	LBP	Along anterior border of spine	Needle Kd meridian
	Tong Yin Mai	Sudden bulging appearance on the back. Pin and needle pain	Needle GB 38 (diversion connection)
	Hui Yin Mai	Sweating over the low back, thirst likes to drink.	Needle Zhi Yang Mai (UB meridian)
	Fei Yang Mai	Sad, fearful, swollen back.	Needle Kd 9, Zhu Bing
	Chang Yang Mai	Chest pain, back hyper-extended when severe, blurry vision, tongue curled up	Needle Kd 7, Fu Liu
Liver	Jue Yin meridian disorder	Low back tight like a stretched string. Cluster of bulged swellings around the spot. Quiet, foggy brain.	Needle Lv 5, Li Gou

When our bodies are exposed to cold, a negative feedback mechanism would likely be initiated to counter the coldness. Numerous mechanisms are responsive in such negative feedback processes – mechanical, neurological or chemical processes or a mixture of them. As a result, events such as muscle twitching or spasm, capillary constriction or localized inflammation and blood capillary enlargement occur. All of these events cause discomfort and pain.

3.4.2.4 Heat Therapy

Since coldness in tissues can be helped by warmth, this type of pain can be helped through application of heat. If not done, it can be a chronic problem and linger on for ever.

Unlike in Western medicine, inflammation caused by coldness is treated as "deficient fire" in TCM. In treatment, herbs that stimulate the body's metabolism, so-called "yang" tonics are used. In acumoxa therapy, moxibustion is preferred to needles. Since infrared radiation can penetrate deep into the body interior, infrared lamps are highly recommended in such applications. They are indispensable devices in acupuncture clinics.

A common misconception about using heat in therapy is that heat will be absorbed and dissipated without any significant physiological effect by circulation. In a healthy body, this is true. With people who have Qi or blood stagnation, it is definitely not true. Circulation itself is the problem, and an external boost is necessary. Without such help, a warm area remains warm and a cold area remains cold. Leaving it alone is the main cause of inflammation.

The effectiveness of heating depends largely on the devices. A heating pad is only good for the surface. A red lamp goes deeper. A popular special kind of IR lamp called "Shen Deng," a "magic" lamp that generates longer IR wavelengths, goes even deeper into tissues.

3.4.3 Key Features of the Nei Jing's LBP Treatment

3.4.3.1 Distal Needling

As is evident from Figure 3.4, the most distinctive feature in the Nei Jing's LPB treatment is the predominant usage of distal acupoints. The only exception is UB 34, Xia Liao, usage, which is at the fourth sacral foramen. It is used when LBP occurs together with tightness of the abdomen and flank. The UB 34 point from the opposite side of the tightness is to be used for this purpose.

The implication of distal acupoint selection is the use of acupoints from the limb's extremities, the so-called five Shu points. As explained earlier, such acupoints are the most influential points in acupuncture.

The book of Ling Shu, which is the Volume 2 of the Nei Jing, also contains treatments for LBP with varieties of associated symptoms: in Chapters 10 and 12, liver (Lv) meridians are used, and in chapter 26, UB and St meridians are used for hot low back. In Chapter 47, the kidney (Kd) meridian is used to treat LBP caused by kidney disorders. No acupoints were specified. However, of all the meridians used, only the UB meridian runs over the low back. To use other meridians, we have to choose acupoints of the legs or feet. Once again, distal acupoints are the main choices for LBP.

3.4.3.2   Use of Multiple Meridians

The second feature is that several meridians are utilized: UB, Lv and GB meridians. The other three foot meridians are used but with less frequency. The Nei Jing's prescription is a holistic approach, utilizing all the meridians on the lower limbs. It would be even better to include the upper limbs.

3.4.3.3   Blood Letting

The third feature is the bleeding of UB 40, Wei Zhong. UB 40s sit at the middle of the popliteal fossa of both legs. Bleeding is done on the bulging veins when there is apparent blood stasis around UB 40.

In TCM, blood stagnation is considered a serious problem for health. It is thought that blood stagnation leads to Qi stagnation. For blood stagnation, bloodletting is a quick and necessary way of restoring order. Unlike Western-style blood letting,[11] only a very small quantity of blood is let. In certain applications, such as in reducing fever, only one drop of blood is needed. The effect on the body is dramatic.

3.5   Modern Day LBP Treatment

After the Nei Jing's publication, acupuncture practice went through a decline in China. Among the reasons is Confucian ideology that the body's integrity, including the hair and skin, should not be breached out of respect for the ancestors. Bright people were encouraged to become governmental officers. Acupuncture was not considered a high-standing profession. Against the preference of the Nei Jing, medical treatments were mostly replaced by herbal medicine.

Revival of acupuncture as a modality of healing started in China in the 1950s. College-level textbooks were published. In Table 3.2, the acupoint selection for LBP is based on one of the standard textbooks on acupuncture.[12] Such an approach is common. For instance, a recent study on LBP treatment adopted this acupoint selection.[13] Other studies have some variation but the approach is basically similar in terms of acupoint selection and local orientation.

Table 3.2 list of acupoint selection by modern text books.

Meridian used	Point selection	Used in all cases	Additional problems
Urinary bladder	UB 23 (Shen Shu)	Yes	
	UB 25 (Da Chang Shu)		Cold dampness
	UB 26 (Guan Yuan Shu)		Cold dampness
	UB 40 (Wei Zhong)	Yes	
	UB 52 (Zhi Shi)		Kd Yin deficient
Hua Tuo para-spinal points	1.5 cm lateral to the lower border of each spinous process, from L1 to L5		
Kidney	Kd 3 (Tai Xi)		Kd Yin deficient
Extra	Yao Yuan (on the low back)		Kd Yang deficient
Du	Du 3 (Yao Yang Guan)	Yes	
	Du 4 (Ming Men)		Deficient Kd Yang
	Du 26 (Ren Zhong)		Traumatic injury

3.5.1 Features of Modern Acupuncture for LBP Treatment

3.5.1.1 Local Point Selection

As Table 3.2 indicates, a distinct feature in the modern approach is the heavy use of acupoints located on the low back and on the sacrum where the pain is most intense. Exceptions are UB 40 and Kd 3 (Tai Xi). Kd 3 is used when pain is associated with a kidney and the pain feels deep and anterior to the spine. It is a kidney problem and a rare incidence of LBP.

Hua Tao's para-spinal points are located on both sides of the spinal column, close to the lumbar spinal processes. They are also considered local points.

3.5.1.2 Ashi as a Point of Choice

In addition to acupoints on the meridians, the use of Ashi points, which are located outside of the meridians, is prominent. Originally, an Ashi point was used to designate a location where, upon palpation, the patient exclaimed "Ah Shi", that is, "Oh, Yes!" in English. Pressing such a point results in a reduction of pain. Needling such a point also often results in pain relief. A well-known physician in the Tang dynasty (618–907 AD), Dr Sun Si Miao, is credited with this practice. He was also the one who proposed the idea of "using pain spots as acupoints." Dr Sun was a very pragmatic physician, and his words should not be taken literally. The practice of Ashi should not be used indiscriminately. However, choosing tender points and treating them as acupoints has become a popular practice in modern times. This practice seems to be encouraged by the popularity of trigger point release theory.[14]

With chronic pain conditions, numerous places on the body become painful to palpation. Needling these spots may or may not relieve the pain. Therefore, acupoints based on pain does not produce consistent results. Contrary to popular belief, trigger point theory is difficult to implement effectively in practice.

The most serious issue with needling painful spots is that it can make the patient feel worse after treatment. Take fibromyalgia, for example: patients experience pain all of the time throughout the whole body, including severe LBP. They most likely are feeling exhausted from such suffering. They are considered to be deficient from a TCM perspective and should not be needled aggressively. Minimal needling and careful insertion should be applied. Painful spots should preferably not be needled at all.

The other situation where painful spots preferably should not be needled at all involves pain caused by hyperextension of muscles. This is also a deficiency situation. Moxibustion should be used instead. More on this subject will be discussed later.

In the last 20 years, there have been numerous published clinical studies on acupuncture therapies for LBP, and some do incorporate limited distal acupoints. The emphasis is always on the selection of local points on the low back. In one study, for the controls, moxibustion was explicitly excluded.[15]

3.5.2   Summary of LBP Treatment Comparisons

A comparison of LBP treatment principles is given in Table 3.3.

As can be seen from Figure 3.4, the selection of acupoints is significantly different between the Nei Jing style and the modern style of acupuncture. It is clear that the Nei Jing's style avoids the common pitfalls of the modern Ashi approach.

Table 3.3   Comparison of LBP treatment principles.

	Nei Jing	*Modern day*
Location of acupoints chosen for LBP	Legs (distal to painful location)	On the low back
Guiding principle	Follow meridians	Follow the diagnosis pattern or Ashi point
Meridian involved	Mostly liver, gall bladder, urinary bladder	Mostly Du (over the spine) and UB (on low back), except for UB 40, Wei Zhong
Differential diagnosis	Pain's impact on body mobility	Mode of injury, body condition

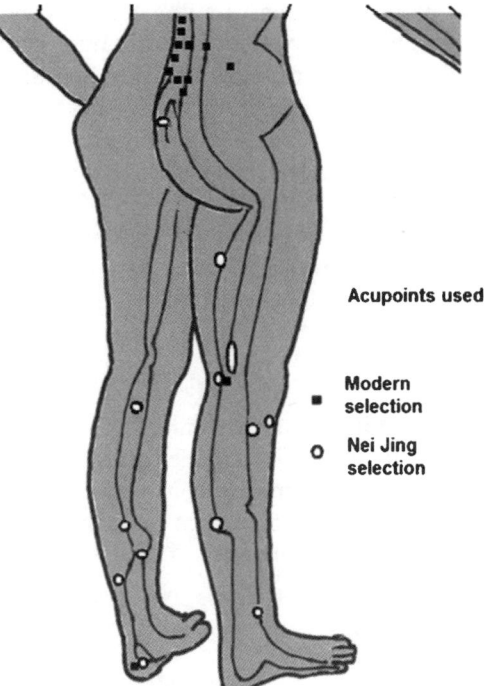

Figure 3.4   Comparison of acupoint selection.

3.6   Scientific Basis of LBP Treatment

3.6.1   Strategies for LBP Treatment

The most crucial issue in treating LBP is the urgency of pain reduction, with the goal of restoring structural normalcy in the end. To accomplish such results, several physiological processes have to be understood and taken into consideration.

3.6.2   The Nature of the Pain

The center of the LBP is where pain is sensed. Without pain, the suffering will disappear. But what is pain? Much research has been done and will continue to be done on pain. We have learned a lot about pain, but as an entity it remains quite illusive. Each new finding only makes pain appear more complicated. It is certain that pain involves psychological, neurological, muscular, biochemical and postural issues.

3.6.2.1   Neurological

3.6.2.1.1   Nociceptors

Nociceptors are specialized nerve endings that are capable of sensing thermal, chemical or mechanical stimulations. These receptors remain dormant in general. When these receptors are sensitized by chemicals released from tissue injury or from the inflammation process, they produce electrical signals that pass on to the brain through neural networks. The sensitizing agents include such chemicals as prostaglandins, nerve growth factor, hydronium ions and bradykinin. Traumatic injury and muscle overuse can sensitize the receptors. The brain interprets the signals it receives. Depending on the status of the mind, pain sensation can vary.

For chronic low back pain, continued mechanical stress may be responsible for sustaining the production of sensitizing agents. Therefore, reducing the mechanical stress should be the focus of therapy.

3.6.2.1.2   Golgi organs

At the termini of muscle spindles are sensors called Golgi organs which keep track of the tension on the muscles. If muscles are stretched passed the safety points of Golgi organs, reflective muscle contraction can occur. The role of Golgi organs in pain has not been studied enough since the stressed and shortened muscles that result in pain mostly do not exhibit any tissue damage.

3.6.2.1.3   Neurological disorders

Nerves themselves also can cause LBP. The pain is produced by lesions on the nerve itself. Such an incidence is rare among LBP sufferers and treatment should be focused on repairing the lesion. Pain relief through soft tissue work may be beneficial.

3.6.2.2   Musculo-Skeletal Issues

3.6.2.2.1   Muscle tension

Pain causes reflective contraction of muscles. Contracted muscle creates mechanical tension to the surrounding tissues, increases pressure on the vascular ducts and decreases blood infusion to the muscle tissues. Muscle contraction consumes energy. Reduced blood supply and increased energy usage can initiate local inflammatory processes which cause discomfort. Here, the muscle as the recipient of the nociceptor reaction becomes the initiator of pain stimulation. It is clear that a vicious circle can easily be formed.

3.6.2.2.2   Trigger points

Researchers have been studying ways to release involuntary muscle contraction. Simon and Travell[16] spent a lifetime studying this subject, focused on trigger points (TPs) that center at the mid-lines of muscles. TPs are focal, discrete and hyper-irritable and are associated with taut muscle bands or nodules that are painful upon palpation. They found that such points can be used to release the muscle spasm through injection of anesthetic solution or stretching after spraying with coolants. Many physicians still practice such techniques.[6]

There are arguments about whether the contraction of muscles constitutes "spasm." For this chapter, muscle contraction that is not under voluntary control is referred to as muscle spasm.

TPs are highly correlated with acupoints.[17] Properly done, inserting acupuncture needles (referred to as "dry needling" by medical acupuncturists) into these TPs can cause the contracted muscle to relax and the associated pain to subside. Conceptually, such a practice is very similar to the use of Ashi points in acupuncture. It is so simple that it is easy to understand why it has become so widespread among acupuncturists.[17]

3.6.2.2.3 Trigger point release

The mechanism which causes an activated TP to be inactivated, resulting in relaxation of contracted muscle, is still not clear. There are other ways that are used to inactivate muscle spasm, including isometric contraction, in which the contracted muscle is voluntarily contracted against a resistance. Others found that by compressing the shortened muscles further, the TPs can be released, resulting in muscle relaxation.[18,19] Done properly, all of these can achieve certain levels of release for the muscle spasm.

Common to all these processes is the proper stimulation to the motor nerve that is partially responsible for the contraction. In TCM, this is called moving the stagnated Qi. In contrast to TP injection, no chemical agents are utilized in these processes. This is possible because no biological damage is involved in the spasm. This is consistent with Headley's study,[4] which suggests that these spasming muscles are functionally shut down and electrically silent.

3.6.2.2.4 Hyperextension of muscles

Whenever one voluntary muscle contracts, called an agonist, another muscle or muscles, called an antagonist, will relax and elongate. Together, they achieve the intended mechanical action. When the contracted muscle relaxes, the role reverses. In TCM, this reflects the co-dependence feature of yin and yang on bodily functions. Both aspects are needed when considering treatment.

If a muscle is under sustained contraction, the antagonist muscle would become hyper-extended. The process is called reciprocal inhibition. The relaxed muscle is not resting as in a normal relaxed state; instead, it is under constant tension. Such a muscle will be sensitive to palpation. The combination of overextension at one muscle and persistent shortening at the opposing muscle may cause the supporting structure to deform. It is important to realize that the inhibition or hyper-extension does not involve apparent tissue injury or damage. Again, it is a Qi stagnation issue. Qi has to be reset.

3.6.2.2.5   Resetting hyper-extended muscles

Hyper-extended muscles also contain tender spots just like the shortened muscles. This is easily overlooked by patients and physicians. The patient would normally not be aware of the existence of these painful spots until palpated. Such muscles also become weakened with time, facilitating the continued contraction of the agonist muscles. For the low back, these weakened muscles include the abdominal muscle group, gluteus medius and gluteus maximus. The contraction–hyperextension combination facilitates the distortion of the postural configuration.[20] Common distortions with LBP are an uneven pelvis and/or internal rotation of the pelvis.

From a therapeutic perspective, it is beneficial to strengthen the hyper-extended muscles. In practice, it is difficult to achieve a good result as long as the agonist muscle remains contracted. In TCM, the muscle is in a "deficient" state, and needs to be strengthened or "tonified." If needling is chosen, the right technique is required such that it is applied slowly and gently. The best approach is to use moxibustion, as it is intrinsically a strengthening process. The moxibustion can be applied over the tender spots on these hyper-extended muscles. Any other therapeutic process that causes pain is counterproductive as it triggers the pain reflex and negates any potential benefit.

3.6.2.2.6   Pain referral

An active TP produces other noticeable painful spots that are at a distance from the TP. Such pain is called referred pain. The locations of referred pains are rarely related to the originating TP through dermatome or peripheral nerve distribution.[16]

Voluntary muscles work with bones to achieve intended activities. When a muscle contracts, it shortens and exerts tension on the attached bones. When the muscle is relaxed through voluntary control, the tension disappears. In the spasm situation, the muscle stays shortened, resisting voluntary control. Mechanically, such shortened muscle creates a constant tension at the attached bones. The tension is transmitted to muscles that are connected to these bones, some with positive and some with negative tension. The magnitude of tension can be estimated using Newtonian mechanics. As the result of continued tension, referred pain develops. When the muscle shortening is severe, the pain pattern can be very extensive over many parts of the body. The expanded distribution of referred pain is mostly absent in published studies, yet it is crucial in understanding and achieving a good result in therapy.

3.6.2.2.7   Propagation of stress

Stressed low back muscles exert tension on their attachments: the pelvis, spinal column and the ribs. The tension extends further outwards in all directions, and analysis is somewhat difficult. Numerous muscles in the low back and the pelvis are present and involved. The muscles are fairly irregular in shape and in distribution. The tension pattern from the interaction may be too complex to be delineated precisely.

However, simplification of the analysis can be achieved. All of the upper leg muscles are attached to the pelvis. When shortening of a low back muscle is severe, corresponding upper leg muscles will be stressed through the pelvic connection. As can be seen from Figure 3.5, such tension will be most apparent

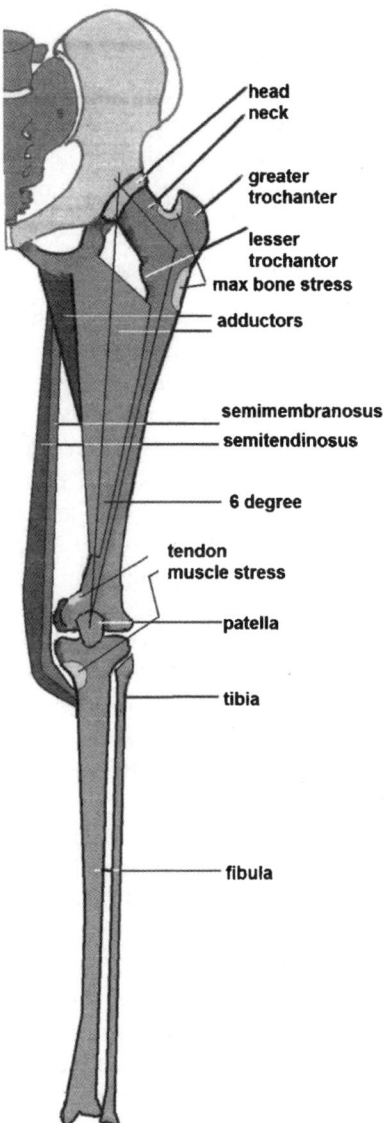

Figure 3.5 Structural support of body weight. Note that the line through the head of the femur and the center of the knee, called the mechanical axis, is about 3° from the gravitational vertical line. The light shaded spots take on the most mechanical stress.

at the muscle insertion sites, the lower medial part of the femur and the upper part of medial surface of the tibia. Such tension renders the knees prone to pain problems. These are the positions of acupoints Lv 7, Xi Guan, and Lv 8, Qu Quan, of the liver (Lv) meridians. The tension can also be transmitted to the lower legs and all the way down to the feet. Referred pain spots and therefore the TPs can be found in many places along the meridians of the Lv or GB. This is one of the reasons for the existence of the meridian theory.

The consequence of stress propagation is significant. Thus, even if the TPs on the low back are successfully released through intervention, the TPs from pain referral to the legs and feet also need be treated in order to have a sustained outcome. On the other hand, if the TPs on the legs and feet are treated, the LBP will be significantly improved through a chain reaction up the meridians.

3.6.2.2.8 Gravitational stress on the liver meridian

Lv 7 and Lv 8 are unique for LBP. There are many reasons that make these areas centers of LBP from a mechanical perspective. The configuration of the human body is intrinsically unstable. Unlike four-legged animals, human babies have to spend many months learning to use the proper muscles to stand up.

To make standing possible, adductor muscles are used partially as structural support muscles and are constantly under stress. To minimize this stress, the neck and the shaft of the femur forms an angle called angle of inclination. This angle varies among people between 90° and 160°, with an average of 125°. The balance of the body becomes more challenging mechanically for individuals with higher deviations from the average, and people with such issues are more prone to have back or knee problems. The head of the femur bears the total weight of the main body. Such weight is directed downwards at the greater trochanter, which makes the neck and the outer edge of the femur below the greater trochanter take on great mechanical stress (see Figure 3.6). Our bodies provide many large muscles that attach to the greater trochanter to provide adequate support.

In an ideal situation, the head of the femur should align vertically with the center of the knee joint so that the gravitational force will be maximally supported by the lower leg. The magnitude of muscle stress on the medial side of the femur will be minimized.

Any deviation from this ideal alignment increases the mechanical stress on the femur. On the lateral side, the pain will be most intense at the hip or the lateral side of the knee. This follows the distribution of the GB meridian. On the medial side, stress will be on the adductors and on the muscles that connect to the head of the fibula, semitendinosus or semimembranosus muscles. This distribution follows the Lv meridian. When the stress level becomes significant, many places along the meridians will be stressed. On the medial side of the femur, the adductors attach to the femur in a fanned out configuration, so the stress on the femoral shaft is quite diffuse. Patients may not be aware of the pain without palpation.

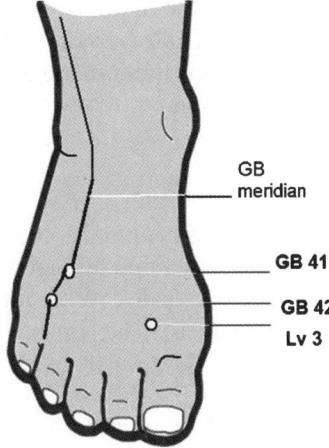

Figure 3.6 GB meridian and selected acupoints: GB 41, GB 42, Lv 3.

It should be noted that when the lateral side of the femur becomes shortened as indicated by lateral knee pain or hip pain, the adductors on the medial side will necessarily be hyper-extended and become tender upon palpation. Without dealing with this hyper-extension issue, improvement is hard to achieve.

Because of variations in an individual's body build, it is not always clear which femur of the two legs will be affected in LBP. Therefore, in therapy, both legs may have to be treated simultaneously.

3.6.2.2.9   Dermatome crossing

The tension around the medial side of the knee will likely increase the tension on the hind side of the calcaneus through the medial head of the gastrocnemius muscle. This either can cause tension at the underside of the foot causing plantar fasciitis or it can increase stress at the feet between the second and fifth toes. GB 41 or GB 42, Di Wu Hui, which are located between the fourth and fifth toes, will be sensitive to touch. Needling this acupoint will regulate the Qi along the meridian up to the low back. In TCM, this is the practice of using the GB meridian to treat Lv meridian disorders.

Application of a meridian pathway helps elucidate the pain distribution beyond what is predicted from a purely dermatome perspective. The LBP pain is not restricted to one or two dermatomes; rather, an extensive array of dermatomes are involved. The pain readily crosses the boundaries of several dermatomes.

3.6.2.2.10   Distribution of low back tension upwards

The tension from LBP on the ribs is normally not as prominent. All the ribs are tied together through inter-coastal tendons. The effect of tension from the low back is greatly dissipated. But closer to the spine, the tension is passed on along the para-spinal muscles up to the upper back, neck and shoulder. Since these are on the passage way of the triple burner meridian (TB) and the small intestine meridians (SI), acupoints on these meridians are useful in resolving LBP that is close to the spinal column.

3.6.2.2.11   Postural adaptation

Adaptation is a multifaceted phenomenon. In this discussion, it refers to structural alteration that accommodates and mitigates the tension. With short-term stress, the alteration recovers easily. With long-term stress, postural adaptation can become semi-permanent. The composition of muscles will also change, becoming more fibrotic and less elastic. At this stage, pain cannot be quickly eliminated.

Restoring the proper posture will be required in order to become pain free. This process may take considerable time and effort as it involves re-establishing the proper musculature. At this stage, the pain may be only mild, but since the body configuration is not correct, the mobility of the body is compromised. Much energy is consumed just to maintain body posture.

3.6.2.2.12   Breaking the interlocked cycle

When a soft tissue lesion in the low back occurs, mechanical stress increases. The stress propagates outwards, impacting the surrounding skeletal system and beyond, particularly into the lower limbs.

The situation is depicted in Figure 3.7. The body's response to the pain eventually develops and renders the body into the interlocked state with

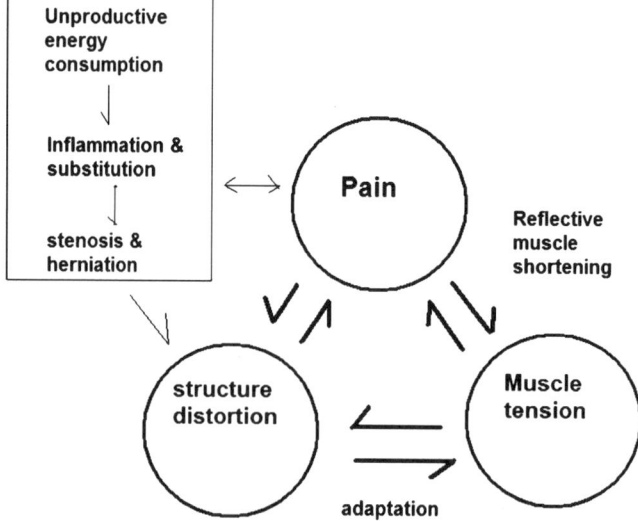

Figure 3.7   Chronic pain interlock cycle.

postural imbalance. If healing is left to the body's own healing process, the pain will not disappear. In therapy, the optimum result will be obtained if all three issues are addressed at the same time. Resolving a single issue makes the result tenuous.

3.6.2.2.13   Reciprocating disorders

Knee pain also causes the surrounding muscles in the upper and lower legs to become tense. Such tension easily becomes transmitted upwards and causes tension in the low back. Therefore, LBP is commonly associated with pain at or around the knees and vice versa.

3.6.2.2.14   Extra benefit of distal acupoint needling

It is well demonstrated that the proper treatment of TPs can result in release of muscle spasm. It is also well known that such release through either injection or acupuncture needles does not last for long. In Simon and Travell's work,[16] this observation was attributed to the presence of other TPs. However, the extent of development of such secondary TPs was not properly recognized in their work. Aggressive work directly on the TPs can also aggravate the muscle spasm through pain and irritation, causing great suffering for the patients.

Distal needling does not aggravate the muscle in question and will not cause further reactive spasming. Furthermore, it works through the chain of stressed muscles. It provides more extended relief to chains of muscles and tendons than the direct TP release approach. Hence the result will be longer lasting.

3.6.2.3   *Spinal Issues*

3.6.2.3.1   Stenosis

Some of LBP cases are related to spinal lesions. Naturally, one-sided LBP will cause a one-sided tension to the spinal column. When the stress is high, the spinal column will be significantly bent. This can result in two serious conditions. The spinal foramen affected will become narrow and squeeze the nerve passing through, and so-called nerve pinching will occur. This will likely cause severe LBP with tingling and numbness sensations traveling along the nerve distribution.

Disc herniation The second issue is that increased tension will increase the pressure on vertebral discs on one side and at the same time decrease disc tension on the opposite side. Such uneven pressure on the disc can be significant enough to result in disc herniation towards the side of less tension. The disc bulges and squeezes the adjacent nerve. The pain of such pinching will be on the opposite side of the stressed muscles.

To resolve both issues, first and foremost, the uneven tensions on both sides of the spinal column have to be eliminated. This can be achieved through general LBP treatment as specified in the treatment procedure session.

3.6.2.3.2 Application of anti-inflammatory agents – NSAIDs, steroids or ice

In a traumatic injury, a classical inflammatory reaction occurs. A chain of chemical processes takes place, resulting in increased blood circulation and enhanced tissue perfusion of the injured tissues. The increased blood flow speeds up the metabolic rate and tissue healing. It also brings along unproductive consequences through swelling of the tissue. Swelling causes physical tension, resulting in increased pain which can significantly immobilize the musculature, hindering the healing process.

Icing A popular practice in dealing with pain issues is icing. Icing is often claimed to reduce inflammation, but this claim can only be valid for a brief period. Cold certainly reduces blood flow, which stops inflammation, but it is not in synchrony with the healing process. The accumulated tissue fluid in the swollen sites needs be removed through vascular conduits. Icing causes vascular constriction and inhibits the removal of the accumulated fluid prolonging the pain. Icing should be used carefully in painful conditions.

Medications It is also common to use anti-inflammatory medications. To the extent that an inflammatory process is causing the pain, this approach can be beneficial. Medications do not provide additional benefit in resolving the muscle tension. For chronic pain, continued use is complicated by the side effects. Steroids such as prednisone are very powerful agents, and conservative application can help some acute LBP cases. Prolonged use creates extensive degenerative consequences.

On the other hand, application of heat to the painful site is almost always a good approach. It enhances circulation in addition to relaxing muscle tension. It effectively provides an anti-inflammatory result. When heating is coupled with mechanical movement such as stretching, the benefit is greatly enhanced towards eventual healing. Contrary to popular belief, the combination of heat plus moving provides the best results in reducing swelling.

Heating can be easily achieved by heating pads or infrared devices which provide deeper penetrating heat. It is obvious that the warming process has its limitations. When the pain is extensive, it is difficult to implement effective warming. Medications may be beneficial in such situations.

3.6.2.4   Emotional

3.6.2.4.1 Brain pain center

Because the brain is actively involved in pain processing, the perception of pain is subject to the psychological and emotional status of the person. This is why the placebo effect plays an important role in medicine. There are at least two aspects to the influences of the brain's participation, positive and negative.

Positive emotions elevate the pain threshold and reduce pain perception. This also reduces the overall tension of the body, which is very beneficial to the reduction of pain. Pain itself increases tension, as does the daily stress of modern living.

Modern medicines take advantage of the effects of emotion on pain. Numerous medicines such as antidepressants are designed to influence a patient's mood and thus reduce pain perception.

3.6.2.4.2 Acupuncture for the mind

Many acupoints are known to relax patients. These include Du 20, Bai Hui, which is at the top of the head, and ear point Shen Men, mind gate as shown in Figure 3.8. These acupoints help calm the mind in treating pain.

There are other acumoxa points that have strong relaxation effects; these include LI 4, He Gu, and Lv 3, Tai Chong. These points have strong effects on promoting endorphin production. Often people attribute the effects of acupuncture to the production of endorphin but this is an over-simplification. Suffice it to say, endorphin notwithstanding, improper acupuncture technique on these points can make a patient's body tense up and elevate blood pressure. The result is just the opposite effect of endorphin production.

3.6.3 Summary of Scientific Principles for LBP Treatment

From the above discussions, it should be clear that LBP involves a very complex set of physiological events. A simplistic approach based on the TP principle falls well short of being effective. Such an approach cannot be expected to provide consistent good outcomes. The approach described in the Nei Jing appears to circumvent these pitfalls and should be seriously considered in order to improve the clinical efficacy of LBP treatment.

3.7 A Rational Approach to LBP Treatment Based on the Nei Jing

This may be a surprise to many acupuncturists. In spite of the Nei Jing's detailed discussion on LBP treatment, it does not specify the acupoints to be used. Most of the specific points were suggested by Wang Bing of the Tang Dynasty, who was credited with the revision of the Nei Jing. The differential criteria for meridian selection in the Nei Jing are also difficult to follow in the clinic. To overcome these difficulties and provide specific instructions in the spirit of the Nei Jing, the guide in this section was devised.

The guide presented below offers steps that are easy to apply. It adheres to the Nei Jing's principle of distal needling for LBP. It also expands the concept, making use of the upper limbs as well. Furthermore, the deficiency issue which is associated with the treatment of hyper-extended muscles is incorporated into the guide.

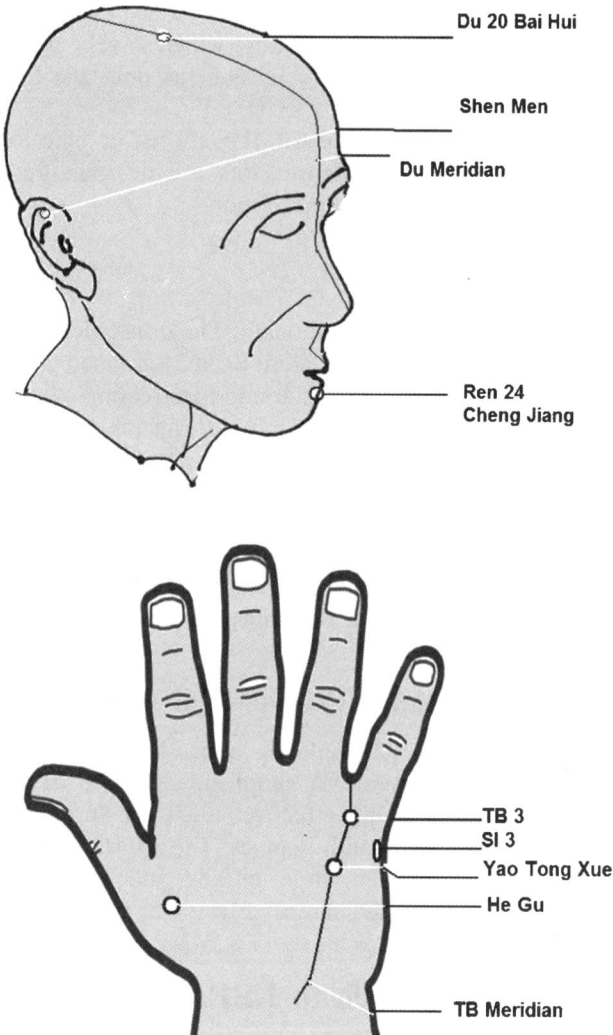

Figure 3.8   Selected acupoints on the hand and head. Also shown are the TB
meridian on the hand and Du meridian on the head. Selected acupoints:
Ear point, Shen Men, Ren 24, Du 20, TB 3, SI 3, LI 4 (He Gu) and Yao
Tong Xue, an extra- meridian acupoint.

3.7.1   Mapping the Low Back for Point Selection

This approach involves dividing the low back into several anatomical zones
and selects the acupoints based on which zone the pain is felt in. This simplifies
the selection considerably.

3.7.2 Treatment Guide for LBP

3.7.2.1 Relax the Patient

First, needle LI 4, He Gu, and Lv 3, Tai Chong. Choose one from each pair. These points help relax the entire body.

3.7.2.2 Identify the Pain Zone

Instruct the patients to lie on their back and ask them to twist or stretch the low back gently. This allows the patient to identify the specific location of the pain. Confirm the diagnosis by palpation of the location. A tense muscle or taut band can usually be felt. Determine the zone involved.

3.7.2.3 Select Acupoints According to the Zone (See Figure 3.9)

- a. Zone S (for spine), along the spinal column. For pain concentrated in Zone S, needle acupoints on the Du or Ren meridian. For example, use Du 20, Bai Hui, or Ren 24, Cheng Jiang.
- b. Zone P (para-spine), parallel to the spinal column. For zone P pain, needle SI 3, Hou Xi, on the same side as the pain. This is the master of the Du meridian, closely linked to spinal column function.
- c. Zone I, pain on the inner line of the UB meridian on the back, about 3–4 cm from the spinal column. Stress from this zone travels up. Use the TB meridian to treat it, e.g., Yao Tong Xue, an empirical acupoint for low back pain.
- d. Zone O (outer UB line), pain on the outer line of the UB meridian on the back, about 6–7 cm from the spine's center line. Use GB acupoints on the feet, such as GB 42, Di Wu Hui.
- e. Zone G (for gluteus) – pain and soreness in the gluteus. Needle GB 34, Yang Ling Quan.

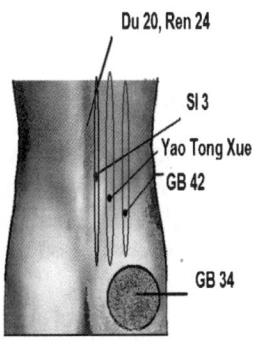

Figure 3.9 Acupoint selection guide showing an example of point selection according to the zones.

- f. Zone L (for liver meridian) – pain along the liver meridian on the leg. Needle or moxibustion on Lv 7 and Lv 8. Moxa tight and tender areas above and below these points. These two acupoints are also good for other cases of LBP.

Once the site of pain has been confirmed, apply needles according to the guide. Needling the same side as the pain is more important than the opposite side, but often both sides are needed.

3.7.2.4   Application of the Proposed Guide

Many people, especially females, experience tenderness to palpation around Lv 7, Xi Guan, and Lv 8, Qu Quan. Many also have tenderness extending up along the mid-lines of the inner thighs. When this pain is treated, moxibustion with or without needles is preferred. The muscles at Lv 8 and the area above it tend to be hyper-extended. Moxibustion is the best choice for therapy.

The acupoints specified in the guide can be substituted with other acupoints from the same meridian. Thus, in Zone I, TB 4 or TB 5 can be used instead. For Zone G, GB 38, Yang Fu, GB 39, Jue Gu, or GB 41, Zu LIN QI, can be used instead. The effectiveness can vary somewhat.

Zone P, because of its location, is particularly useful for people with spinal stenosis or disc herniation. It is also very helpful for people who have back pain after back surgery. For any surgery, moxibustion over the surgical scar is often crucial for continued success in treatment.

3.7.2.5   Shifting of the Pain Location

Patients often report shifting of pain as the treatment gets under way. This situation is indicative of a wide distribution of pain on the body. Under such circumstances, people can only report the most severe pain location initially. Additional needles have to be applied based on the shifting situation. Often, therapy involves all of the zones.

3.7.2.6   Supplemental Treatment

In the event the result is not totally satisfactory, one can add UB 40, Wei Zhong, or explore Ashi locations. Adding heat therapy or electrical stimulation with acupuncture on the low back may also enhance the result. One reason for doing this is to address the concerns of certain patients who believe that the low back must be needled.

In severe situations, the pelvis can be significantly distorted. This is called tilt pelvis. The Dai meridian may be involved. Palpate GB 28 for tensed muscle. Apply acupuncture or moxibustion on this acupoint if painful.

3.7.2.7   Dao Qi, Moving Qi

Generating and moving proper Qi to the target site is a delicate skill that has to be harnessed. This is especially critical in distal point acupuncture. The effectiveness of therapy depends on it. Often the therapy is not satisfactory, not because of the point selection, but rather it fails due to insufficient Qi movement.

Once a needle has been inserted, there are many different ways to manipulate the needle to assure that proper Qi is obtained. These include tapping or scraping the needle handle, lifting and reinserting the needle and warming the needle. In a difficult case, multiple needles may be required to enhance Qi. These are the standard techniques learned from acupuncture education.

3.7.2.8   Advanced Technique: Combination of Needling with Stretching

After needles have been inserted in selected locations on the feet, hands or legs, the patient is instructed to stretch the low back in a way that would have provoked pain prior to the needle insertion. Manipulate the needles more if not enough Qi is obtained. With correct diagnosis, the patient will instantly experience an increase in the range of motion and a reduction of pain and stiffness with each stretching.

This technique provides the feedback of "De Qi", obtaining Qi, for the acupuncturist. Although there are many known signs and signals that indicate the possibility of "De Qi," none is reliable. By watching the patient's stretching, whether or not proper Qi has been obtained can clearly be gauged. Adjustments can be made accordingly.

Intrinsic to this process is stretching the stressed muscles, a very crucial step in muscle rehabilitation. Most importantly, the stretching is accomplished without provoking pain and causing backlash. The result is therefore much more pronounced in comparison with the forced stretching implemented in physical rehabilitation protocols.

3.8   Conclusion

LBP is a medical issue that costs society enormously in terms of suffering, money and quality of life. It is a major contributor to immobility in the elderly through its influence on the knees. Current medical practices have yet to provide a satisfactory way of handling this issue.

The ineffectiveness of treating LBP is due to the narrow focus of treatment. To improve the efficacy, the treatment should include other parts of the body, especially the legs and the feet. Such an emphasis was clearly exemplified in Huang Di Nei Jing.

One of the most thorough clinical studies on the effectiveness of acupuncture for LBP to date was conducted in Germany.[17] Acupuncture was clearly shown to be much more effective than the combination of medicine and physical therapy. The study also discovered a puzzling conclusion, that sham acupuncture performed better than "traditional" acupuncture. This challenges the validity of current acupuncture education.

As argued in this chapter, the modern approach of "traditional" acupuncture to LBP is flawed. It is not the acupuncture approach promulgated in the Nei Jing. Instead, the modern practice is in danger of becoming a modified TP therapy. The text books used in educational institutions further enforce such practices. This approach cannot be expected to fulfill high expectations. As the result, a seasoned acupuncturist has to develop his or her own style of "real-world acupuncture" to achieve better results. This sentiment is reflected in a recent article.[21] This situation is not acceptable.

A systematic guide is presented in this chapter to remedy this status. It is simple to implement. It is based on principles laid out in the Nei Jing with some expansion. The logic behind the approach is explained in terms of human physiology and science. The discussion can also be applied to the treatment of other disorders. The effectiveness for LBP treatment has been borne out in clinics. Hopefully, future studies on the efficacy of acupuncture in treating LBP will not repeat the same mistakes.

References

1. R. A. Deyo, S. K. Mirza and B. I. Martin, *Spine*, 2006, **31**, 2724.
2. T. S. Carey, A. T. Evans, N. M. Hadler, G. Lieberman, W. D. Kalsbeek and A. M. Jackman, *Spine*, 1996, **21**, 339.
3. K. Von Korff and K. Saunders, *Spine*, 1996, **21**, 2833.
4. B. J. Headley, *Phys. Ther. Forum*, 1993, **1**, 24.
5. N. Nathan, in *Early Chinese Texts: a Bibliographical Guide*, ed. M. Loewe, University of California Press, Berkeley, CA, 1993, p. 196.
6. D. J. Alvarez and P.G. Rockwell, *Am. Fam. Physician*, 2002, **65**, 653.
7. D. C. Cherkin, R. A. Deyo, K. Wheeler and M. A. Ciol, *Arthritis Rheum.*, 1994, **37**, 15.
8. G. F. Kominski, K. C. Heslin, H. l. Morgenstern, E. L. Hurwitz and P. I. Harber, *Med. Care*, 2005, **43**, 428.
9. D.B. Helmann and J.B. Imboden Jr, in *Current Medical Diagnosis and Treatment 2010*, 49th edn., ed. S. J. McPhee and M. A. Papadakis, McGraw-Hill Medical, New York, 2010, p. 738.
10. B. H. Pomeranz, *APS J.*, 1994, **3**(2), 96.
11. A. Hirsch, *Allgemeine Deutsche Biographie*, Historical Commission of the Bavarian Academy of Sciences, Duncker & Humblot, Leipzig, 1879, Vol. 9, p. 57.
12. X. N. Cheng, *Chinese Acupuncture and Moxibustion*, Foreign Language Press, Beijing, 1987, p. 437.

13. B. M. Berman, H. M. Langevin, C. M. Witt and R. Dubner, *N. Engl. J. Med.*, 2010, **363**, 454.
14. M. Belgrade, *APS J.*, 1994, **3**(2), 73.
15. B. Brinkhaus, C. M. Witt, S. Jena, K. Linde, A. Streng, S. Wagenpfeil, D. Irnich, H.-U. Walther, D. Melchart and S. N. Willich, *Arch. Intern. Med.*, 2006, **166**, 450.
16. D. G. Simon and J. G. Travell, *Myofascial Pain and Dysfunction. The Trigger Point Manual*, 2nd edn., Lippincott, Williams & Wilkins, New York, 1999, p. 19.
17. M. Haake, H.-H. Müller, C. Schade-Brittinger, H. D. Basler, H. Schäfer, C. Maier, H. Endres, H. G. Endres, H. J. Trampisch and A. Molsberger, *Arch. Intern. Med.*, 2007, **167**, 1892.
18. L. H. Jones, *Strain and Counter Strain*, American Academy of Osteopathy, Newark, OH, 1981.
19. W. A. Kuchera and M. L. Kuchera, *Osteopathic Principles in Practice*, 2nd edn, Greydon Press, Columbus, OH, 1994.
20. V. Janda, *J. Manual Med.*, 1991, **6**, 136.
21. M. D. Bauer, *Acupuncture Today*, 2012, **13**, 1.

CHAPTER 4

Understanding Ch'i: The Life-Force Energy that Determines Vitality, Health and Wellbeing

MARCIA R. BABCOCK

Anatomy and Cell Biology; MBA; T'ai Chi Ch'uan Instructor, Reiki Master Teacher, Longmont Recreation Center, Longmont CO 80501, USA
E-mail: marcia@bodyenergetic.com

4.1 The Origin of the Universe: the Cosmology of Ch'i and Its Manifest Forms as Yin and Yang

In man's attempt to understand the universe and his place in it, he hypothesized that the origins of all there is came from a great formless void, the Wuji. This great formless void is ch'i[1] in a dispersed state. Ch'i is the undifferentiated material from which all things are made. According to legend, in the beginning there was no thing, no movement; this formless silence is called the Wuji. Then, movement occurred and in the same way that the Big Bang gave rise to the fundamental particles of the Universe in a huge expansion of matter and anti-matter, light and sound, the Wuji gave rise to two fundamental states of Being, the states the Chinese call yin and yang. The yin and yang are complementary opposites that form a whole. These states create light and darkness, soft and hard, Earth and the Heavens, female and male. From the Wuji, the yin and yang came into Being and hence all manifest reality was created. If all things of the Universe are made from the same ch'i or life

RSC Drug Discovery Series No. 31
Traditional Chinese Medicine: Scientific Basis for Its Use
Edited by James David Adams, Jr. and Eric J. Lien
© The Royal Society of Chemistry 2013
Published by the Royal Society of Chemistry, www.rsc.org

force, then all physical aspects of the Universe are one great body.[2,3] Hence ch'i forms the basis for the philosophical statement "We are all One."

My T'ai Chi Ch'uan teacher, Xu, Tingsen, explained Chinese cosmology to me using the analogy of two fish. This is represented in the Chinese yin/yang symbol or Taijitu that looks like a white fish with a black eye circling a black fish with a white eye. In the beginning, there was no thing, that is a blank circle. The creation of complementary aspects of ch'i as yin and yang is expressed as a symbol of the two fish, a white fish and a black fish. Within each fish developed an eye of the opposite color. Each eye symbolizes that each aspect is transforming into the other, yin is always becoming yang and yang is always becoming yin. In eternity, the white fish with a black eye and the black fish with the white eye swim around each other and movement is actually the eternal transformation of one into the other.[4] Examples of yin and yang are given in Table 4.1.

It is believed that this energy flows in a balanced manner in healthy bodies, and in disease states the flow is disrupted or stagnant. This life force can be cultivated through righteous thinking and action.[5]

In practical terms, this means that life force or vital energy can be enhanced through physical, mental and emotional training. These concepts form the basis of several techniques to develop, maintain or restore health and wellbeing with healthy ch'i flow. These techniques include traditional Chinese medicine (TCM) methods of acupuncture, herbology and massage (Tui na), movement forms that include Chi Kung, internal martial arts such at T'ai Chi Ch'uan, external martial arts such as Gung Fu and mirroring the harmony of ch'i in the Universe through the placement of objects in one's environment as described in the geomancy practices of I Ching and Feng Shui.

Although these are widespread practices in the East and gaining prominence in the West, there is distrust in conventional Western society of these concepts owing to the lack of scientific evidence to support their use. Despite this fact,

Table 4.1 Examples of yin and yang.

Examples of yin	Examples of yang
Slow	Fast
Soft	Hard
Yielding	Solid
Diffuse	Focused
Cold	Hot
Wet	Dry
Passive	Aggressive
Water	Fire
Earth	Sky
Moon	Sun
Feminine	Masculine
Night	Day
Broken line (I Ching)	Solid line (I Ching)

these practices have proven invaluable as tools to promote healing and wellbeing for millions of people. Maybe the lack of scientific evidence is a reflection of the lack of appropriate tools of measurement for the phenomenon. Evidence comes not from the direct measurement of ch'i, it comes from observing the effects of ch'i building or restoring practices.

4.2   Ch'i as Life Force Is Described in Many Cultures

This life force is believed to permeate all matter and the flow of this energy creates a combination of matter and energy that we recognize as manifested reality in what would be conventionally called "inert" as well as "living" matter. Ch'i is present in the Five Elements or Phases of Earth, Air, Fire, Water and Ether that make up all manifested reality.[4,6]

The concept of ch'i is present in some way in many Earth cultures. It would be easy to say that these were primitive peoples trying to understand the nature of reality; however, its pervasive extent suggests that a universal phenomenon was present and observed. Each culture has a healing system based on something like ch'i and many have components that incorporate treatment, exercise movement, diet and lifestyle (Table 4.2).

4.3   Perceiving and Measuring Ch'i

If ch'i is the Vital Energy of the Universe, how would it be measured? In general, the perception of ch'i is subjective, as a sensation most often described as heat or cold or vibration that is felt most easily in the palms of the hands. The ability to perceive ch'i is a skill that can be learned and cultivated. This cultivation is the basis of a variety of health-building modalities such as T'ai

Table 4.2   Universal healing systems based on ch'i.

Country/culture	Energy concept	Healing modality
Japan	Ki	Reiki[7–9]
India/Vedic	Prana	Ayurveda, Marma Point therapy[10,11]
Ancient Egyptians	Ka	Ka Shen Sekhem[12]
Ancient Greeks	Pneuma	Greek or Ionian medicine, similar to Ayruveda and Chinese medicine[13]
Native Americans	Great spirit	Dance, Sweat Lodges, Shamanism[14]
Africa	Ashe	Akan, Yoruba[15]
Hawaii	Ha	Huna[16]
China	Ch'i	T'ai Chi Ch'uan, Ch'i Kung, acupuncture, herbology[4,6]
Tibetan Buddhism	Lung	Gso-wa Rig-pa[17]
Western philosophy	Vital energy	Subtle energy and vibrational therapies[18,19]

Chi Ch'uan and Ch'i-kung. For practitioners of acupuncture, skill is developed in reading the "pulses" of the body and the appearance of the tongue,[4,6] very similar to Ayurvedic practitioners' reliance on these methods for diagnosis.[10,11] Owing to the inability to develop an objective method comparable to Western formats for measuring ch'i, conventional medicine has been unable to reconcile the benefits of ch'i-building modalities with conventional methods of medical treatment. It is possible, however, to document the effects of these alternative healing systems on a variety of well-known parameters such as blood pressure, blood flow, physical strength, agility and balance. Research efforts are ongoing to compare and contrast conventional Western medicine with TCM.

4.4 Historical Measurement of Vital Energy Using Scientific Methods

That being said, others have used a variety of techniques to measure the vital energy of plants, animals and the human body. These techniques are a marriage of technology and heuristic methods. These may be useful here in that the results reflect the basic theories of yin and yang found in Chinese culture.

Vital[18] energy was described in Western literature as early as 500 BC by Pythagoreans as a luminous body that could be perceived and could be cultivated for various uses including healing from disease. Early twelfth century scholars saw that humans have the ability to affect each other positively or negatively through an interaction with an energy field. Paracelsus recognized this force in the Middle Ages. Helmont in the 1800s visualized a vital energy that pervades all Nature.[18] A biological electromagnetic or "odic field" was studied by Von Reichenbach in the mid-1800s.[20] The properties of this field were both particulate and wave-like. His studies in the human body discerned a polarity along the major axes of the body, with the left side the negative pole and the right side the positive pole. These terms reflect the yin and yang described by the Chinese almost 2000 years earlier.

4.5 Modern Western Doctors Interested in the Energy Field Surrounding Human Beings

Throughout the twentieth century, doctors from a variety of disciplines have created devices to measure the energy field of the human body.[18] In 1911, Dr. William Kilner used colored filters to detect three distinct zones that he called the aura.[21] In 1948, Dr. Wilhelm Reich used various electronic and medical instruments including microscopy to detect "orgone" energy flows in living organisms. In the 1950s, Dr. George De La Warr and Dr. Ruth Drown built instruments to detect energy radiating from living tissues. In the resulting photographs, tumors could be distinguished from the surrounding healthy tissue.[22]

Objective measurement of the frequency, amplitude and latency of human energy field transmissions was determined by Dr Valerie Hunt, a researcher at UCLA. She initially used electromyographic (EMG) recording electrodes to measure skilled and pathological muscular activity. She found that during movement with an associated state of altered consciousness, subjects displayed unusual EMG recordings. Later, using ordinary bipolar surface electrodes on the skin, she detected increased energy flows in response to alternative healing techniques. Further studies showed that mathematical analysis of these and similar experiments showed consistent wave forms and frequencies that corresponded with subjective reports of increased muscular performance and "energy flow." Dr Hunt learned that areas corresponding to Vedic Chakras experienced the greatest changes in measured frequency, amplitude and latency.[18,23]

A study in 2008 used functional magnetic resonance imaging (fMRI) and positron emission tomography (PET) to investigate pain relief due to acupuncture. The results showed acupuncture-related changes in the regions of the brain governing sensory and affective aspects of pain perception. This study provided some evidence of the neural mechanisms of acupuncture.[24]

4.6   Comparison of TCM and Ayurveda

The TCM system has many similarities to the Indian Vedic system of Ayurveda. Both systems recognize subtle energy bodies of the human bodies. Both systems employ herbs, energy movement, massage and physical movement to bring balance to energy flows and enhance health, vitality and personal growth through the cultivation of ch'i (TCM, Chinese) or Prana (Ayurveda, Vedic or Indian)[11,25] (Table 4.3).

4.7   Cultivation of Ch'i: Improving the Quality and Strength of Ch'i Flows

Flow of ch'i can be directed either through movement practices or through interventions such as acupuncture, moxibustion, herbal medicines or massage.

Table 4.3   Comparison of TCM and Ayurveda.

System	Chinese	Indian
Exercise	Martial arts, including Tai Ch'i and Ch'i-kung	Yoga
Energy organization	Meridians and acupuncture points	Nadis and Marma points
Diagnosis method	Pulses taken at the wrist, observation of the tongue	Pulses taken at the wrist, observation of the tongue

Movement practices that focus on the correct and efficient execution of movements are a matter of purposeful intention during the action. Purposeful movement charges the practitioner with improved ch'i flow during the action. Repetition of the movements creates a personal practice wherein the cultivation of ch'i is achieved. This cultivation can be witnessed over time as increased health, wellbeing, recovery from disease, increased immune function, lower heart rate, achieving optimal body weight, improved mood, increased concentration and improved cognitive abilities.[26] So although ch'i itself is not measured directly in this case, the effects and positive benefits have been widely documented in both the West and East.

4.8   Energy Healing Practices Across the World

The concept of ch'i is pervasive across the globe and likewise so are healing practices that take advantage of the ability to manipulate ch'i flows as described in Table 4.4.

4.9   Chinese Practices to Balance and Facilitate Ch'i Flow

4.9.1   Traditional Chinese Medicine

TCM is a broad range of medicine practices sharing common theoretical concepts which were developed in China and are based on a tradition of more than 2000 years, including various forms of acupuncture, herbal medicine, massage (Tui na), exercise/movement and dietary therapy.

The doctrines of Chinese medicine are rooted in books such as the *Yellow Emperor's Inner Canon* and the *Treatise on Cold Damage*, and also in cosmological theories such as yin and yang and the five elements or five phases. Starting in the 1950s, these precepts were modernized in the People's Republic of China so as to integrate many anatomical and pathological aspects of scientific medicine. However, many of its precepts, including the model of the body or the concept of disease, are not supported by Western medicine.

TCM's view of the body places little emphasis on anatomical structures, and is mainly concerned with the identification of functional entities which regulate

Table 4.4   Energy healing practices.

Method	Modality	Origin
Acupuncture	Needles	Chinese
Reiki	Touch or at a distance	Japanese
Marma point therapy	Touch	Ayurvedic/Indian
Healing touch	Touch	American
Yuen technique	Body as a biocomputer	Shaolin/Chinese[27]

digestion, breathing, aging, etc. While health is perceived as the harmonious interaction of these entities and the outside world, disease is interpreted as a disharmony in interaction. TCM diagnosis consists in tracing symptoms of the basic disharmonies of the body through patient reporting, reading the pulse and inspecting the tongue.[6,28]

This said, a number of clinical trials have been performed that revealed acupuncture to be efficacious in inducing analgesia, protecting the body against infection and regulating various physiological functions. In the case of analgesia, acupuncture has been shown to be as efficacious as morphine in some cases. Acupuncture is a leading long-term treatment due to its lack of side effects such as dependency that is evidenced with leading pharmaceutical agents.[29,30]

4.9.2   What Can Happen to the Ch'i?

Ch'i can be in excess or deficient. Excess ch'i occurs when flow is blocked so that the flow is stagnant. Treatment is based on dispersing the blockage so that the ch'i flow is re-established and ch'i can flow in a balanced manner in a meridian.

Ch'i can be deficient, and one can squander or lose ch'i through one of the excesses such as emotions, sex/reproductive activity, exercise/taxation of the body or severe illness. Treatment is to engender or increase ch'i through herbs, meditation or energetic practices such as Ch'i Kung. Ch'i can also become deficient through blockage of ch'i that causes downstream deficiency at another location within the body. The treatment practice is to disperse the blockage as described above.[4,6,28]

4.9.3   TCM: Acupuncture and Moxibustion

Acupuncture is a type of TCM that involves the insertion of needles into the skin, subcutaneous tissue and muscles at specific acupuncture points to facilitate the proper balance of ch'i and ch'i flow through channels called *meridians*. Although Western methods have not identified anatomical or physiological correlates for ch'i meridians and acupuncture points, Western research methods have found that acupuncture is effective for relief from pain and nausea. When administered by well-trained practitioners using sterile needles, acupuncture has a very low risk of serious adverse effects. Acupuncture is often performed in concert with moxibustion, a treatment that involves burning mugwort on or near the skin of an acupuncture point.[4,6,28]

4.9.4   TCM: Herbal Medicine

Herbs have yin and yang natures that affect ch'i. Herbs are chosen for a decoction/prescription according to their yin and yang properties. For example, the most commonly used herb is RenShen (ginseng, *Panax ginseng*). It is the strongest herb for engendering ch'i, and is considered the ch'i tonifier.[6]

In TCM, there are roughly 13,000 medicinals used and over 100,000 medicinal recipes recorded. Plant, animal and mineral elements and extracts are used. A prescription typically contains a decoction of 9–18 substances.

4.9.5 TCM: Massage (Tui Na)

Tui na is a form of massage similar to acupressure. Oriental massage is typically administered with the patient fully clothed, without the application of grease or oils. The massage involves thumb presses, rubbing, percussion and stretches. The movements tonify, engender and increase ch'i in an area or disperse stagnant ch'i out of an area where it has been blocked.[6]

4.10 Movement/Exercise

The use of ch'i to generate power is cultivated in the *internal martial arts* (neija gong). The three main martial arts of the Wudang school are T'ai Chi Ch'uan, Hsing I and Pa Kua Chang. These arts include realistic and sublime fighting skills that are developed through the application of proper body mechanics and the infusion of internal ch'i through the linking of body, mind, emotions and spirit. To advance truly, assiduous practice leads the practitioner to sense ch'i moving through the body as a physical sensation akin to biofeedback. Once this level of sensitivity has been attained, the practitioner advances from a state of learning how to move correctly, breathe correctly and stand correctly to a state of directly sensing energetic ch'i flow allowing immediate and accurate self-correction. The practitioner can now "feel" whether the moves are being executed with the greatest efficiency and total body balance. This allows the practitioner to move to an advanced state of martial arts practice.[31,32] The following classical T'ai Chi expression sums this up: "Four ounces defeat a thousand pounds."

4.10.1 T'ai Chi Ch'uan

T'ai Chi Ch'uan employs the *Grand Terminus*, a subtle system of Chinese philosophy. When the Grand Terminus acts, it forms *yang*. When activity reaches the extreme point, it becomes inactivity and forms *yin*. Extreme inactivity returns to become activity. Thus, there is always movement from *yang to yin* and from *yin to yang* in all matters of the Universe.

In the practice of T'ai Chi Ch'uan, one of the essential points is distinguishing full and empty. Fullness is weight bearing, empty is without weight on the limb. During the form set, the practitioner is always moving, from full to empty and from empty to full. Like the Grand Terminus, there is never a time when the full and empty are completely separated or equally balanced because this would stop the movement of the Grand Terminus.

Fullness contains emptiness and emptiness contains fullness. This is expressed as continuous movement in the practice of T'ai Chi Ch'uan.

As a practice, T'ai Chi Ch'uan is an ancient Chinese martial art that combines fluid movements of the upper and lower body to produce a subtle and powerful full-body exercise. The movements are uninterrupted like a flowing stream. They are performed by the coordination of the hands, eyes, trunk and limbs with the legs as a base and the waist as an axis. Each part of the body is in constant motion and the mind is tranquil and alert. Most common styles are Yang Styles that are derived from Chen Style, Sun Style and Wu/Hao Style. T'ai Chi specialties include Push Hands and Weaponry (Staff, Sword and Saber).[33–37]

Regular practice of T'ai Chi Ch'uan enables one to develop coordination, strength and flexibility, reduce stress and promote inner strength. T'ai Chi Ch'uan has proven effects on pulmonary function, cardiovascular health (especially blood pressure) and balance. It was the first Chinese self-healing art to appear in the *Journal of the American Medical Association* and other peer-reviewed journals.[26,38]

4.10.2   Pa Kua Chang

Pa Kua Chang trains the body and hands (Chang) to move in circular patterns (Pa Kua). It is generally considered the most mysterious of the inner martial arts. The founder of Pa Kua Chang was nursed back to health by Taoist priests who taught him their healing and martial art. Pa Kua Chang is a superb art for cultivating flexibility and rooted strength. When applied to self-defense, the Pa Kua boxer imagines that the attacker is the center of his or her circle. The boxer whirls around the attacker with a combination of ingenious locks, throws and strikes. Pa Kua Chang literally means "eight trigram palm" referring to the trigrams of the I Ching, one of the canons of Taoism.[31,32]

4.10.3   Hsing I

Hsing I (Body Mind Boxing) is based on five linear strikes, each related to one of the Five Elements of Chinese philosophy:

- Splitting moves like an ax chopping wood (Metal Element) and benefits the lungs.
- Crushing darts out like a wooden arrow (Wood Element) and benefits the liver.
- Drilling coils like a meandering stream (Water Element) and stimulates the kidneys.
- Pounding explodes like a canon ball (Fire Element) and is related to the heart.
- Crossing trains diagonal footwork (Earth Element) and benefits the spleen.

There are 12 Animal Movements: Dragon, Tiger, Monkey, Chicken, Sparrow, Hawk, Lizard, Horse, Phoenix, Snake, Eagle and Bear.

As a martial art, Hsing I Ch'uan is the opposite of and complement to T'ai Chi Ch'uan and Pa Kua Ch'uan. The practitioner never retreats. He or she drills into the opponent, defending and counterattacking at the same time.[31,32]

4.11 Ch'i Kung

Ch'i Kung is a practice involving coordinated breathing, movement and awareness, traditionally viewed as a practice to cultivate and balance ch'i. With roots in traditional Chinese medicine, philosophy and martial arts, Ch'i Kung is now practiced worldwide for exercise, healing, meditation and training for martial arts. Typically, a Ch'i Kung practice involves rhythmic breathing coordinated with slow stylized movement, a calm mindful state and visualization of guiding ch'i as well as external ch'i healing.[32,39]

4.12 Geomancy

4.12.1 Geomancy: The application of ch'i to divination

The I Ching, or The Book of Changes, is an attempt to use balance of yin and yang energies of the Universe to answer questions of mundane and moral context. Thus, the I Ching is a method for asking for inspiration from the universal energies of creation to guide man in his quest for understanding and growth through the resolution of human life experience.

In the I Ching, yin and yang are depicted with broken (yin) or solid (yang) lines. The lines are arranged in columns of six trigrams called a hexagram. Each trigram is derived by choosing yarrow stalks or throwing three bronze Chinese coins to determine the combination of yin and yang aspects of each trigram. The yin side of the coin is valued 2 and the yang side of the coin is valued 3. Thus, each trigram contains a series of three yin or yang components, so that three coins all showing broken lines would be an "old" yin trigram and three coins showing all solid lines would be an "old" yang trigram. A trigram with two yang components and one yin component would yield a young yin line and value 8. The hexagram is built from the bottom up. The hexagram is interpreted as a whole and each line is interpreted according to its proper placement and whether it lends strength or weakness to the hexagram. There are 64 possible hexagrams.

	Value of the trigrams	Lines
Old Yin	2 + 2 + 2 = 6	—— ——
Young Yang	2 + 2 + 3 = 7	————
Young Yin	3 + 3 + 2 = 8	—— ——
Old Yang	3 + 3 + 3 = 9	————

A hexagram composed of the numbers 7, 9, 6, 8, 8, 8 would be interpreted as trigram 19. Lin/The Approach and the "age" of each line of the hexagram would give more detail about the situation in question. This trigram would be depicted as follows:

Young Yin	3 + 3 + 2 = 8	—— ——
Young Yin	3 + 3 + 2 = 8	—— ——
Young Yin	3 + 3 + 2 = 8	—— ——
Old Yang	3 + 3 + 3 = 9	————
Young Yang	2 + 2 + 3 = 7	————

The meaning of this trigram is "becoming great" and this is a function of the two strong lines that are coming up from below and the light-giving power of the upper lines allows these strong lines to expand.[40]

4.12.2   Geomancy: Feng Shui

Originally, Feng Shui was used to orient objects in space in an energetically auspicious orientation based on calculating the balance of ch'i, the interactions between the five elements, yin and yang and other factors. The location of each item in a space affects the flow of ch'i by slowing, redirecting or accelerating ch'i. The flow of ch'i is believed to affect the health, wealth, energy level, luck and many other aspects of the occupants of the space. Following the close of the Cultural Revolution, Feng Shui has increased in popularity among Easterners and Westerners alike. The ch'i of a building would reflect the orientation of the building, its age, its interaction with its environment such as placement in relation to hills, water, roads and vegetation. Modern Feng Shui has developed principles allowing the practitioner to determine the most energetically favorable locations for the locations of rooms and objects within a house, office or any building of a specific function. Each room has a designated function according its location in a building, and each room is benefited by certain colors, objects and functions. Inauspicious placements can be mitigated through the addition of specific tools such as a mirror.[41–43] Wikipedia has several write ups that are helpful in this regard.[44]

References

1. Throughout this chapter, Chinese terms are romanized using the Wade–Giles system. The reason is that for most terms used here, Wade–Giles appears phonetically more like the word is pronounced. For comparison, the following are some examples in Wade–Giles/Pin-Yin: T'ai Chi/Taiji, Pa Kua/Ba Gua, Hsing I/Xing Yi, Tao/Dao, Kung Fu/Gong Fu.
2. Y.-L. Fung, *A Short History of Chinese Philosophy*, Free Press/Macmillan Publishing, New York, 1948.
3. T.-H. Jou, *The Tao of Tai-Chi Chuan, Way to Rejuvenation*. Tai Chi Foundation, Warwick, 1981, p. 87.

4. H. Beinfield and E. Korngold, *Between Heaven and Earth, a Guide to Chinese Medicine*, Ballantine Books, New York, 1991.

5. W.-T. Chan, *A Source Book in Chinese Philosophy*, Princeton University Press, Princeton, NJ, 1963.

6. T. J. Kaptchuk, *The Web That Has No Weaver, Understanding Chinese Medicine*, Contemporary Books, Chicago, 2000.

7. D. Stein, *Essential Reiki*, Crossing Press, Berkeley, CA, 1995.

8. M. Usui and F. A. Petter, *The Original Reiki Handbook of Dr. Mikao Usui*, Lotus Press, Twin Lakes, WI, 1998.

9. F. A. Petter, T. Yamaguchi and C. Hayashi, *The Hayashi Reiki Manual*, Lotus Press, Twin Lakes, WI, 2003.

10. V. Lad, *Ayurveda, the Science of Self-Healing*, Lotus Press, Twin Lakes, WI, 1984.

11. V. Lad and A. Durve, *Marma Points of Ayurveda*, Ayurvedic Press, Albuquerque, NM, 2008.

12. P. Zeigler, Ka Shen Sekhem, http://www.vibrational-alchemy.com/kashen/intro.htm (last accessed 14 September 2012).

13. D. K. Osborn, Greek Medicine: Pneuma and Ignis, http://www.greekmedicine.net/b_p/Pneuma_and_Ignis.html (last accessed 14 September 2012).

14. K. Cohen, *Honoring the Medicine, the Essential Guide to Native American Healing*, Ballantine Books, New York, 2006.

15. K. B. Konadu, *Indigenous Medicine and Knowledge in African Society*, Routledge, New York, 2007.

16. R. A. Morrell, *The Sacred Power of Huna: Spirituality and Shamanism in Hawai'i*, Inner Traditions – Bear & Company, Rochester, VT, 2005.

17. T. W. Rinpoche, *Healing with Form, Energy and Light: the Five Elements in Tibetan Shamanism, Tanta and Dzogchen*, Snow Lion Publications/Shambala, Ithaca, NY, 2002.

18. B. A. Brennan, *Hands of Light, a Guide to Healing Through the Human Energy Field*, Bantam Books, New York, 1987.

19. R. Gerber, *Vibrational Medicine*, Inner Traditions – Bear & Company, Rochester, VT, 2001.

20. C. Von Reichenbach, *Physio-physiological Researches on the Dynamics of Magnetism, Electricity, Heat, Light, Crystallization and Chemism in Their Relation to the Vital Force*, Clinton-Hall, New York, 1851.

21. W. J. Kilner, *The Human Aura*, University Books, New Hyde Park, NY, 1965.

22. G. De La Warr, *Matter in the Making*, Vincent Stuart, London, 1966.

23. V. V. Hunt, *Infinite Mind, Science of the Human Vibrations of Consciousness*, Malibu Publishing Company, Malibu, CA, 1989.

24. D. D. Dougherty, J. Kong, M. Webb, A. Bonab, A. Fischman and R. Gollub, A combined [^{11}C]diprenorphine PET study and fMRI study of acupuncture analgesia, *Behav. Brain Res.*, 2008, **193**, 63.

25. D. Frawley, S. Ranade and A. Lele, *Ayurveda and Marma Therapy, Energy Points in Yogic Healing*, Lotus Press, Twin Lakes, WI, 2003.

26. R. Jahnke, L. Larkey, C. Rogers, J. Etnier and F. Lin, A comprehensive review of health benefits of qigong and tai chi, *Am. J. Health Promotion*, 2010, **24**(6), 1.
27. K. Yuen, *Yuen Energetics, the Power of Instant Healing*, Yuen Energetics/ Shaolin Press, Los Angeles, 2001.
28. X. Cheng, *Chinese Acupuncture and Moxibustion*, Foreign Languages Press, Beijing, 1987.
29. Office of the Director, *Acupuncture, NIH Consensus Statement*, 1997, **15**(5), 1.
30. World Health Organization, *Viewpoint on Acupuncture*, World Health Organization, Geneva, 1979.
31. J. Bracy and X.-H. Liu, *Ba Gua, Hidden Knowledge in the Taoist Internal Martial Art*, Blue Snake Books, Berkeley, CA, 1998.
32. K. S. Cohen, *The Way of Qigong, the Science and Art of Chinese Energy Healing*, Ballantine Books, New York, 1997.
33. Z. Yang, *Yang Style Taijiquan*, Morning Glory Publishers, Beijing, 1991.
34. *Chen Style Taijiquan*, Zhaohua Publishing House and Hai Feng Publishing Company, Hong Kong, 1984.
35. Y. C. Chen, *T'ai Chi Ch'uan, Its Effects and Practical Applications*, Millington, Shanghai, 1947.
36. W. Liao, *T'ai Chi Classics*, Shambala Books, Boston, MA, 1990.
37. D. Wile, *Tai Chi Touchstones: Yang Family Secret Transmissions*, Sweet Ch'i Press, Brooklyn, NY, 1983.
38. S. L. Wolf, N. Kutner, R. Green and E. McNeely, The Atlanta FICSIT study: two exercise interventions to reduce frailty in elders, *J. Am. Geriatr. Soc.*, 1993, **41**(3), 329.
39. W. K. Kit, *The Art of Chi Kung, Making the Most of Your Vital Energy*, Element Books, Shaftesbury, 1993.
40. R. Wilhelm, *The I Ching or Book of Changes*, Princeton University Press, Princeton, NJ, 1950.
41. K. R. Carter, *Move Your Stuff, Change Your Life, How to Use Feng Shui to Get Love, Money, Respect and Happiness*, Fireside Press, New York, 2000.
42. J. Butler-Biggs, *Feng Shui in 10 Simple Lessons*, Watson-Guptill Publications, New York, 1999.
43. L. Too, *The Complete Illustrated Guide to Feng Shui, How to Apply the Secrets of Chinese Wisdom for Health, Wealth and Happiness*, Element Books, Shaftesbury, 1996.
44. http://en.wikipedia.org/ articles on Qi, Traditional Chinese Medicine, Tai Chi Chuan and Feng Shui.

CHAPTER 5

When Modern Computational Systems Biology Meets Traditional Chinese Medicine

CALVIN YU-CHIAN CHEN[a,b,c,d,e,f,g]

[a] Laboratory of Computational and Systems Biology, China Medical University, Taichung 40402, Taiwan; [b] Department of Medical Research, China Medical University Hospital, Taichung 40402, Taiwan; [c] Department of Biotechnology, Asia University, Taichung 41354, Taiwan; [d] Department of Bioinformatics, Asia University, Taichung 41354, Taiwan; [e] China Medical University Beigang Hospital, Yunlin, Taiwan; [f] Department of Systems Biology, Harvard Medical School, Boston, MA 02115, USA; [g] Computational and Systems Biology, Massachusetts Institute of Technology, Cambridge, MA 02139, USA
E-mail: ycc929@mit.edu

5.1 Introduction

Traditional Chinese medicine (TCM) and the practice of TCM are separate but often confused subjects. TCM practice focuses on *yin-yang* and involves concepts such as *wuxing*, *qi xue* and *jing luo*. TCM, on the other hand, refers to pharmaceuticals documented in ancient Chinese texts such as *ShenNong's Herbal* and *The Compendium of Materia Medica*. The pharmacological effects of TCM are categorized under terms corresponding to those used by TCM practitioners to identify *patterns* (*i.e.*, the outward expression of physical imbalance brought on by illness, environment and individual physical

RSC Drug Discovery Series No. 31
Traditional Chinese Medicine: Scientific Basis for Its Use
Edited by James David Adams, Jr. and Eric J. Lien
© The Royal Society of Chemistry 2013
Published by the Royal Society of Chemistry, www.rsc.org

conditions). Based on the identified patterns, TCM formulas are combined from multiple TCMs with different pharmacological effects to maximize effectiveness in restoring bodily balance. Many instances of recovery by TCM have been reported although evidence-based validation remains elusive. The ancient history of TCM combined with its seclusion in Asia due to language barriers has long shrouded TCM and TCM practice in a mysterious light. Here, we discuss the integration of computational and systems biology methods to provide a scientific basis for the workings of TCM and how it might serve as a possible step forward for personalized medicine.

5.2 Modernizing Traditional Chinese Medicine for a Global Stage

TCM, with its ancient history, is a rich accumulation of experiences in administrating minerals, small molecular organic compounds and biomacro-molecules with therapeutic effects.[1] Although similarly rooted in observation-based substance usages, evidence-based scientific methods later emerged as the golden rule of Western medicine. TCM and its related philosophies were viewed as folk remedies. Owing to the recent focus shift in medical practice from disease treatment to a more holistic treatment approach,[2] the resourcefulness of TCM is gradually gaining attention due to its "natural" nature and its ability to treat the bodily "system" as a whole.[3] However, before the medical potential of TCM can be tapped into, TCM must be modernized in order to create a common platform to facilitate information exchange with mainstream Western medical knowledge.

TCM modernization is an extensive systematic task ranging from the identification of active ingredients to elucidating underlying rules of complex TCM formulas and investigating pharmacodynamic mechanisms using modern scientific concepts and technologies. Yin et al.[4] summarized an interesting four-tiered research approach which highlights the need for integrating active component studies with pharmacological studies, elucidating active component(s) of TCM, encompassing active compounds, effective parts and medicinal material or TCM formulas in TCM research and investigation of medicinal mechanisms at the molecular, cell, organism and whole host levels. It is clear that such modernization will require the integration of multiple disciplines and extensive work to cover the great deal of experience-based TCM recordings accumulated over the years. With the increasing demand for natural medicine,[5] much progress in TCM research has been made. Most notably, various active compounds from TCM have been identified, with anticancer, antioxidant, antibacterial, antiflu and neuropro-tective properties, to name just a few.[6–15] In addition, several attempts to elucidate the pharmacokinetic mechanisms of TCM formulas have been successfully reported.[16–18] Lukman et al.[19] also reported interesting attempts at modernizing TCM and TCM formulas.

5.3 TCM in Drug Discovery: Application from the Molecular Level

With the explosion of scientific advances in the 1990s, particularly in the fields of sequencing and protein structure analysis, there was an evolution from conventional trial-and-error to rational drug design. Rational drug design refers to the rational design of receptor-specific drug compounds with interdisciplinary incorporation of biochemistry, enzymology, molecular biology, genetics, bioinformatics and computational biology. The concept of rational drug design revolutionized modern drug development, with computer-aided drug design (CADD) techniques playing a crucial role in this advancement. A comprehensive review on CADD can be found elsewhere.[20–21]

The "lock-and key" theory is the dominant drug development paradigm adopted in the pharmaceutical industry in past decades. The introduction of computer simulation techniques in CADD allowed the *in silico* modeling of ligand docking to the receptor of interest. Major developments in CADD now utilize dynamic ligand–receptor binding models that consider the plasticity of both ligands and proteins and allow flexible docking under various solutions. Within this drug development paradigm, the expansive TCM resources present an extensive database of chemical compounds that might be exploited for bottom-up drug discovery. The affinity of TCM compounds with a specified target protein may be calculated *in silico* and can greatly reduce the time and labor associated with biological "wet" experiments. Parallel to candidate identification using affinity or structural compatibility, ligand-based bioactivity prediction models may be constructed on compounds of unknown bioactivity through the quantitative structure–activity relationship analysis of compounds known to affect the selected protein target. Potential drug leads from TCM for cancer, hypertension, influenza, weight control, stroke, *etc.* have been identified through these procedures.[22–37]

5.4 Futuristic Drug Discovery: The Top-Down and Bottom-Up Systematic View

Although the single drug–single protein theory has served medical research well in the past, increasing evidence now reveals that such a reductionist approach is overly simplified.[38] Support for this is seen in the multiplex of genetic and epigenetic factors involved in complex multifactorial diseases. For example, mutations of at least 189 genes have been associated with human breast and colorectal cancer,[39] and the actual number may possibly be higher since non-coding regulatory regions of coding genes may also contribute.[40] This multitude of factors involved in a single disease is the perfect illustration of the intricacies and protective mechanisms of biological systems. Redundancy, in which different proteins have similar functions, is a critical protection mechanism against shutdown of the entire system by a single factor.

This inherent mechanism, although important for maintaining proper biological functions, makes the control of an illness by targeting a single target protein a challenging process.

Better understanding of human physiological inner workings and disease networks has pushed for a re-evaluation of drug development principles. One concept that has emerged is the systems biology-based approach which integrates genomics, proteomics and metabolomics to enhance our understanding of the effects of a lead compound on whole pathways and networks rather than isolated protein targets, thereby increasing the success rate and decreasing the adverse effects of drugs.[41,42] Interestingly, although the concept of viewing an illness as the collective expression of multiple imbalanced factors and using various treatments to "restore" balance to alleviate illness is relatively new to practitioners of modern medicine, it is the central idea for TCM and the guiding concept for TCM formulations.

A TCM formula is the essence of TCM medication and formulation is based on four functional roles of the compositions, termed *sovereign, minister, assistant* and *envoy*, according to the ancient TCM text Shen Nong Ben Cao Jing.[43] The *sovereign* component targets the root cause and primary symptoms and is the primary composition of the formula. The *minister* acts synergistically to the *sovereign*. The *assistant* relives secondary symptoms or neutralizes toxicities of the major components to reduce side effects. The *envoy* serves as a homing agent leading the active components to the source of the illness and/or balancing the effect of all components.

The effectiveness of a TCM formula might be affected by the interaction of different components and is largely dependent on the formula composition. For example, *Ramulus cinnamomi*, when used with *Ephedra sinica* Stapf., induces sweating, but reduces perspiration when combined with *Paeonia lactiflora*. Concerted application of multiple TCM components can induce synergistic drug effects or reduce adverse side effects. When *Rheum officinale* is used with mirabilite, the laxative effects become more pronounced. In other instances, a composite combination leads to the reduction of adverse side effects. For example, the toxicity of *Pinellia ternata* (Thunb.) Breit. can be reduced when used in conjunction with *Zingiber officinale* Rosc., permitting the use of *Pinellia ternata* (Thunb.) Breit in limiting vomit and dispelling phlegm.

The composition of a TCM formula follows set principles and has certain applicability. It is the practical adaptation of differential treatment. Therefore, its clinical application is not rigid and should change according to the progression of the illness, physical conditions and age. When the compositions of the primary medicinals are changed, the function of the TCM formula changes. As an example, the range of functions of *Astragalus membranaceus* (Fisch.) Bge. in different TCM formulas are shown in Table 5.1.

Similarly, concentration changes of a TCM compound will also affect the pharmacological effect of the TCM formula. Both Zhizhu decoction and Zhizhu pills are composed of *Citrus aurantium* and *Atractylodes macrocephala* Koidz. The concentration of *Citrus aurantium* is higher than that of

Table 5.1 Pharmacological effects of *Astragalus membranaceus* (Fisch.) Bge. in TCM formulas formed from different supporting compositions.

Primary component	Supporting compositions	TCM formula	Function
Astragalus membranaceus (Fisch.) Bge.	*Panax ginseng* *Bupleurum chinense* DC. *Atractylodes macrocephala* Koidz. *Cimicifuga foetida* L.	Bu Zhong Yi Qi Tang	Boosts qi, with which its vacuity is associated with general physical weakness
	Angelica sinensis (Oliv.) Diels	Danggui Buxue Tang	Boosts blood; addresses conditions associated with blood vacuity
	A. macrocephala Koidz. *Stephania tetrandra* S. Moore	Farng Jii Hwang Chyi Tang	Facilitates the release of water; relieves conditions such as swelling and diarrhea
	A. macrocephala Koidz. *Saposhnikovia divaricata* (Tutez.) Schischk.	Yupingfeng powder	Reduces copious sweating leading to exterior vacuity; relieves sweating, heart palpitations, fatigue, dry lips, etc.

Atractylodes macrocephala Koidz in Zhizhu decoction and functions to eliminate stagnation. In Zhizhu pills, *Atractylodes macrocephala* Koidz. is at a higher concentration than *Citrus aurantium* and has therapeutic effects in invigorating the spleen. From this example, we can see that despite identical medicinal components, a change in concentration will lead to changes in the primary and secondary effects of the formula and affect the therapeutic usage of the formula.

As described previously, there seems to be much similarity between TCM and systems biology. Both differ from the conventional bottom-up research approach by taking into account the entire biological system. Systems biology may be applied to "translate" a TCM formula into logical, scientific information that can be communicated and generally applied. The question is, how are we going to do this?

5.5 Unleashing the Potential of TCM Through Systems Biology

Rational investigation of TCM using a systems biology approach requires molecular analysis of action mechanisms in biological networks. Needless to say, detailed information regarding kinetics, localization and biochemical dynamics of specific signaling, metabolic and transcriptional pathways must first be established in order to build robust proteomic, genomic and metabolomics models. However, even more fundamental is the need for a digitalized database of all available TCM compounds to facilitate subsequent computer modeling work characteristic of omic work and drug design.

Prior to 2010, the greatest issue with regard to the computational analysis of TCM was the lack of a TCM-specific compound database. Many websites contained compiled information on TCM sources, usages and traditional processing procedures, but very few contained molecular information on TCM constituents. The TCMGeneDIT database[44] is an effective search engine for TCM-related literature, but information on TCM constituents is not well organized. Other databases, such as the Chinese Traditional Medicinal Herbs Database,[45] Traditional Chinese Medicine Information Database (TCM-ID)[46] and TCMD,[47] provide general information and 3D structures of some TCM constituents. In the past decade, increasing interest in TCM has spurred the development of dedicated TCM compound databases. In 2012, the Chemical Database of Traditional Chinese Medicine (Chem-TCM) was launched through the collaborative efforts of King's College and the Shanghai Institute of Materia Medica (http://www.chemtcm.com/database.html).[48] The database contains chemical and source information on 12,070 TCM compounds and offers powerful functions to predict the pharmacological properties of each compound. Nonetheless, the commercial nature of Chem-TCM greatly limits the audience that can have access to this database.

An alternative to Chem-TCM is the open-access, web-based TCM Database@Taiwan (http://tcm.cmu.edu.tw/)[49] (Figure 5.1). With continuous updating efforts since its launch in 2011, TCM Database@Taiwan currently houses nearly 42,000 TCM compound structures and provides traceback information to its TCM source. The search options allow users to search for TCM compounds using chemical structures or by TCM categories. Not only can one search for a specified TCM compound, but one may also conduct virtual screening against a given protein target for potential leads either by downloading the compound directory directly or by conducting virtual screening online through its cloud computing portal TCM intelligent screening system (Figure 5.2) (iScreen; http://iscreen.cmu.edu.tw/).[50] Flexibility is designed into the TCM Database@Taiwan, allowing users to specify the pool of compounds used for screening through a user-friendly checklist. Most notable is the bilingual nature of the websites, allowing researchers of both Western and Eastern origins to access TCM resources. ZINC,[51] the largest open-access 3D molecule database known, serves as an entry portal to many

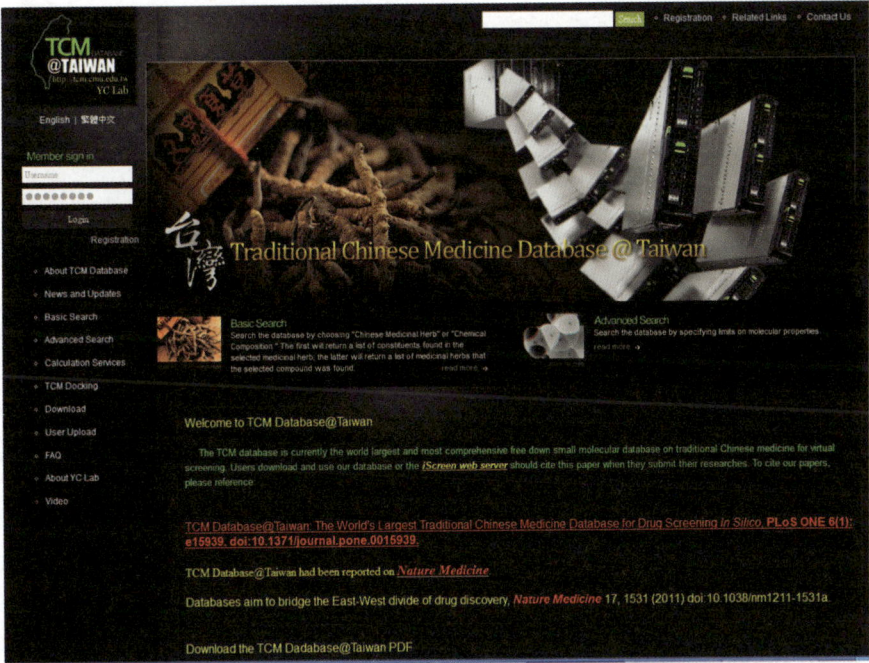

Figure 5.1 TCM Database@Taiwan,[49] the comprehensive traditional Chinese medicine database for drug screening *in silico*. This database maintains over 42 000 TCM compounds based on TCM classification. Simple molecular properties and the molecular structure of each compound are recorded and open to browsing and searching. The screenshot was taken from TCM Database@Taiwan[49] (http://tcm.cmu.edu.tw/) with permission.

Figure 5.2   iScreen, the innovative cloud computing web server for TCM drug
screening and *de novo* design.[50] Docking settings may be customized to
user specifications and protein preparation tools are implemented to
reduce errors caused by excessive information in the raw input data.
Users may conduct screening on all TCM compounds or select categories
of interest from the checklist. The screenshot was taken from iScreen[50]
(http://iscreen.cmu.edu.tw/) with permission.

structural databases. ZINC primarily adopts TCM Database@Taiwan for its
TCM resources and its retailer links are an important function for researchers
to locate a specific compound.

To take a further initiative in TCM drug development, a cloud computing
system dubbed iScreen was developed based on TCM Database@Taiwan.
iScreen is an excellent option for researchers who have limited computational
resources for conducting TCM docking and *de novo* drug design. A range of
customizable docking modes, including standard, water, pH docking and
flexible docking, are available for users to select from in order to achieve the
most relevant docking simulation. As with TCM Database@Taiwan, iScreen
allows users to specify the pool of TCM compounds with which to conduct the

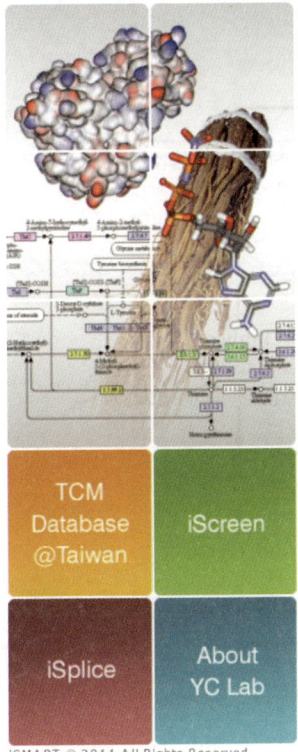

iSMART integrated SysteMs Biology Associated Research with TCM

iSMART is a multi-funcitonal cloud-computing web system that comprised with three components:
(1) TCM Database@Taiwan, (2) iScreen, and (3) iSplice.

TCM Database@Taiwan
The TCM database is currently the world largest and most comprehensive free down small molecular database on traditional Chinese medicine for virtual screening.

iScreen
To take initiative in the next generation of drug development, we constructed a docking and screening web-server, iScreen, based on our world's largest TCM database, the TCM Database@Taiwan.

iSplice
iSplice is an online tool combining gapped-dinucleotide pattern probability with logarithmic odds (GO algorithm) to assess the likelihood of the activation of a cryptic 5' splice site, which competes with its paired authentic 5' splice site.

Future Works
The framework of iSMART system has been built and relevant development has been conducted to fabricate multiple analysis components prior the pre-clinical trial. More comprehensive analysis would be conducted with the establishment of complementary functions in iSMART.

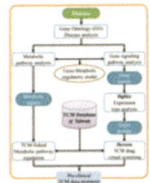

iSMART © 2011 All Rights Reserved.

Figure 5.3   The title page of iSMART. iSMART is a pioneering system for bridging TCM studies to systems biology research. Reproduced from Ref. 52 with permission.

docking. Users interested in a specific TCM category may select compounds in that category or related categories to conduct docking. Users may download the top 200 TCM compounds ranked by docking and utilize them for further biological studies or further chemical modification. Integrated SysteMs biology Associated Research with TCM (iSMART; http://ismart.cmu.edu.tw)[52] is the all-in-one entry domain that currently contains TCM Database@Taiwan and iScreen (Figure 5.3). In addition, an online tool termed iSplice is also included and enables users to assess activation probability of a cryptic 5 splice site. Further expansion of iSMART aims to include various analysis components of drug design prior to pre-clinical trial. The proposed integration of iSMART systems biology research is shown in Figure 5.4.

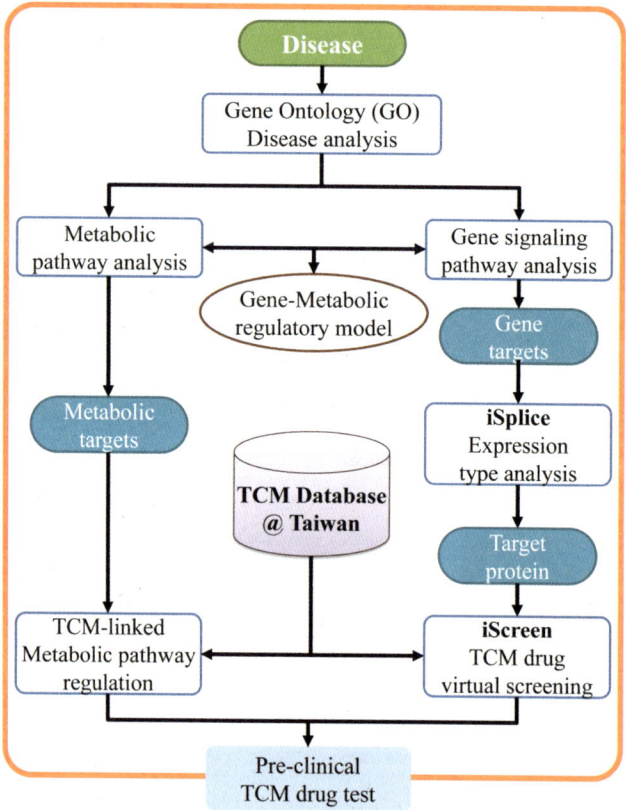

Figure 5.4   The schematic standard operating procedure of iSMART in systems
biology research.[52] A disease would be categorized, its associated
signaling and metabolic pathways defined, genetic defects, such as
cryptic 5 splice site activation, identified and relevant targets extracted
for drug design. Therapeutic targets along the metabolic pathway would
also be identified accordingly. Reproduced from Ref. 52 with permission.

5.6   An Example of a Systems Approach to TCM Drug Development

In this section, we use stroke medication development as an example of the
possible integration of computational systems biology and TCM towards
personalized medication.

The integration of CADD and systems biology to construct a stroke model
and understand its underlying mechanisms for drug development are outlined
in Figure 5.5 and CADD procedures to conduct this investigation are
illustrated in Figure 5.6. The CADD procedure can be roughly divided into
virtual screening and validations using structure-based and ligand-based drug
design approaches.

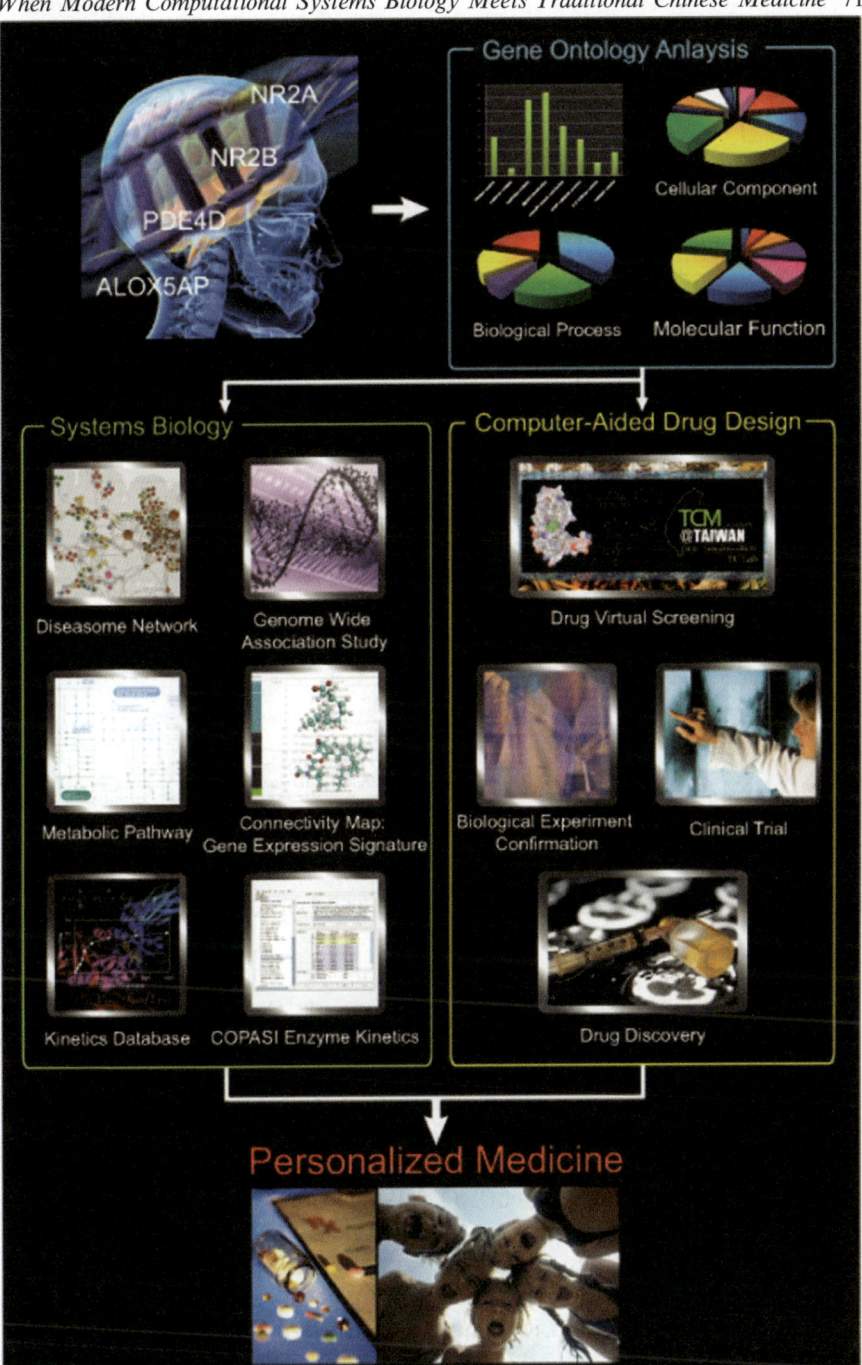

Figure 5.5 Integration of computational and systems biology to target stroke-inducing genes and metabolic pathways to develop personalized stroke medicine from TCM origins.

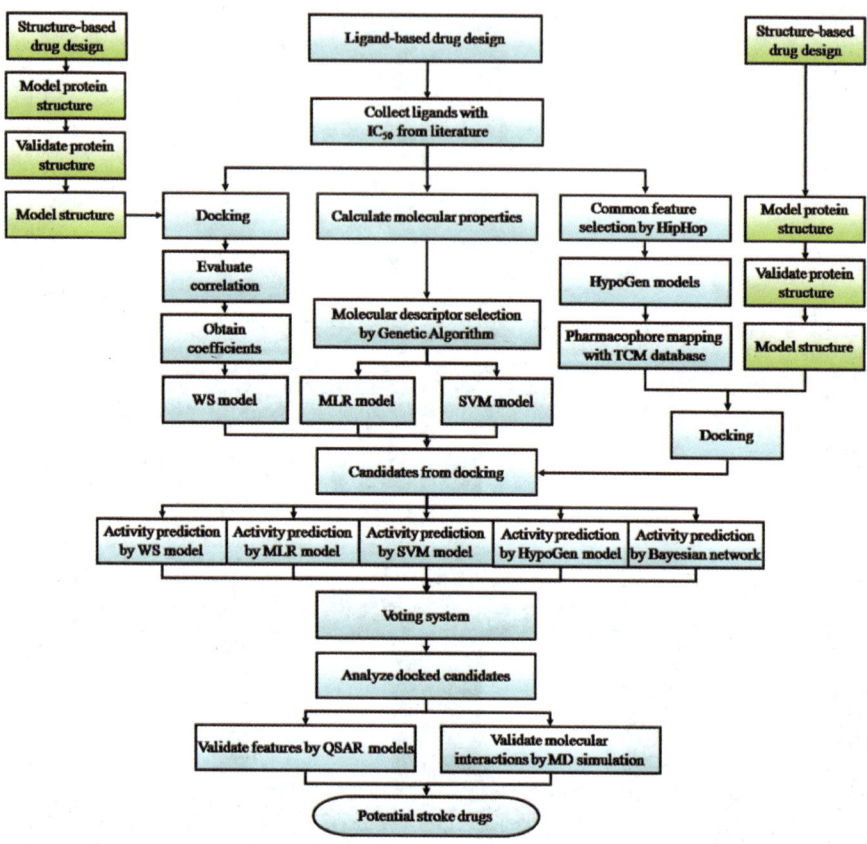

Figure 5.6   CADD procedure for investigating TCM stroke medicine.

Structure-based drug design methods utilize docking and molecular dynamics (MD) simulation. First, simulation and validation of the target protein for stroke are conducted and then TCM Database@Taiwan and Zinc are employed to conduct virtual screening via docking. Subsequently, MD is used to analyze time-dependent interaction changes between the drug compound and the target protein to achieve a stable final complex structure. In-depth analysis of the dynamic motions of the drug and target protein can facilitate new drug development. The Weight Score[53] integrative scoring system has been proven to be more accurate than the Consensus Score.

Ligand-based drug design methods utilize different models, including Hypogen,[54] machine-learning methods of support vector machine (SVM),[55–56] multiple linear regression (MLR)[57] and Bayesian networks,[58–59] *etc.* Drug models are constructed based on the pIC_{50}, molecular characteristics and fingerprints of known drugs and can be used to predict the pIC_{50} of new drugs. Data on the single nucleotide polymorphism (SNP) of healthy *versus* stroke individuals are used to construct a SVM model to predict the risk of having a stroke.

Then, we connect to the Gene Ontology (GO) database,[60] Kinetic Data of Bio-molecular Interactions (KDBI) database[61] and CellNetAnalyzer[62] to obtain information on different levels of the biological system (Figure 5.5). The GO database allows the construction of static models that describe the drug–genes and genes–pathways interactions (Figure 5.7); KDBI allows the calculation of kinetic simulation parameters. These parameters are required to conduct dynamic simulations critical to building a drug model that is close to real life.

Once the kinetic model has been built, biological experiments must be conducted to acquire complete kinetic parameters describing protein–protein, protein–RNA, protein–DNA, protein–ligand, RNA–ligand and DNA–ligand interactions. CellNetAnalyzer is used to calculate signal transduction networks within the cell and to understand the transduction mechanisms of stroke drugs. From real-time patient microarray analysis, the effect of drugs on gene expression can be observed and utilized to construct stroke-related gene networks using Genetic Network Analyzer (GNA).[63] This dynamic gene network can be integrated with previous dynamic metabolic networks to investigate the comprehensive interaction mechanisms of stroke drugs and patient cells. Since complicated relationships often exist between different illnesses, particularly in multifactorial chronic diseases such as stroke, Human Disease Network, Diseasome and Disease Gene Network[64] were constructed to investigate the relationship of stroke with other illnesses. These networks can provide information on protein–protein interactions, transcription factor–promoter interactions and metabolic reactions. As a final step, connectivity maps[65] are constructed to provide information on stroke drug dosages *versus* gene expression.

5.7 Integration of TCM with Systems Biology: A Long Way to Go

The rapid development of TCM databases has contributed phenomenally to advancing TCM drug design. Although databases are convenient, caution should be exercised when selecting a suitable database that fits one's research needs.

5.7.1 Science Should Be Open Access

A public database not only invites a wider audience, but through user feedback one can understand upgrade needs from a user's perspective. For example, users of TCM Database@Taiwan have been a fundamental force in our continuous efforts to update database records. Currently the database includes a total of 122,249 compounds of natural origin, including 41,490 compounds specific to TCM, and another update of 23,000 compounds is scheduled during 2012.

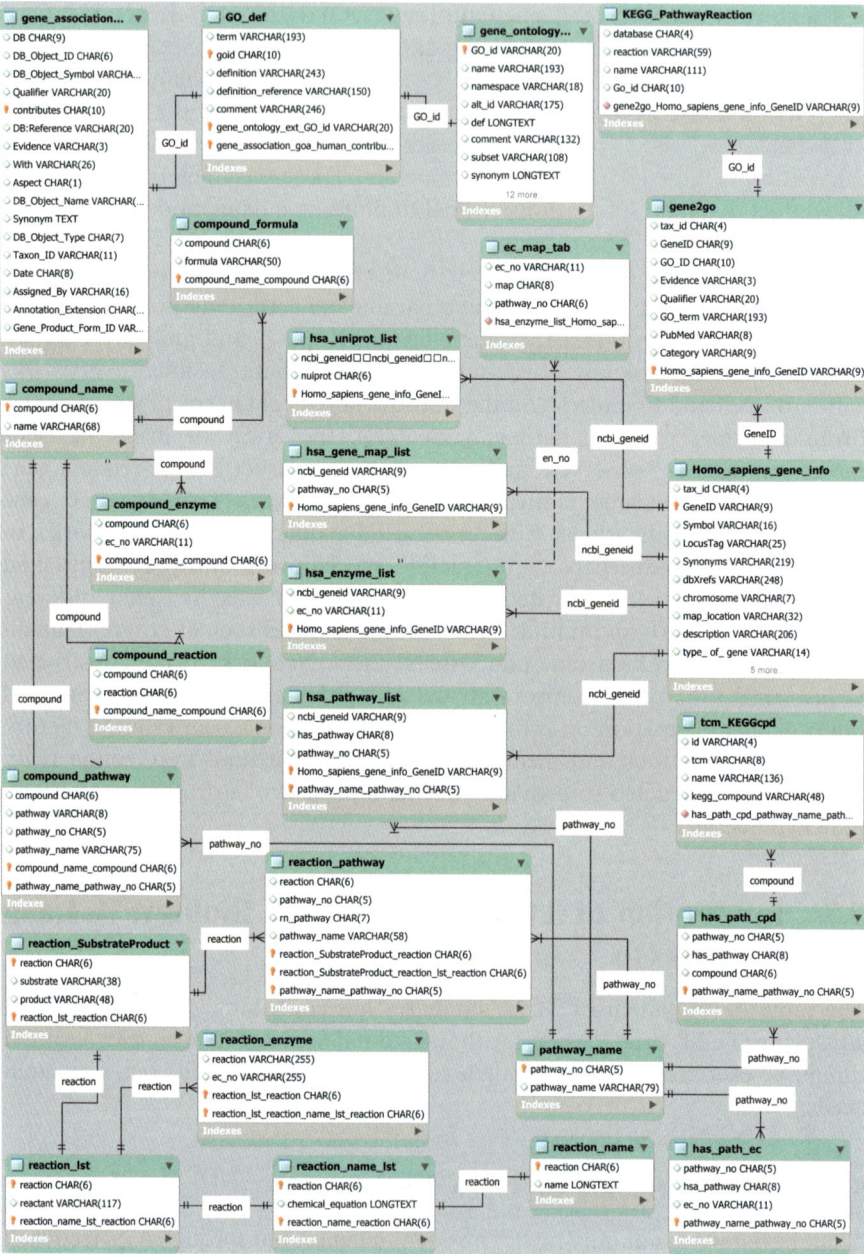

Figure 5.7   Interaction between TCM compounds, metabolic pathways, disease genes and drug reactions determined by GO analysis.

5.7.2 Pharmacological Data Have Limited Applicability in Investigating Novel Leads

One of the newest features introduced in structure databases is the co-listing of pharmacological activity data with chemical structure. Although attractive, it may be of limited significance since drug discovery is primarily focused on new drugs for new targets. Bioactivity information may be more useful during the drug modification stage. In fact, a simple online literature search can provide sufficient pharmacological information.

5.7.3 Bigger Is Not Better in Terms of TCM Drug Compound Screening

Although it is true that larger databases are better for identifying potent drug leads, screening of super databases containing millions of compounds through the docking method is almost impossible to complete unless conducted through pharmacophore mapping. From our experience, screening 20,000 compounds via flexible docking requires ~ 10 days of parallel computing by high-performance 192 core CPUs (24 workstations with 8 core CPU). Needless to say, it would be extremely time consuming for millions of compounds. In addition, super databases are depositories of diverse chemical compounds. Screening results may contain a wide variety of chemical compounds, many of which are not suitable for pharmaceutical use. Although TCM-specific databases may not cover the wide chemical diversity offered by comprehensive databases,[66] the compounds are naturally present and less likely than synthesized chemicals to provoke intense body reactions leading to serious adverse side effects. In addition, TCM includes medicinal herbs that have been used for thousands of years and compounds derived from TCM are naturally drug-like. Screening from specialized TCM databases offer advantages of efficient screening and outputs of natural drug-like compounds. To improve practicability, some websites, such as TCM Database@Taiwan, offer TCM compounds and natural compounds which have passed absorption, distribution, metabolism, and excretion - toxicity (ADME-Tox).

5.7.4 Traceback Information Is Critical

One of the most attractive features of ZINC is that it provides users with links to compound suppliers. This link-forward information is limited for TCM compounds. In the absence of commercial availability, traceback information to plant origins and TCM categories of these compounds is extremely important. Plant origins of the compounds can provide researchers with clues on how to isolate the compound of interest. The ancient TCM categories under which the plants are filed also provide rough guides to expected effects of the compounds. These categories are extremely helpful when designing drugs for a specific action. Very limited numbers of databases organize their compound catalog according to TCM categories, but the importance of these fundamental elements should not be underestimated. Although there is no convenient way to obtain TCM compounds, providing sufficient traceback information can help minimize guesswork and promote practical applications of TCM.

5.7.5 Modernization of TCM Formulas

Much effort has been placed on elucidating the scientific basis of TCM formula.[67–73] We have previously attempted to utilize KEGG (Kyoto Encyclopedia of Genes and Genomes)[74] and algorithms to establish relationship maps of TCM components in several TCM formulas and to establish a standard research procedure for integrating TCM with systems biology. However, our efforts have met with major difficulties. First, establishing coherent relationships between TCM components is hard. Among the different TCM formulas we have tried, only the one in Figure 5.8 shows an acceptable correlation between the formula constituents. Identifying correlations between illnesses and TCM formulas through signal pathways is also a major hurdle that remains to be overcome.

Although the integration of TCM and systems biology is an intriguing concept, many hurdles remain to be resolved. First, the definition of "systems" is vague. Does it include symptom expression or merely cell expression? In addition to the complexity of the human body, limitations in current

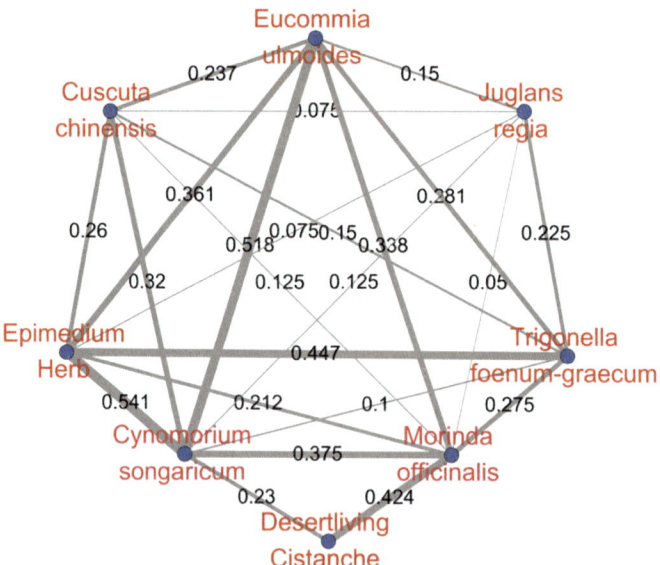

Figure 5.8   Correlation network of eight TCM drugs categorized under the yang-tonifying class of TCM medicine.[52] Correlations between pairs of TCM drugs are indicated by a line and the width of the line indicates the correlation strength of the compounds. Values given are calculated correlations. Reproduced from Ref. 52 with permission.

technologies are also difficult to overcome. A stable protein folding in molecular dynamics drug design requires stabilization exceeding the microsecond timescale, but current computer capabilities can hardly support such a huge calculation capacity. Both TCM and Western medicine still require specialists in physics, engineering, electrical engineering, chemistry, computer engineering and mathematics to make cutting edge discoveries. These improvements, accompanied by the newest sequence analysis techniques, may facilitate the use of individual protein sequences in personalized medication and bridge the current gap between TCM and systems biology.

Second, the spirit of TCM lies in TCM formulas and not as individual pure compounds. We have spent more than 10 years modernizing TCM and building a comprehensive TCM database (http://tcm.cmu.edu.tw/) for small molecule screening. However, this is the study of TCM components using single protein–single drug research methods and differs vastly from TCM formulas. In our attempt to remain true to the spirit of TCM, we analyzed TCM formulas through the KEGG network and COPASI-enzyme kinetics and assessed the overall effect of TCM formulas on human physiology. The results were not coherent enough to reach a solid conclusion. The lack of metabolic data on TCM compounds within the human body and the ability of enzymatic kinetics to truly reflect complex protein–protein interactions within the body add to the complex network of difficulties. When a TCM formula is ingested, what is the interrelationship between the numerous TCM compounds and the various proteins? While TCM formulas may address illnesses from a systems biology-like standpoint, there is still a long way to go before TCM formulas can be established as a science.

References

1. W. F. Li, J. G. Jiang and J. Chen, *Arch. Med. Res.*, 2008, **39**, 246.
2. V. Floriani and C. Kennedy, *Curr. Opin. Pediatr.*, 2007, **19**, 99.
3. F. Yu, T. Takahashi, J. Moriya, K. Kawaura, J. Yamakawa, K. Kusaka, T. Itoh, S. Morimoto, N. Yamaguchi and T. Kanda, *J. Int. Med. Res.*, 2006, **34**, 231.
4. W. Yin, M. Li and S. Liu, *Chin. J. Exp. Tradit. Med. Formulae*, 2006, **12**, 60.
5. G. A. Cordell and M. D. Colvard, *J. Ethnopharmacol.*, 2005, **100**, 5.
6. Y. Tu, *Nat. Med.*, 2011, **17**, 1217.
7. D. Z. Wong, H. A. Kadir and S. K. Ling, *J. Ethnopharmacol.*, 2012, **139**, 256.
8. L. W. Wang, B. G. Xu, J. Y. Wang, Z. Z. Su, F. C. Lin, C. L. Zhang and C. P. Kubicek, *Appl. Microbiol. Biotechnol.*, 2012, **93**, 1231.
9. T. B. Ng, J. Liu, J. H. Wong, X. Ye, S. C. Wing Sze, Y. Tong and K. Y. Zhang, *Appl. Microbiol. Biotechnol.*, 2012, **93**, 1795.
10. X. C. Tang and Y. F. Han, *CNS Drug Rev.*, 1999, **5**, 281.

11. Z. F. Chen, Y. C. Liu, Y. Peng, X. Hong, H. H. Wang, M. M. Zhang and H. Liang, *J. Biol. Inorg. Chem.*, 2012, **17**, 247.

12. N. Zhao, Y. Y. Liu, B. H. Hu, K. Sun, X. Chang, C. S. Pan, J. Y. Fan, X. H. Wei, X. Li, C. S. Wang, Z. X. Guo and J. Y. Han, *Am. J. Physiol. Heart Circ. Physiol.*, 2010, **298**, H1166.

13. H. L. Xu, W. Tang, G. H. Du and N. Kokudo, *Drug Discov. Ther.*, 2011, **5**, 202.

14. G. Q. Chen, X. J. Shi, W. Tang, S. M. Xiong, J. Zhu, X. Cai, Z. G. Han, J. H. Ni, G. Y. Shi, P. M. Jia, M. M. Liu, K. L. He, C. Niu, J. Ma, P. Zhang, T. D. Zhang, P. Paul, T. Naoe, K. Kitamura, W. Miller, S. Waxman, Z. Y. Wang, H. de The, S. J. Chen and Z. Chen, *Blood*, 1997, **89**, 3345.

15. H. Ge, Y. F. Wang, J. Xu, Q. Gu, H. B. Liu, P. G. Xiao, J. Zhou, Y. Liu, Z. Yang and H. Su, *Nat. Prod. Rep.*, 2010, **27**, 1758.

16. Y. C. Lo, Y. T. Shih, Y. T. Tseng and H. T. Hsu, *Evid. Based Complement. Alternat. Med.*, 2012, **2012**, 501032.

17. Y. Liu, Q. Hua, H. Lei and P. Li, *J. Ethnopharmacol.*, 2011, **137**, 1035.

18. H. W. Wang, K. T. Liou, Y. H. Wang, C. K. Lu, Y. L. Lin, I. J. Lee, S. T. Huang, Y. H. Tsai, Y. C. Cheng, H. J. Lin and Y. C. Shen, *J. Ethnopharmacol.*, 2011, **138**, 22.

19. S. Lukman, Y. He and S. C. Hui, *Comput. Methods Programs Biomed.*, 2007, **88**, 283.

20. G. Schneider and U. Fechner, *Nat. Rev. Drug Discov.*, 2005, **4**, 649.

21. G. M. Keseru and G. M. Makara, *Drug Discov. Today,*, 2006, **11**, 741.

22. W. I. Tou and C. Y. C. Chen, *PLoS One*, 2012, **7**, e33728.

23. K. C. Chen, M. F. Sun, S. C. Yang, S. S. Chang, H. Y. Chen, F. J. Tsai and C. Y. C. Chen, *Chem. Biol. Drug Des.*, 2011, **78**, 679.

24. U. Grienke, J. Mihaly-Bison, D. Schuster, T. Afonyushkin, M. Binder, S. H. Guan, C. R. Cheng, G. Wolber, H. Stuppner, D. A. Guo, V. N. Bochkov and J. M. Rollinger, *Bioorg. Med. Chem.*, 2011, **19**, 6779.

25. S. C. Yang, S. S. Chang, H. Y. Chen and C. Y. C. Chen, *PLoS Comput. Biol.*, 2011, **7**, e1002189.

26. S. D. Zhang, W. Q. Lu, X. F. Liu, Y. Y. Diao, F. Bai, L. Y. Wang, L. Shan, J. Huang, H. L. Li and W. D. Zhang, *MedChemComm*, 2011, **2**, 471.

27. T. M. Ehrman, D. J. Barlow and P. J. Hylands, *Curr. Pharm. Des.*, 2010, **16**, 1785.

28. X.-J. Li, D.-X. Kong and H.-Y. Zhang, *Curr. Drug Discov. Technol.*, 2010, **7**, 22.

29. K. C. Chen and C. Y. C. Chen, *Soft Matter*, 2011, **7**, 4001.

30. S. Zhang, W. Lu, X. Liu, Y. Diao, F. Bai, L. Wang, L. Shan, J. Huang, H. Li and W. Zhang, *MedChemComm*, 2011, **2**, 471.

31. K. Y. Chen, S. S. Chang and C. Y. C. Chen, *PLoS ONE*, 2012, **7**, e43932.

32. W. I. Tou and C. Y. C. Chen, *PLoS ONE*, 2012, **7**, e33728.

33. S. S. Chang, H. J. Huang and C. Y. C. Chen, *PLoS Comput. Biol.*, 2011, **7**, e1002315.

34. S. C. Yang, S. S. Chang and C. Y. C. Chen, *PLoS ONE*, 2011, **6**, e28793.

35. T. T. Chang, K. C. Chen, K. W. Chang, H. Y. Chen, F. J. Tsai, M. F. Sun and C. Y. C. Chen, *Mol. Biosyst.*, 2011, **7**, 2702.
36. K. C. Chen, K. W. Chang, H. Y. Chen and C. Y. C. Chen, *Mol. Biosyst.*, 2011, **7**, 2711.
37. S. S. Chang, H. J. Huang, H. Y. Chen and C. Y. C. Chen, *Mol. Biosyst.*, 2011, **7**, 3366.
38. A. Schrattenholz and V. Soskic, *Curr. Med. Chem.*, 2008, **15**, 1520.
39. T. Sjoblom, S. Jones, L. D. Wood, D. W. Parsons, J. Lin, T. D. Barber, D. Mandelker, R. J. Leary, J. Ptak, N. Silliman, S. Szabo, P. Buckhaults, C. Farrell, P. Meeh, S. D. Markowitz, J. Willis, D. Dawson, J. K. Willson, A. F. Gazdar, J. Hartigan, L. Wu, C. Liu, G. Parmigiani, B. H. Park, K. E. Bachman, N. Papadopoulos, B. Vogelstein, K. W. Kinzler and V. E. Velculescu, *Science*, 2006, **314**, 268.
40. R. N. Venkatesan, J. H. Bielas and L. A. Loeb, *DNA Repair (Amst.)*, 2006, **5**, 294.
41. E. J. Davidov, J. M. Holland, E. W. Marple and S. Naylor, *Drug Discov. Today*, 2003, **8**, 175.
42. A. Bugrim, T. Nikolskaya and Y. Nikolsky, *Drug Discov. Today*, 2004, **9**, 127.
43. Anonymous, *Shen Nong Ben Cao Jing*, Fan-I Publishing, Hsin-Chu, Taiwan.
44. Y. C. Fang, H. C. Huang, H. H. Chen and H. F. Juan, *BMC Complement. Altern. Med.*, 2008, **8**, 58.
45. X. Qiao, T. Hou, W. Zhang, S. Guo and X. Xu, *J. Chem. Inf. Comput. Sci.*, 2002, **42**, 481.
46. X. Chen, H. Zhou, Y. B. Liu, J. F. Wang, H. Li, C. Y. Ung, L. Y. Han, Z. W. Cao and Y. Z. Chen, *Br. J. Pharmacol.*, 2006, **149**, 1092.
47. M. He, X. Yan, J. Zhou and G. Xie, *J. Chem. Inf. Comput. Sci.*, 2001, **41**, 273.
48. K. Sanderson, *Nat. Med.*, 2011, **17**, 1531.
49. C. Y. C. Chen, *PLoS One*, 2011, **6**, e15939.
50. T. Y. Tsai, K. W. Chang and C. Y. C. Chen, *J. Comput. Aided Mol. Des.*, 2011, **25**, 525.
51. J. J. Irwin, T. Sterling, M. M. Mysinger, E. S. Bolstad and R. G. Coleman, *J. Chem. Inf. Model.*, 2012, **52**, 1757.
52. K. W. Chang, T. Y. Tsai, K. C. Chen, S. C. Yang, H. J. Huang, T. T. Chang, M. F. Sun, H. Y. Chen, F. J. Tsai and C. Y. C. Chen, *J. Biomol. Struct. Dyn.*, 2011, **29**, 243.
53. C. Y. C. Chen, *J Biomol. Struct. Dyn.*, 2009, **27**, 271.
54. H. Li, J. Sutter and R. Hoffman, *Pharmacophore, Perception, Development and Use in Drug Design*, O. Guner (Ed.), 2000, International University Line: La Jolla, CA.
55. V. Vapnik, *Statistical Learning Theory*, 1998, Wiley, New York.
56. A. Ben-Hur, C. S. Ong, S. Sonnenburg, B. Schölkopf and G. Rätsch, *PLoS Comput. Biol.*, 2008, **4**, e1000173.

57. D. A. Konovalov, L. E. Llewellyn, Y. V. Heyden and D. Coomans, *J. Chem. Inf. Model.*, 2008, **48**, 2081.
58. Z. Tang, M. J. Taylor, P. Lisboa and M. Dyas, *Drug Discov. Today*, 2005, **10**, 1520.
59. N. Friedman, M. Linial, I. Nachman and D. Pe'er, *J. Comput. Biol.*, 2000, **7**, 601.
60. The Gene Ontology Consortium, *Nat. Genet.*, 2000, **25**, 25.
61. Z. L. Ji, X. Chen, C. J. Zhen, L. X. Yao, L. Y. Han, W. K. Yeo, P. C. Chung, H. S. Puy, Y. T. Tay, A. Muhammad and Y. Z. Chen, *Nucleic Acids Res.*, 2003, **31**, 255.
62. S. Klamt, J. Saez-Rodriguez and E. D. Gilles, *BMC Syst. Biol.*, 2007, **1**, 2.
63. H. de Jong, J. Geiselmann, C. Hernandez and M. Page, *Bioinformatics*, 2003, **19**, 336.
64. K. I. Goh, M. E. Cusick, D. Valle, B. Childs, M. Vidal and A. L. Barabasi, *Proc. Natl. Acad. Sci. U. S. A.*, 2007, **104**, 8685.
65. J. Lamb, E. D. Crawford, D. Peck, J. W. Modell, I. C. Blat, M. J. Wrobel, J. Lerner, J. P. Brunet, A. Subramanian, K. N. Ross, M. Reich, H. Hieronymus, G. Wei, S. A. Armstrong, S. J. Haggarty, P. A. Clemons, R. Wei, S. A. Carr, E. S. Lander and T. R. Golub, *Science*, 2006, **313**, 1929.
66. F. Lopez-Vallejo, M. A. Giulianotti, R. A. Houghten and J. L. Medina-Franco, *Drug Discov. Today*, 2012, **17**, 718.
67. S. Zhao and S. Li, *Bioinformatics*, 2012, **28**, 955.
68. G. Luo, Y. Wang, Q. Liang and Q. Liu, *Systems Biology for Traditional Chinese Medicine*, 2012, John Wiley & Sons.
69. A. Buriania, M. L. Garcia-Bermejod, E. Bosisiof, Q. Xu, H. Li, X. Dong, M. S. J. Simmonds, M. Carrara, N. Tejedor, J. Lucio-Cazanae and P. J. Hylands, *J. Ethnopharmacol.*, 2012, **140**, 535.
70. Q.-L. Liang, X.-P. Liang, Y.-M. Wang, Y.-Y. Xie, R.-L. Zhang, X. Chen, R. Gao, Y.-J. Cheng, J. Wu, Q.-B. Xu, Q.-Z. Xiao, X. Li, S.-F. Lv, X.-M. Fan, H.-Y. Zhang, Q.-L. Zhang and G.-A. Luo, *J. Transl. Med.*, 2012, **10**, 26.
71. P. H. Chiu, H. Y. Hsieh and S. C. Wang, *PLoS ONE*, 2012, **7**, e31648.
72. J. Zhao, P. Jiang and W. Zhang, *Brief Bioinform.*, 2010, **11**, 417.
73. P. Tian, Nature, 2011, **480**, S84.
74. M. Kanehisa, *Novartis Found Symp.*, 2002, **247**, 91.

CHAPTER 6

Modern Drug Discovery from Chinese Materia Medica Used in Traditional Chinese Medicine

KUO-HSIUNG LEE*[a,b], SUSAN L. MORRIS-NATSCHKE[a], JOSHUA BRATTLIE[a], JINGXI XIE[c] AND EILEEN BELDING[a]

[a] Natural Products Research Laboratories, UNC Eshelman School of Pharmacy, University of North Carolina, Chapel Hill, NC 27599, USA; [b] Chinese Medicine Research and Development Center, China Medical University and Hospital, Taichung, Taiwan; [c] Department of Medicinal Chemistry, Institute of Materia Medica (IMM), Chinese Academy of Medical Sciences and Peking Union Medical College, 1 Xian Nong Tan Street, Beijing 100050, China
*E-mail: khlee@unc.edu

6.1 Introduction to Drug Discovery Based on Traditional Chinese Medicine (TCM)

Traditional Chinese medicine (TCM) presents a valuable opportunity for drug discovery and development with a high probability of success in the end. Approximately 500 plant species are commonly prescribed and documented by doctors of TCM as Chinese *materia medica* or traditional drugs (Chung Yao), and these medicinal plants have been used for thousands of years to maintain health and treat disease. Although these millennial-scale trials are not the same as modern-day clinical trials, they still provide a vast body of data that may be

RSC Drug Discovery Series No. 31
Traditional Chinese Medicine: Scientific Basis for Its Use
Edited by James David Adams, Jr. and Eric J. Lien
© The Royal Society of Chemistry 2013
Published by the Royal Society of Chemistry, www.rsc.org

usefully mined for relevance to twenty-first century medicine. In TCM, treatments are chosen based on careful observations and systematic principles and most often involve multicomponent processed herbal formulations. Thus, tradition is combined with diagnostic and scientific evaluations to generate a well-established medical system. In the previous century, the application of modern scientific technology to TCM herbal products led to many new and effective "single-component" drugs. In this century, medicinal chemistry approaches will continue to provide clinical trial candidates from TCM, in addition to adjunct therapies to current Western medicine. Mechanism of action studies on active components, active extracts and effective formulas of TCM will provide the scientific proof and future direction for new drug research. This chapter presents 10 case studies of new drug discovery from Chinese M*ateria* M*edica* used in TCM: ephedrine – pseudoephedrine from *Ephedra sinica* (Ma Huang), podophyllotoxin – etoposide from *Podophyllum emodi* (Gui Jiu), tanshinone IIA from *Salvia miltiorrhiza* (Tanshen or Danshen), anisodine – anisodamine from *Anisodus tanguticus* (Shan Lang Dang), camptothecin – irinotecan from *Camptotheca acuminata* (Xi Shu), schisandrin C – dimethyl dicarboxylate biphenyl (DDB) from *Schisandra chinensis* (Wu Wei Zi), artemisinin from *Artemisia annua* (Qing Hao), indirubin from *Indigofera tinctoria* (Qing Dai – Dang Gui Lu Hui Wan), huperzine A from *Huperzia serrata* (Qian Ceng Ta) and PG2 from *Astragalus membranaceus* (Huang Qi). Most often, the final clinical success relates in great part to the original documented use in TCM. By combining modern science with traditional practices, research may be guided to a greater probability of success with a more efficient allocation of effort and resources.

6.2 *Ephedra sinica*/Ephedrine – Pseudoephedrine

6.2.1 Introduction

Ephedra sinica (Ma Huang, Chinese ephedra) has been used as part of TCM for thousands of years, primarily to treat respiratory ailments such as colds and asthma. This traditional use was validated by the discovery of the constituent alkaloids, from which pseudoephedrine is now a common ingredient in many over-the-counter cold or allergy formulations, including Claritin, Contac and various Sudafed products.[1] However, it is now being widely replaced by often less effective alternatives, such as phenylephrine.[2] This latter action resulted mainly from the consequences of *Ephedra* moving "from mainstream medicine … into the twilight zone of street drugs and nutritional supplements."[3] Several adverse effects followed from the uncontrolled use of *Ephedra* and eventually led to a ban on its inclusion in dietary supplements in the USA. Thus, the history of *Ephedra* is a testament, and also a contrast, to the benefits and consequences of controlled (TCM) versus uncontrolled use of herbal medicine.[4]

6.2.2 Source Plant and Traditional Uses

Ephedra sinica (Family: Ephedraceae) is the most common among the *Ephedra* species in China. The plant has long, thin stems, a few tiny leaves, small yellow flowers and red berry-like fruit and grows in the arid areas of western China. Other species are found in different areas of the world, including India and Pakistan (also used medicinally for respiratory ailments), southern Europe and northern Africa and arid parts of North and South America. However, the chemical composition varies and the species in the Americas do not generally contain significant amounts of the ephedrine alkaloids, the major therapeutic constituents found in the stems and leaves of many Eurasian *Ephedra* species.[5]

Ephedra was recorded in the Chinese herbal classic *Shen Nong Ben Cao Jing* (*The Divine Husbandman's Classic of the Materia Medica*), written in the first century AD, for use to induce perspiration and to counter symptoms of allergies.[6] In TCM, both historically and currently, *E. sinensis* is generally used as a primary component in multi-herbal formulas to treat asthma, bronchitis, colds and flu, particularly the associated symptoms of nasal congestion, coughs, chills, aching joints, fever, headaches and wheezing.[6,7] For instance, *Ephedra* decoction, which contains *Ephedra* stem, cinnamon twig, apricot seed and licorice, and San-ao decoction, which contains ginger rhizome rather than cinnamon, are popular TCM prescriptions for cold/flu and asthma.[8,9] However, *Ephedra*'s constituent alkaloids, can also act as potent stimulants on the central nervous system, which likely accounts for the traditional stories of Zen monks and Mongolian soldiers using Ma Huang to stay awake and alert for long periods of time.[6]

6.2.3 Chemistry

The best-known constituents of *Ephedra* species are optically active, structurally relatively simple alkaloids. The major component is (–)-ephedrine [(*R,S*-2-methylamino-1-phenylpropan-1-ol], with minor amounts of (+)-pseudoephedrine [(*S,S*-2-methylamino-1-phenylpropan-1-ol] and the corresponding nor-(demethyl) and *N,N*-dimethyl derivatives of both compounds.[10] Ephedrine and pseudoephedrine vary only in the configuration (three-dimensional arrangement) at one carbon atom. Stereochemically, (–)-ephedrine has the (1*R,2S*)-configuration and (+)-pseudoephedrine has the (1*S,2S*)-configuration. Almost all commercial applications of *Ephedra* derive from these *N*-methylated alkaloids.

(-)-Ephedrine   (+)-Pseudoephedrine

6.2.4   Pharmacology

Ephedrine and pseudoephedrine are also structurally similar to adrenaline (otherwise known as epinephrine), an important hormone/neurotransmitter, and are classified physiologically as adrenergics (adrenaline mimics) or sympathomimetic agents. They can have direct agonist effects at adrenergic receptors (both α- and β-),[11] indirect effects by causing the release of endogenous noradrenaline (norepinephrine) from sympathetic neurons or effects at other unknown receptors.[12] Ephedrine causes central serotonin and dopamine release,[13] which might contribute to its suspected addictive nature.[14]

(–)-L-Adrenaline
or -Epinephrine

Many effects occur in the body as a result of adrenergic receptor activation. The decongestant action of pseudoephedrine is a result of vasoconstriction of blood vessels in the nasal membranes, causing less fluid to leave the vessels, less mucus to be produced and nasal membranes to become less inflamed. However, bronchial muscles are relaxed, leading to bronchodilation and easier breathing, which relieves shortness of breath in asthma. Respiratory rate and heart rate, in addition to blood pressure, are increased. These stimulant actions are similar to those caused by adrenaline in the flight-or-fight response.

6.2.5   Modern History

In 1953, the history of the *Ephedra* alkaloids was described in the monograph series *The Alkaloids*, Volume III.[15] In 1885, the Japanese chemist N. Nagai first isolated, characterized and recognized ephedrine as a bioactive compound from *Ephedra* species. Early syntheses of ephedrine were performed by Nagai in 1911, Eberhard in 1917 and Späth and Goring in 1920.[15] In the 1920s, Chen and Schmidt renewed interest in ephedrine,[16] leading to detailed analyses of the pharmacological effects of it and related compounds.[17,18] Freudenberg *et al.* established the stereochemistry of ephedrine and pseudoephedrine in 1932.[19] More recently, a history of *Ephedra* (Ma Huang) was published in 2011.[3]

During the 1990s, *Ephedra*-containing dietary supplements, which often also contained caffeine,[20] were widely promoted and used as diet aids or sports enhancers. The latter use, especially, could have stemmed from *Ephedra*'s stimulation of the central nervous system leading to increased alertness or faster reactions. However, adverse cardiological or neurological problems, including hypertension, heart palpitations, convulsions and even heart failure, led to serious safety concerns.[21–23] Consequently, many associations and businesses banned the use of such supplements and on 12 April 2004, the US Food and Drug Administration (FDA) enacted a ban on the unregulated, uncontrolled sale of *Ephedra* supplements.[24,25]

The ban on *Ephedra* sales and the move to replace pseudoephedrine with alternative agents in over-the-counter cold medications also reflect the fact that ephedrine and pseudoephedrine have similar chemical structures to that of methamphetamine (or "speed") and can be used in the illegal synthesis of this street drug.[26] Internationally, *Ephedra* alkaloids are listed by the International Narcotics Control Board as Table I chemicals frequently used in the illicit manufacture of narcotic drugs and psychotropic substances.[27]

It should be stressed that, as used according to TCM practice (particularly in multi-herbal formulations, rather than alone), *Ephedra* is safe.[4] *Ephedra*, as other herbal plants, must be used with appropriate care and monitoring, most often with other ingredients, *e.g.*, herbal expectorants. Additional processing may also be involved in TCM practice to modify and reduce adverse effects.[28]

The ephedrine alkaloids and numerous analogs still have important pharmaceutical applications, including their use to treat attention deficit hyperactivity disorder (ADHD)[1] and to control hypotension during spinal anesthesia for cesarean delivery.[29] Thus, advances in the industrial fermentation-based commercial preparation of *Ephedra* alkaloids are still being avidly pursued.[1]

6.3 *Podophyllum emodi*/Podophyllotoxin, Etoposide and Teniposide

6.3.1 Introduction

Podophyllum emodi (syn. *P. hexandrum*) (Family: Berberidaceae) (Gui Jiu or Kuei Chiu in Chinese) is native to the Himalayan region and is used in TCM as a contact cathartic.[30–32] Pa-chiao-lien in TCM is documented in *Pen Tsao Kang Mu Shih I*, written by Chao Hsueh-ming in 1765 AD.[33] Created from the rhizomes of *P. pleianthum*, it was traditionally used to treat carbuncles and acute tonsillitis, and also trauma, both internal and external.[33] In North America, native people historically used the roots of the related species *P. peltatum* (American mandrake or May apple) to treat snakebite and as a purgative, anthelmintic and cathartic agent.[34] It continued to enjoy medicinal use as a cathartic agent and for other purposes following the establishment of the European colonies and it appeared in the American Pharmacopeia from 1820 until 1942, when it was withdrawn for reasons of toxicity.[34]

Podophyllotoxin, the major chemical constituent of the alcoholic extract of *Podophyllum* rhizomes, is an aryltetralinlactone cyclolignan.[32] First isolated by Podwyssotzki in 1880, its empirical formula was determined by Borsche and Niemann in 1932.[32] Continuing investigations by various research teams culminated in the confirmation of the structure and configuration of the compound and later its total synthesis, by Gensler and Gatsonis in the 1960s.[34,35]

In 1942, the same year that it was withdrawn from the American Pharmacopeia, Kaplan reported the successful application of podophyllin (resin from the alcoholic extract of the *Podophyllum* rhizome) on *Condylomata acuminata* or venereal warts.[34] This marked the beginning of a new phase of study of podophyllotoxin as a potential modern anticancer drug.

6.3.2   Drug Discovery

Following the work of Kaplan, King and Sullivan reported in 1946 that upon topical application, podophyllin was observed to exhibit a mechanism of action similar to that of the suspected antimitotic agent colchicine.[36] Later work by Wilson and Friedkin connected the antimitotic effect of both colchicine and podophyllotoxin on their ability to bind with a high molecular weight "species A," speculated to be the basic subunit of the mitotic apparatus (and subsequently confirmed to be tubulin).[37]

Interest continued in the anticancer potential of podophyllin and, in 1963, scientists at Sandoz commercialized a *Podophyllum* extract derivative with reduced toxicity, which they named SP-G.[38] Inspired by successes with *Digitalis* glycosides for heart therapy, they had extracted the glucoside fraction of *Podophyllum* and condensed it with benzaldehyde to create SP-G. Through subsequent studies of SP-G over several years, which included tests on leukemic mice, an unusually potent pure compound was isolated from the mixture. This compound, demethylepipodophyllotoxin benzylidene glucoside (DEPBG), was observed to prevent cells from entering mitosis, a finding sufficiently distinct from the metaphase action of tubulin inhibitors to suggest a completely new mechanism of action. Soon after, other products of aldehyde condensation with DEPG (the unaltered *Podophyllum* glucoside) were evaluated as drug candidates, leading to the discovery and development of the clinically-successful drugs etoposide (VP-16 or VePesid) and teniposide (VM-26 or Vumon). Etoposide was approved in the USA for treatment of testicular cancer in 1983.[38] Teniposide was approved in the USA for treatment of pediatric acute lymphoblastic leukemia (ALL) in 1992.[39] Both drugs are effective against a fairly broad range of different types of cancer.[40]

Etoposide: R = CH$_3$

Teniposide: R =

6.3.3 Mechanism of Action

As mentioned above, podophyllotoxin achieves its antimitotic effect through inhibition of tubulin, the fundamental protein subunit that combines to form microtubules or the so-called mitotic spindle. Tubulin inhibition results in destabilization of microtubules and cell cycle arrest in metaphase.[41]

In contrast, at the time of its isolation from SP-G in 1965, DEPBG was observed to prevent cells from entering mitosis in cultures of chick embryo fibroblasts.[38] This suggested a new mechanism of action, a mechanism which the subsequent aldehyde condensation products, etoposide and teniposide, were also found to exhibit. In 1976, Loike and Horwitz reported observing a change in sedimentation of DNA from HeLa cells treated with etoposide.[42] They interpreted this to indicate that single-stranded DNA breaks resulted from etoposide treatment. Since purified HeLa cell DNA did not appear to be cleaved by etoposide, they speculated that the fragmentation was mediated by either unspecified cellular endonucleases or products of cellular metabolism of etoposide (or both). This discovery led to a period of active research into the nature of the interaction of etoposide and related compounds with DNA. It was ultimately resolved that etoposide affects both single- and double-stranded DNA breaks by stabilizing the complex formed by topoisomerase II and DNA ("the cleavable complex"); cytotoxic activity was attributed to double-stranded breaks.[34] The publications of Long and Minocha (1983) and Ross *et al.* (1984) represent significant events in the efforts to understand the etoposide/teniposide mechanism of action and the role of DNA topoisomerase II.[43,44]

The discovery of the DNA topoisomerase II inhibitor class of cytotoxic agent was arguably the most significant event in the history of the use of podophyllotoxin-related compounds against cancer. Discussions that point to additional possible mechanisms of action for podophyllotoxin analogs and derivatives may be found in the literature, including induction of apoptosis through various pathways[41] and other mechanisms that appear to be independent of either tubulin inhibition or DNA topoisomerase II inhibition.[45,46] Work continues in this area across a range of disciplines.

6.3.4 New Drug Design

Although the toxic side effects of podophyllotoxin, in particular gastro-intestinal toxicity, limited its usefulness as an anticancer drug in its own right, modifications of the podophyllotoxin structure led to the less toxic derivatives etoposide and teniposide. Despite being improvements over the parent compound from the perspective of toxicity, these derivatives nevertheless suffer from various shortcomings, including limited water solubility, metabolic inactivation and acquired drug resistance.[32] Myelosuppression is commonly associated with their use,[47] and both parent compound and derivatives demonstrate an undesirable lack of selectivity to the target cancer cells in their cytotoxic effect.[40]

Accordingly, a significant body of work has been devoted to creating podophyllotoxin derivatives and analogs to overcome these problems. This work has been reviewed by Gordaliza *et al.* (2000),[40] You (2005)[48] and Xu *et al.* (2009),[41] among others. In general:

- Modification of the A- and B-rings has led to few promising new anticancer agents.
- C-ring modification has focused on C7. This led to the glycosidic derivative NK-611 and also the 7β-alkylated TOP-53. Both are topoisomerase II inhibitors.
- D-ring modification has been attempted in order to address metabolic inactivation problems. These efforts resulted in both topoisomerase II inhibitors and tubulin inhibitors. C9-oxidized analogs demonstrated tubulin inhibiting activity.
- E-ring modification has shown that 4′-demethylation increases topoisomerase II inhibition. Etopophos (from Bristol Myers), a water-soluble prodrug of etoposide, is the 4′-phosphate ester of etoposide.[32]

GL-331, discovered in the present authors' laboratories, is a non-glycosidic derivative based on modification at C7. A DNA topoisomerase II inhibitor with activity superior to etoposide both *in vitro* and *in vivo*, this compound overcomes multidrug resistance (MDR) in many cell lines and can therefore be used for the treatment of any tumor that expresses MDR.[49] GL-331 advanced to Phase II clinical trials as an anticancer drug.

Azatoxin and tafluposide are two other noteworthy derivatives of podophyllotoxin to emerge over the past several decades. Azatoxin was rationally designed from a composite pharmacophore model of topoisomerase II inhibitors.[50] It was subsequently found also to inhibit tubulin at lower concentrations.[51] Tafluposide, or F 11782, is a dual inhibitor of topoisomerase I and II with a mechanism of action distinct from that of other topoisomerase inhibitors.[52,53]

GL-331   Etopophos   Azatoxin

Various structure–activity relationships (SARs) have been described for podophyllotoxin- and etoposide-like compounds.[41,49,54] Among the discussions, a 7β configuration is agreed to be important for topoisomerase II inhibitory activity, as is demethylation at C4′. The *trans*-lactone favors activity but is not essential to it and free rotation of the E-ring is needed.

MacDonald *et al.* (1991) considered the nature of the ternary drug–DNA–enzyme complex in formulating a general SAR model for drugs with topoisomerase II activity.[55] Their hypothetical pharmacophore featured three domains: a planar, polycyclic "intercalation" domain, a pendant "groove-binding" ring and a variable substituent region. This model contributed to the design of azatoxin and continued to be relevant to the design of new topoisomerase II inhibitors for over a decade.[54] The recent publication of the observed etoposide ternary cleavable complex crystal structure is also expected to assist new drug discovery efforts in going forward.[56]

6.4   *Salvia miltiorrhiza*/Tanshinone IIA

6.4.1   Introduction

Salvia miltiorrhiza has been used extensively in Asian medicine for cardiovascular disorders, but the distinct mechanisms by which it exerts its pharmacological actions still remain elusive. Structural determination of the various chemical constituents has allowed further insight into the effects, but more conclusive research is needed. Tanshinone IIA is a pharmacologically important constituent of *S. miltiorrhiza*. The design and development of novel derivatives of tanshinone IIA, and also other chemical constituents of this plant (such as *neo*-tanshinlactone in the authors' laboratories), is a promising area of natural product research and could provide significant benefits in medicine.

6.4.2   Source Plant and Traditional Uses

The rhizome of *Salvia miltiorrhiza* (Family: Labiatae), Tanshen or Danshen in Chinese, has been used in TCM for treating various disorders of the heart and blood vessels, including atherosclerosis or blood clotting abnormalities. Tanshen has also been used to treat hemorrhage, dysmenorrhea, miscarriage, swelling, inflammation, chronic hepatitis and insomnia.[57,58] Although used infrequently in ancient Chinese medicine, Tanshen has become increasingly popular in modern Chinese medicine.[59]

6.4.3   Chemistry

The more than 70 constituents isolated from *S. miltiorrhiza* have been divided into two groups based on physical and chemical properties and structural characteristics. Water-soluble compounds such as salvianolic acids and lithospermic acids along with other phenolic acids fall into the hydrophilic group. Abietane diterpenes fall into the lipophilic group, which contains more than 40 compounds. The tanshinones are abietane diterpenes with a common *o*- or *p*-naphthoquinone chromophore and are found exclusively in the *Salvia* genus. They were first isolated from the roots of *S. miltiorrhiza* by Nakao in 1930[60] and have been studied extensively since then.[61] The three constituents that account for the bulk of the total tanshinone content are tanshinone I, tanshinone IIA and cryptotanshinone.

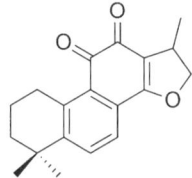

Tanshinone I   **Tanshinone IIA:** R = H   **Cryptotanshinone**
Sodium tanshinone IIA sulfate:
R = SO₃Na

Tanshinone IIA is the most abundant tanshinone and has been extensively studied to determine its role in the pharmacological actions of Tanshen.[59,60] Tanshinone IIA is not well absorbed through the intestinal walls owing to its high lipophilicity, so a salt derivative was developed utilizing a sodium sulfate group on ring D to increase the bioavailability.[62]

6.4.4 Pharmacology

Tanshen is associated with cardioprotective properties and is useful in angina pectoris, myocardial infarction, stroke and atherosclerosis, acting by improving blood circulation, decreasing blood stasis or exerting free radical scavenging, anti-inflammatory activity and other actions.[62–66] Owing to the abundance of tanshinone IIA in Tanshen, many studies have been carried out to elucidate its mechanism(s) of action.

6.4.4.1 *Angina Pectoris and Myocardial Infarction*

Both angina pectoris and myocardial infarction result from insufficient blood and oxygen supply to the heart and are commonly caused by atherosclerosis. Gao *et al.* discussed how tanshinone IIA can slow the progression and development of atherosclerosis via mechanisms involving free radical scavenging, transcription factor regulation and anti-inflammatory activities.[63] They determined that tanshinone IIA decreased low-density lipoprotein (LDL) oxidation and also lipid peroxidation by reducing the formation of reactive oxygen species (ROS). High levels of oxidized LDL (oxLDL) have been shown to correlate positively with increased risk of coronary heart disease and, therefore, decreased formation of oxLDL would be beneficial in preventing angina pectoris and myocardial infarction. Yang *et al.* demonstrated that sodium tanshinone IIA sulfate was able to protect cardiomyocytes from oxidative stress-induced apoptosis by preventing the formation of ROS through inhibition of JNK phosphorylation.[67] Fang *et al.* also provided evidence that tanshinone IIA decreases matrix metalloproteinase-2 (MMP-2) and MMP-9 expression and activity present in atherosclerotic plaques.[68] Decreasing the activity of both enzymes could potentially stabilize atherosclerotic plaque and thus prevent the ruptures that lead to vasculature occlusions. The cardioprotective effects of tanshinone IIA after myocardial infarction could be due to inhibition of nuclear factor kappa B (NF-κB) and subsequent anti-inflammatory activity.[63]

6.4.4.2 *Other Uses*

Sodium tanshinone IIA sulfate has been studied for its ability to protect against cardiotoxicity induced by doxorubicin. Doxorubicin is a broad spectrum antibiotic used in cancer chemotherapy, but it exhibits dose-related cardiotoxicity believed to be due to an increase in oxidative stress. Both *in vitro* and *in vivo* results have shown that the scavenging and antioxidant properties of sodium tanshinone IIA sulfate can decrease the overall cardiotoxicity of doxorubicin, therefore alleviating its dose-dependent limitation.[69,70] Tanshinone IIA and sodium tanshinone IIA sulfate have both been studied to determine whether they exhibit antihypertensive properties,[66,71] or have uses in cardiac hypertrophy, myocarditis in children and myocardial ischemia.[72-75]

While most studies on tanshinone IIA have addressed its effects on the heart and its vasculature, significant research has also been performed on whether tanshinone IIA exhibits antitumor activity.[60,76] In addition to determining tanshinone IIA's antitumor profile, efforts have also been focused on the mechanism by which the antitumor activity occurs. Possible mechanisms are induction of apoptosis, inhibition of invasion and metastasis, inhibition of angiogenesis and regulation of growth factor receptors.[60,77]

Other studies have explored the effects of tanshinone IIA in obesity, postmenopausal syndrome, hepatitis, osteoclast formation and psoriasis.[78-82]

6.4.5   Use as a Lead for New Drug Design

Tanshinone IIA has also been used as a lead compound in the development of novel derivatives that aim to increase therapeutic activity or increase selectivity for a specific pharmaceutical action. Liu *et al.* explored how modifications of ring A of tanshinone IIA affects its antiprostate cancer activity. They concluded that ring A has an important role in the antiandrogen activity and that further optimization would be beneficial in the development of novel antiprostate cancer agents.[83] Novel tanshinone IIA derivatives have also been developed and assessed as vasodilative agents.[84] Bi *et al.* developed several novel derivatives and determined their actions on contractile force of vascular thoracic aorta smooth muscle.[84] They concluded that the newly developed derivatives exhibited good activity, but further optimization would be necessary. Other constituents of Tanshen have also been explored as leads for drug development. One example is *neo*-tanshinlactone, which contains a lactone ring C in place of the *o*-quinone ring of tanshinone IIA. Lee and co-workers have carried out extensive work on the design and optimization of *neo*-tanshinlactone derivatives.[85] *neo*-Tanshinlactone exhibits potent and selective activity against several breast cancer cell lines (10-fold more potent and 20-fold more selective than tamoxifen citrate, a widely used selective estrogen receptor modulator). Chemical modification led to 4-ethyl-*neo*-tanshinlactone, which showed potent *in vivo* activity and was selective for a subset of breast cancer-derived cell lines, but notably less active against normal breast-derived tissue.[86] Other structural modification studies have investigated new analogs with simplified skeletons.[87-92] Overall, these novel analogs display significant antitumor activity, high selectivity compared with normal cell lines and different tumor-tissue type selectivity. Hence they show significant promise for development as candidates in clinical trials for breast cancer.[60]

Neo-tanshinlactone: R = CH$_3$
4-Ethyl neo-tanshinlactone: R = CH$_2$CH$_3$

6.4.6   Use as a Chemical Probe

Owing to the ambiguity of tanshinone IIA's mechanism of action, the drug development process would benefit from the design of chemical probes to elucidate which proteins play a role in the compound's activity and, thus, provide important targets for drug development. Lee *et al.* designed and synthesized novel chemical probes with tanshinone IIA as the backbone.[93] They selected trifluoromethylaryldiazirine as the probe's photophore owing to its chemical stability, photoactivation under long-wavelength ultraviolet light and generation of a reactive carbene that can react with the proteins with little intramolecular rearrangement. It was concluded that the functionalized tanshinone–photophore probe maintained the bioactivity of the natural product and therefore it could be used in the determination of key target proteins involved in osteoclastogenesis.

6.5   *Anisodus tanguticus*/Anisodamine and Anisodine

6.5.1   Introduction

Anisodus tanguticus (also known as *Scopolia tanguticus*) (Family: Solanaceae) is one of the 50 fundamental time-honored herbs used in TCM. However, *A. tanguticus* looks physically very much like *Phytolacca esculenta*, which was frequently used to treat edema. Consequently, roots of *A. tanguticus* were mistakenly exported to various Chinese provinces in the 1960s for use as *P. esculenta*. Because of this misidentification, detrimental effects, including mydriasis, hallucination and sleep/mental disturbances, occurred.[94,95] This circumstance shows that careful identification and quality control must be maintained in the use of herbal medicines. However, the mistake also led to detailed studies of the chemical constituents of *A. tanguticus* and to the identification of anisodamine and anisodine as active principles.[95]

Subsequently, anisodamine was found to be a preferred agent to treat septic shock in children resulting from epidemic toxic dysentery and meningitis in China during the 1960s.

6.5.2   Source Plant and Traditional Uses

A. tanguticus is a perennial flowering plant native to China, Tibet, India, Bhutan and Nepal. It is found widely in the Qinghai–Tibetan Plateau at altitudes of 2800–4200 m and, because of the topography, is often located in isolated areas, leading to a high level of genetic differentiation.[96] Its widespread use as a medicinal herb has led to a declining supply in the wild.

In China, *A. tanguticus* is known medicinally as Shān Làng Dàng or Zang Qie and locally in the Qinghai–Xizang Plateau as Zhangliushen. The plant roots were used traditionally to treat toothache, stomachache and other pains.[94,95]

6.5.3   Chemistry

The medicinal effects of *A. tanguticus* are mainly related to natural alkaloids, especially tropane alkaloids. Chemically, all tropane alkaloids contain a core tropane ring, a seven-membered ring with a methylated nitrogen bridging between C1 and C5. Alternatively, the tropane system can be described as two fused heterocyclic rings, one six-membered piperidine ring (C1–C2–C3–C4–C5–N8) and one five-membered pyrrolidine ring (C1–N8–C5–C6–C7), with two carbons and the nitrogen (C1–N8–C5) common to both rings.

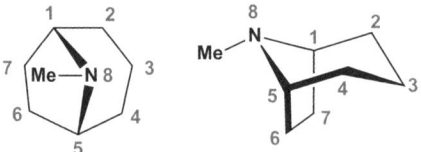

Structure of Tropane Ring System

Chemically, anisodamine is 6β-hydroxy-1αH,5αH-tropan-3α-ol tropate. Its tropane ring is hydroxylated on C3 of the piperidine ring and C6 of the pyrrolidine ring. Furthermore, the 3-hydroxy group is esterified with the carboxylic acid moiety of tropic acid (3-hydroxy-2-phenylpropanoic acid).

Anisodine is structurally similar to anisodamine. However, instead of a 6-OH group, an epoxy group is present between C6 and C7 on the pyrrolidine ring. In addition, the C3 acyl moiety contains an additional hydroxy group at C2′ (2-hydroxytropic acid). The structure of (–)-anisodine was assigned by Xie *et al.* as 6β,7β-epoxy-1αH,5αH-tropan-3α-ol (*S*)-2′-phenyl-2′,3′-hydroxypropionate based on chemical correlation and spectroscopic analysis.[95]

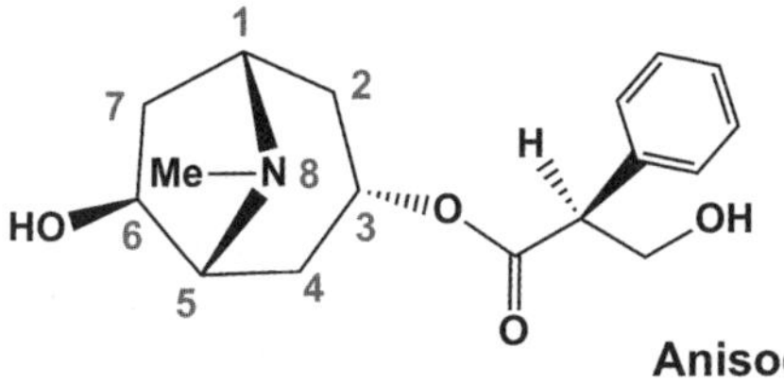

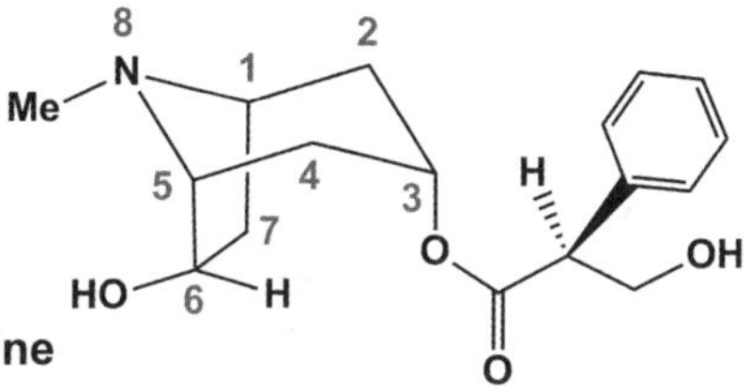

Anisodamine

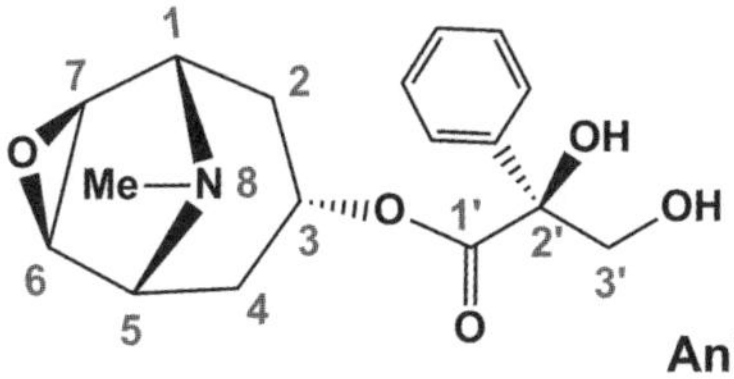

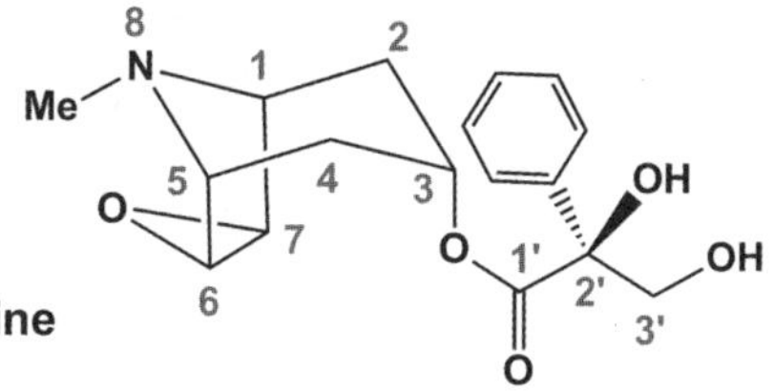

Anisodine

Anisodamine and anisodine are structurally related to other tropane alkaloids, such as hyoscyamine (which lacks the 6-hydroxy group found in anisodamine) and scopolamine (which lacks the 2′ hydroxy group found in anisodine). Atropine [the racemic mixture of (*R,S*)-hyoscyamine] is best known for its use as a topical ophthalmic mydriatic to dilate the pupil of the eye, and scopolamine is frequently used in a transdermal patch to treat motion sickness.

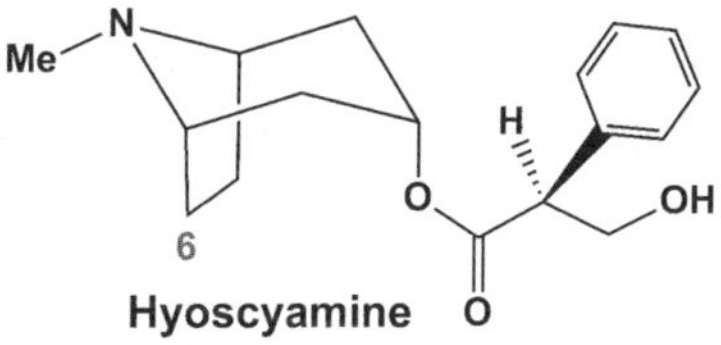

Hyoscyamine

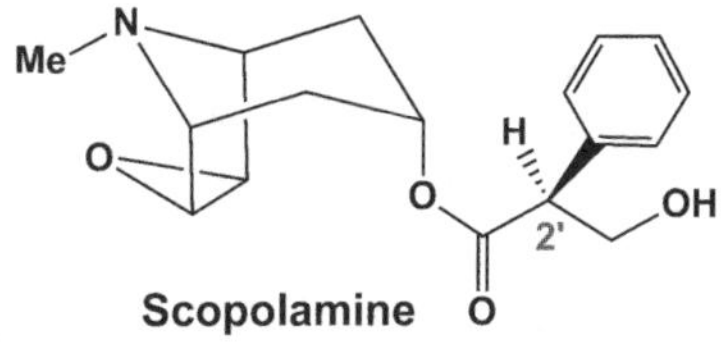

Scopolamine

6.5.4 Pharmacology

Mechanistically, anisodamine and anisodine (and also atropine and scopolamine) affect the parasympathetic nervous system and act as non-specific muscarinic acetylcholine receptor antagonists (also known as cholinoceptor antagonists) and α_1-adrenergic receptor agonists.[97] Thus, their major actions are due to antimuscarinic and anticholinergic effects. The actions of these compounds result in various physiological effects in the body, some of which are listed below:

- *antispasmolytic action:* inhibits gastrointestinal motility/propulsion and bladder contractions;
- *antisecretory action:* inhibits gastrointestinal, salivary and sweat secretions;
- *mydriatic action:* dilates the pupil of the eye and can cause cycloplegia (paralysis of ciliary muscle);
- *bronchodilating action:* dilates bronchi and bronchioles by reducing smooth muscle contraction and respiratory mucus secretion.

Because these four tropane alkaloids have similar pharmacological actions, their potencies and efficacies are often compared in scientific studies. A review on anisodamine by Poupko *et al.* in 2007 compared its potency relative to atropine (also scopolamine and anisodine) with respect to many pharmacological effects.[97] Early experimental and clinical studies in China reported that anisodamine had essentially the same spasmolytic activity as atropine, but was a better spasmolytic agent owing to its lesser effects on the CNS, salivary glands and pupils.[98,99] Anisodine exhibits stronger CNS action than anisodamine and has a lower LD_{50} than atropine, scopolamine and anisodamine.[1,100]

6.5.5 Drug Development

The therapeutic use of anisodamine has not extended much beyond Asia. However, in China, anisodamine has been used clinically to treat circulatory disorders, such as septic shock, disseminated intravascular coagulation (DIC), thrombophlebitis and coronary artery disease,[100] rheumatoid arthritis, eclampsia and lung edema[100] and morphine addiction.[101] Anisodine is a good ganglion-blocking agent, *i.e.*, blocks nerve impulses at autonomic ganglionic synapses, and is used clinically to treat motion sickness, migraine and vascular spasm in the interior of the eye.[94,95]

6.5.5.1 Septic Shock

The major therapeutic avenue for anisodamine is to treat septic shock. In China during the 1960s, large doses of tropane alkaloids were used successfully to decrease the number of deaths resulting from acute circulatory and septic shock caused by various infectious diseases, including toxic dysentery and meningitis, particularly in children.[102] In 1966, a study reported particularly good results using anisodamine to treat epidemic fulminant meningitis.[103] Mortality rates dropped from 20 to 0.5% in cases of shock from toxic dysentery and from 60 to 14% with shock from meningitis. Anisodamine was the preferred agent, because atropine had a very strong effect on salivary gland secretion, and the resulting dryness of the throat could interfere with intubation. The beneficial effects of the tropane alkaloids on shock were attributed to relaxation of vascular smooth muscle with vasodilation. Consequently, microcirculation and blood flow would be improved, without harmful decreases in blood pressure.[103–106] However, more recent studies have indicated that vasodilation is not the only mechanism and the beneficial effects of these compounds against shock may result from complex mechanisms.[97,107–109]

Xie and co-workers accomplished the synthesis of anisodamine[110,111] and anisodine,[94,95,112] and also quaternary ammonium salts.[113] The synthetic racemic anisodamine was named 654-2 to distinguish it from the natural pure *levo* form 654, but the two compounds had very similar spectroscopic (IR and UV) and pharmacological properties. The laboratory-scale synthesis was extended to industrial-scale production in hundreds of kilogram quantities and the drug is still in production industrially[99] by Beijing Pharmaceutical and Hangzhou Mingshen Pharmaceutical Companies.

Anisodamine passed clinical evaluation by the Chinese Ministry of Health in 1993, has been included as raceanisodamine in the Chinese Pharmacopeia since 1995 and was approved by the Chinese State Drug Administration for both injection and tablet forms in 2002.[114] Grynkiewicz and Gadzikowska listed anisodamine as a new drug for septic shock (although not registered beyond Asia) in a review of tropane alkaloids as medicinally useful natural products and synthetic derivatives as new drugs.[115] Future research in this area is highly warranted and will likely continue.

6.5.5.2 Other Research

Tropane alkaloids are also under investigation as antidotes for organophosphorus poisoning,[97,116,117] to treat snakebite,[118] and to prevent or treat diabetic retinopathy or related eye disorders.[119] Anisodamine has also been found to have preventive and therapeutic effects on various pulmonary conditions, including acute respiratory distress syndrome[120] and allergic asthma.[121] It should also be noted that tiotropium bromide,[115] the active agent of Spiriva, which is successfully marketed for maintenance treatment of chronic obstructive pulmonary disease (COPD), is a synthetic quaternary ammonium tropane analog. It has a tropane alkaloid core structure and an epoxide group like anisodine, but instead has a dimethylated quaternized amine and is esterified at the C3 hydroxyl with 2-hydroxy-2,2-di(thiophen-2-yl)acetic acid. Its antimuscarinic action leads to reduced muscle contraction and mucus secretion and, thus, causes bronchodilation.

Tiotropium bromide

6.6 *Camptotheca acuminata*/Camptothecin

6.6.1. Introduction

Camptotheca acuminata is a deciduous tree that is native to areas of China and Tibet.[122] Called Xi Shu ("tree of joy"), it was first documented in TCM in *Chih Su Ming Shih Tu Kao*.[123] The dried root and fruit were used traditionally for gastric, rectal and cystic carcinoma, and also chronic lymphemia and hepatosplenomegaly due to schistosomiasis.[123] *C. acuminata* became relevant to modern drug discovery following the 1966 publication of an article by Wall *et al.*[124] The article reported a constituent with potent anticancer activity in animal models was isolated from the *C. acuminata* tree stem wood. This was the first step in a chain of events that ultimately led to the discovery and development of the cancer chemotherapy drugs topotecan (Hycamtin) and irinotecan (Camptosar).[125]

6.6.2 Drug Discovery

In 1950, Monroe Wall was selected to lead a US Department of Agriculture (USDA) program to identify plants that could be useful sources of steroids.[126] In the course of collecting data for the program, Wall, in collaboration with Jonathan Hartwell of the National Cancer Institute (NCI), discovered that an extract from *C. acuminata* demonstrated high activity in one of the NCI's *in vivo* assay systems. Wall pursued fractionation and isolation of the active principle, camptothecin, following his establishment of the Natural Products Laboratory at the Research Triangle Institute in 1960.

The earlier article[124] described a novel pentacyclic alkaloid that resulted in life prolongation as high as 100% when tested against leukemia L1210 in mice administered 0.25–1.0 mg kg^{-1} of the compound daily. Against the KB cell culture, the ED$_{50}$ was reported to be 0.07 µg ml^{-1}. However, clinical trials of the sodium salt of camptothecin were unsuccessful owing to limited efficacy and severe toxicity.[122]

Interest in the compound dwindled until 1985, when Hsiang *et al.* reported that camptothecin achieves its cytotoxic effect by inhibiting DNA topoisomerase I,[127] a nuclear enzyme that participates in DNA replication and transcription. It was ultimately resolved that camptothecin binds with the topoisomerase I–DNA complex to form a cleavable complex.[128] Upon interaction with the replication fork, double-stranded DNA breaks result, which in turn cause cell death.

The breakthrough in understanding of the mechanism of action led to further efforts to develop candidates for clinical trials from the camptothecin lead. These efforts eventually resulted in the introduction of topotecan and irinotecan into clinical use. Topotecan (NSC-603071) is approved for treatment of metastatic carcinoma of the ovary and small cell lung cancer[129] and irinotecan (CPT-11) is approved for treatment of metastatic colon cancer.[129]

Topotecan   Irinotecan

More recently, Staker *et al.* in 2002 reported observations of the X-ray crystal structure of the ternary complex formed by topoisomerase I, DNA and topotecan.[130] The planar, polycyclic ring system of topotecan was observed to mimic a DNA base pair. Consistent with SAR observations, topotecan's C7, C9 and C10 were observed to be oriented into the DNA major groove, which would allow a wide range of substitutions without steric hindrance. The lone hydrogen bond between the enzyme and topotecan was observed between Asp533 and the 20(*S*)-hydroxyl, also consistent with SAR observations of the importance of the 20(*S*)-hydroxyl. Enzyme mutations at key observed points of contact (Asp533, Arg364, Asn722) would be predicted to result in drug resistance; such resistance has been reported in the literature. Of course, it is unknown to what extent the crystal structure reflects the biological system; nevertheless, these observations should guide future efforts to design improved camptothecin analogs as anticancer drugs.

6.6.3 Lead for New Drug Design

Without modification, camptothecin is unsuitable as a drug because of limited water solubility, toxicity and metabolic inactivation of the lactone ring.[131] These limitations motivated continuing work to discover new drugs from the camptothecin lead. From the cumulative experience in attempting a wide range of structural changes, the following general conclusions may be drawn regarding the SAR of camptothecin analogs:

- The E-ring α-hydroxylactone is required for activity.
- The naturally occurring 20(S) stereoisomer is active, whereas the 20(R) form is inactive.
- The pentacyclic ring structure is essential for activity.
- The pyridine moiety in the D-ring is also needed for activity.
- Planarity of A-ring substituents is required.[126]

The literature should be consulted for detailed descriptions regarding the effects of specific modifications on bioactivity.

Rahier *et al.* presented a discussion of camptothecin analogs either in clinical use or in clinical trials as of 2011.[125] In addition to topotecan and irinotecan, belotecan has been approved for clinical use in South Korea. Belotecan is a water-soluble camptothecin analog with an isopropylaminoethyl group at C7. Among the drugs in clinical trials are water-soluble analogs (Lurtotecan, OSI-211, DRF-1042, Namitecan, Elomotecan), water-insoluble analogs (Cositecan, AR-67, Gimatecan, Diflomotecan) and camptothecin conjugates (CZ-48, TP-300, PEG-CPT, EZN-2208 and others). Novel strategies are also being explored for camptothecin drug delivery, including microparticulate carriers and liposomes.[131]

6.7 *Schisandra chinensis*/DDB

6.7.1 Introduction

Schisandra chinensis is used as a TCM to treat and protect against liver disorders, including viral and chemical hepatitis.[132] Dimethyl dicarboxylate biphenyl (DDB) was developed as a synthetic intermediate derivative of the natural bioactive lignans found in *S. chinensis*. The successful discovery of this antihepatitis drug was a major achievement that explored and built upon the rich heritage of TCM and is a milestone in drug discovery history.

6.7.2 Source Plant and Traditional Uses

Schisandra chinensis (Family: Schisandraceae) is indigenous to forests of northern and northeastern China and far eastern Russia. It is a woody vine with clusters of small, red berries. All five basic flavors (salty, sweet, sour, savory and bitter) can be tasted in the berries, suggestive of the plant's Chinese name Wu Wei Zi, meaning "five flavor berry." *S. chinensis* is more specifically called Bei-Wu Wei Zi, translated as "northern five flavor berry," to differentiate it from similar plants, including *S. sphenanthera*, called Nan-Wu Wei Zi. Since 2000, these two plants have been regarded as separate crude drugs.[133]

S. chinensis, one of the 50 fundamental herbs of TCM, was recorded in *Shen Nong Ben Cao Jing* and used to treat chronic cough and labored breathing, diarrhea, night sweats, wasting disorders, irritability, palpitations and sleep disorders.[134] In Japan, the Ainu people used *S. chinensis*, or repnihat, as a folkloric remedy for colds and sea-sickness.[135] Most research has been performed on its hepatoprotective, anti-inflammatory and adaptogenic effects.[136]

In Chinese TCM, the berries of *S. chinensis* are most often dried and then boiled to make a tea. In Korea, the berries are known as omija and the tea made from the berries is called omija cha. In Japan, the berries are known as gomishi. In Russia, the berries are widely used in commercial juices, wines, fruit extracts and sweets.

6.7.3 Chemistry

The primary hepatoprotective and immuno-modulating constituents in *S. chinensis* are lignans, found primarily in the seeds of the fruit. Dibenzocyclooctadiene lignans account for over 80% of the lignan content. They are characterized by a biphenyl ring system linked by a four-carbon bridge, creating a cyclooctadiene ring. The compounds have either an (*R*)-configuration or (*S*)-biphenyl configuration, referring to the three-dimensional twist of the biphenyl rings. Although schisandrin B and gomisin N would look the same when drawn in only two dimensions, they are different in three dimensions. The former compound has an (*R*)- and the latter an (*S*)-biphenyl configuration. Detailed NMR and mass spectrometric data, combined with X-ray crystallography, have recently been reported on gomisin N.[137]

Bicyclooctadiene Lignan Stereochemistry

R-Biphenyl Configuration **2D-only Representation** **S-Biphenyl Configuration**

Schisandrin B   Gomisin N

Some dibenzocyclooctadiene lignans with an (*R*)-biphenyl configuration are schisandrin, schisandrin A (deoxyschisandrin), schisandrin B (γ-schisandrin), gomisin A and gomisin K2. Compounds with an (*S*)-configuration are gomisin N, schisandrin C, schisantherin A (gomisin C), schisantherin B (gomisin B), gomisin S and angeloylgomisin Q.[138]

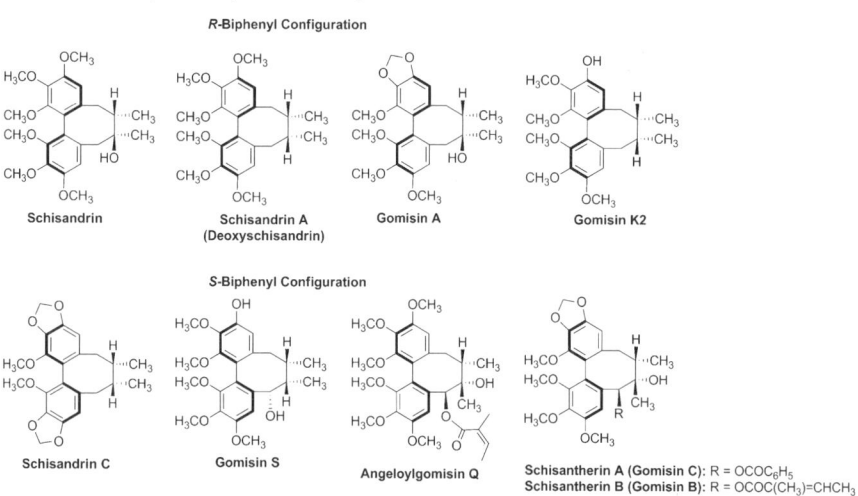

Examples of Bicyclooctadiene Lignans in *S. chinensis*

Two reviews have been published on the chemical composition of *S. chinensis*, focusing on the bioactivities and also isolation and analytical methods, including fingerprint techniques.[139,140] Quantitative determination methods for the lignans have also been summarized, particularly regarding quality control on the crude drug and its preparations.[141,142] Analytical methods used to distinguish between *S. chinensis* (Bei-Wuweizi) and *S. sphenanthera* (Nan-Wuweizi) have also been described.[133]

6.7.4   Pharmacology

6.7.4.1   Adaptogenic Effects

Adaptogens are defined as compounds that increase the ability of an organism to adapt to environmental factors and to avoid damage from such factors.[143,144] In 1961, *S. chinensis* was recognized as a restorative adaptogen in the official USSR Pharmacopea.[145] Extensive studies were documented by Panossian and co-workers,[146,147] including an extensive review in 2008, describing pharmacological studies in animals, on isolated organs, cells and enzymes, on healthy human volunteers and in clinical trials.[145] *S. chinensis* is an adaptogenic herb and has health benefits that have been described.[148]

6.7.4.2. Antioxidant and Immunomodulating Effects

Schisandra promotes activity of antioxidant enzymes, such as superoxide dismutase (SOD) and glutathione peroxidase (GPX), which reduce cell-damaging oxygen free radicals.[149] The major bioactive lignan schisandrin exhibited antioxidative effects both *in vitro* and *in vivo* in mice,[150] and the related schisandrin B directly scavenged active oxygen radicals[151] and prevented mitochondrial dysfunction in oxidatively stressed brain.[152] *S. chinensis* may be therapeutically beneficial in promoting the body's antibody- and cell-mediated immune responses, as schisandrin and gomisin A can induce the release of various cytokines.[153]

6.7.4.3 Anti-inflammatory Effects

Various lignans, including gomisin C (schisantherin A), gomisin N, gomisin J, schisandrin and schisandrin C, exhibit anti-inflammatory effects, such as inhibiting lipopolysaccharide (LPS)-induced nitric oxide production, prostaglandin E2 (PGE2) release and cyclooxygenase-2 (COX-2) expression. The compounds also reduce levels of the pro-inflammatory cytokines tumor necrosis factor alpha (TNF-α) and IL-6.[150,154,155]

6.7.4.4 Hepatoprotective and Detoxifying Effects

Cyong *et al.* reported positive results when several patients with chronic hepatitis C were treated with a multicomponent herbal medicine containing *S. chinensis*.[156] Many other studies attest to the liver protective/detoxification effects of *S. chinensis* and its constituents, although questions remain about the study protocol.[157] *S. chinensis* seed extract can enhance antioxidation and detoxification[132] by increasing and restoring drug metabolism in damaged liver.[158,159] In addition, various bicyclooctadiene lignans, including schisandrin B and C and gomisin A, have also demonstrated important *in vitro* and *in vivo* effects on liver protection and repair.[160] The pharmacological effects and molecular mechanisms[161] include inducing action on liver cytochrome P-450,[162] inhibiting lipid peroxidation,[163] reducing alanine aminotransferase (ALT) levels and death rates in mice[164] and, particularly, elevating the level of hepatic mitochondrial glutathione (GSH).[165] A proteomics study indicated that schisandrin B down-regulates Raf kinase inhibitor protein (RKIP), which may be a key regulatory protein in the development of hepatotoxin-induced cell damage.[166]

6.7.5 Drug Development: Discovery of DDB

Studies in China in the early 1970s showed that *Schisandra* extracts could provide beneficial effects on markers of liver damage and reduce histopathological liver damage.[167] When *Schisandra* was used as an adjuvant treatment for the fatigue common to infectious hepatitis patients, elevated levels of serum glutamic pyruvic transaminase (SGPT), a marker for infective hepatitis and other liver disorders, could be lowered quickly and without obvious toxicity. Subsequently, seven bicyclooctadiene lignans were isolated from an ethanol extract of the seeds of *S. chinensis* by Chen *et al.*,[168] and several compounds were linked to the therapeutic and hepatoprotective effects in liver diseases.[169,170] Schisandrin C most potently lowered SGPT and improved inflammatory liver lesions and necrosis.[171] Although it was selected as the lead compound, the natural supply (the plant content was below 0.08%) could not meet the needs for further study. A collaborative effort, including scientists from Shanghai Zhonghua Pharmaceuticals Co. and the 15th Shanghai Pharmaceuticals Co., was initiated to accomplish the total synthesis of schisandrin C.[172,173] During the synthesis of the four isomers of schisandrin C, the *trans–cis* relationship of the two methyl groups in the cyclooctadiene ring and the relative positions of methoxy and methylenedioxy groups on the biphenyl core were determined.[174] A thorough description of the synthesis of schisandrin C, including spectroscopic analyses, was published in 2009.[175]

DDB

Although the synthesis of schisandrin C was successful, it was relatively long, resulted in low yields and had some technical challenges. Therefore, it was difficult to supply a sufficient amount of the compound for pharmacology studies. Although the biphenyl unit of the bicyclooctadiene lignans is critical for their pharmacological activities, the cyclooctadiene ring is not and, accordingly, simplified active schisandrin analogs were pursued.[176,177] Among more than 30 synthetic analogs, dimethyl 4,4'-dimethoxy-5,6,5',6'-dimethyle-nedioxybiphenyl-2,2'-dicarboxylate (DDB) was designated as a promising drug candidate. It was not as active as schisandrin C, but it was less toxic and more bioavailable.[157] DDB exhibited SGPT-lowering activity, could improve liver function and also protect the liver. A reliable, efficient process was developed for large-scale preparation and sufficient material was produced for toxicity studies. After continued good results, DDB was approved as an antihepatitis drug and was commercialized at the end of 1980. It was rapidly adopted for use in China and hundreds of thousands of hepatitis patients were treated effectively and cured successfully.[178]

The discovery of DDB was highly recognized, winning the First Level Accomplishment Award from the Ministry of Health and the Third Level State Invention Award in China. The drug has been exported to many countries and has been used to treat chronic viral hepatitis B patients in China for more than 20 years, and also in Korea and Egypt for more than 10 years, without any significant adverse effects.[179,180]

It should be noted that different acronyms for DDB, including diphenyl dimethyl bicarboxylate, biphenyl dimethyl dicarboxylate (BDD)[181] and diphenyl dimethyl dicarboxylate (PMC), are used in the literature.[182]

6.8 *Artemisia annua*/Artemisinin

6.8.1 Introduction

The 1971 discovery by Chinese scientists of the antimalarial compound artemisinin from *Artemisia annua* (Family: Asteraceae) was a major break-through in research on TCM for new drug discovery. This event was of global significance and impact, as it led to critically-needed treatment for millions of people suffering from the parasitic disease. Also, it was the highest profile instance of a discovery of a modern drug from TCM by scientists working within China.

6.8.2 Drug Discovery

Malaria is a disease caused by the parasites *Plasmodium falciparum* and *P. vivax*, transmitted among human populations by infected mosquitoes. Symptoms of the disease appear 10–15 days following infection and include fever, chills, sweating, headache, nausea and vomiting. Left untreated, malaria can be fatal. The World Health Organization (WHO) reported 216 million episodes of malaria worldwide in 2010 (655 000 deaths), with the disease endemic to 106 countries.[183]

In the West, the natural product quinine was the first widely available antimalarial drug. An alkaloid occurring in the bark of the Peruvian *Cinchona* tree, quinine was used through World War II, after which it was largely superseded by the newer, more economical synthetic antimalarials, most notably chloroquine.[184,185] Decades of use of quinine and the early synthetic antimalarials resulted in widespread resistance of *P. falciparum* and *P. vivax* to these agents. Hence, in the late 1960s, the Chinese government launched an initiative to find new drugs to combat the malaria epidemic.[186]

As a result, in 1971 a consortium of Chinese scientists discovered that the flowering annual *A. annua* contained a potent antimalarial constituent, subsequently named qinghaosu or artemisinin.[187] Use of *A. annua* had been documented in TCM in *Recipes for 52 Kinds of Diseases* of 168 BC, Ge Hang's *Handbook of Prescriptions for Emergency Treatment* of 340 AD and Li Shihzen's *Compendium of Materia Medica* of 1596,[185] and also *Shen Nong Ben Cao Jing* of the first century AD and *Wenbing Tiaobian* of 1798.[188] The newly discovered active principle artemisinin was noteworthy not only for its efficacy against malaria, but also for its unusual chemical structure, featuring an endoperoxide bridge within a sesquiterpene lactone.[189,190]

Youyou Tu was awarded the 2011 Lasker–DeBakey Clinical Medical Research Award for her contributions to the discovery of artemisinin and has described the events that led to the breakthrough.[187] As part of the national Project 523, a group under Tu's leadership screened more than 2000 Chinese herb preparations, with more than 380 promising extracts then evaluated in a murine model of malaria. On 4 October 1971, an extract was obtained with 100% effectiveness in treating malaria-infected mice and monkeys. Subsequent clinical application confirmed the efficacy against both *P. falciparum* and *P. vivax* in human patients.

The discovery of artemisinin led – both in China and in the West – to the identification of many derivatives with greater solubility and greater potency. These first-generation derivatives included oil-soluble artemether and arteether, water-soluble (but less stable) artesunate and dihydroartemisinin (DHA), to which artemisinin is metabolized in the body.[185,191,192] Like artemisinin itself, the derivatives share a short metabolic half-life in relation to other antimalarial drugs along with their unusual efficacy.[193] Solubility limitations and concerns about the potential for neurotoxicity, particularly with artemether and arteether, motivated the search for additional artemisinin derivatives suitable for clinical use. [185,190,191]

Artemether: R = β-OMe
Arteether: R = β-OEt
Sodium artesunate: R = α-OC(O)CH₂CH₂CO₂Na
Dihydroartemisinin: R = α + β OH

6.8.3 Major Uses and Mechanism of Action

Currently, the WHO recommends artemisinin-based combination therapies (ACTs) for uncomplicated *P. falciparum* malaria, the most lethal form of malaria.[183] These therapies pair a shorter-half-life artemisinin derivative with a longer-half-life partner antimalarial to maximize therapeutic efficacy and minimize the chances of resistance emerging.[192,193] The ACTs currently in clinical use include artemether–lumefantrine (Coartem), artesunate–amodiaquine, artesunate–mefloquine and artesunate–sulfadoxine–pyrimethamine combinations; other combinations continue to be studied.[192,193] The ACTs are generally more expensive than earlier antimalarial drugs ($1.40 for a course of artemether–lumefantrine *versus* $0.10 for a course of chloroquine).[183,192] Safety and quality are growing concerns.

Three major theories have been proposed regarding the mechanism of action of artemisinin and its derivatives:

1. *Cleavage of the peroxide bond by heme-associated iron, followed by formation of carbon-centered radicals*

   This theory was inspired by analogy to a Fenton-like reaction.[194] Heme, the by-product of parasite digestion of hemoglobin, was thought to present labile divalent iron that would catalyze the cleavage of the peroxide bond in artemisinin, thereby yielding alkoxyl radicals that quickly transformed to carbon-centered radicals.[195] The subsequent targets of the carbon-centered radicals were never resolved (or observed); however, they were generally considered to be parasite proteins.[196]

2. *Disruption of mitochondrial function*

   The possible mitochondrial role in the mode of action of artemisinin has been most clearly indicated by studies of the growth of yeast on non-fermentable media, along with selected animal studies.[190,196] In the former case, yeasts, which require mitochondrial participation in the metabolism of non-fermentable carbon sources, failed to grow in the presence of such carbon sources and artemisinin.[197] This inhibition was hypothesized to result from depolarization of the mitochondrial membrane by reactive oxygen species (ROS). With respect to the animal studies, parasite mitochondrial swelling was reported after administration of artemisinin in studies of *P. inui*-infected monkeys and *P. berghei*-infected mice.[198] Impairment of oxidative phosphorylation was speculated to be one of various possible causes. Recent work by Wang *et al.* further considered the role of ROS in impairment of mitochondrial function.[199]

3. *Interference with sarcolendoplasmic reticulum Ca^{2+}-ATPase (SERCA)*

   Eckstein-Ludwig *et al.*, in 2003, originally presented the compelling argument that artemisinin achieves its activity by inhibiting SERCA.[200] They expressed PfATP6 in *Xenopus laevis* oocytes, finding not only that artemisinin had an inhibitory effect, but also that it was antagonized by thapsigargin, a known SERCA inhibitor with a structure similar to that of artemisinin.

6.8.4 Lead for New Drug Design

The efforts to design new artemisinin derivatives have been aimed at improving upon the solubility, pharmacokinetic profile and efficacy of the parent compound (and first-generation derivatives) while reducing the potential for neurotoxicity. Following the strategy used to design the first-generation derivatives, many new compounds have been reported based upon modification at the C-10 position of artemisinin; these have been broadly classified as C10 acetyl or non-acetyl derivatives.[201] Modifications at C15 (according to CAS numbering, alternatively referred to as C9 and C14 in the literature) have also been attempted in designing new compounds with improved drug profiles.[189,191,202] A class of 11-azaartemisinins has been reported,[202] as have fluorinated artemisinins.[191] Other recent work includes efforts to design C10 carba and aryl analogs of DHA.[201]

The successes and failures in these new drug discovery efforts have, over time, clarified the structure–activity characteristics of artemisinin, with the peroxide established as the critical pharmacophore.[189,190] The third, non-peroxidic oxygen atom has been demonstrated to enhance antimalarial activity.[189] The B- and D-rings, and also the lactone function, have been found unnecessary for antimalarial activity.[203]

These observations have guided efforts to design synthetic or semisynthetic eroxide and trioxane analogs that would, in principle, be more economical to produce on a large scale than artemisinin-based compounds.[204] To date, the most promising and widely discussed result of this work has been OZ277, reported by Vennerstrom *et al.* in 2004.[205] They used a medicinal chemistry approach, integrating antimalarial, pharmacokinetic, metabolism, physico-chemical and toxicity data to identify optimal compounds for study. OZ277, also called arterolane maleate or RBx11160, is currently in Phase III clinical trials in combination with piperaquine phosphate through Ranbaxy Research Laboratories, India.[206] More recently, the structurally-related OZ439 has been presented as a potential single-dose alternative for treating uncomplicated malaria.[207]

OZ277

Future work is expected to focus on the search for new artemisinin derivatives with improved drug profiles; better resolution of the mechanism of action of artemisinin will inform this work. While efforts are ongoing to formulate biochemical and molecular strategies to make artemisinin production more economical, a best-case scenario is the convergence of economical synthesis or semisynthesis approaches with the identification of new analogs with optimized efficacy in treating this neglected disease.

6.9 *Indigofera tinctoria*/Indirubin

6.9.1 Introduction

In TCM, the herbal formula Dang Gui Lu Hui Wan (or Danggui Longhui Wan) (DLW) is used to treat "excess heat or fire in the liver and gall bladder." Thus, DLW could be used to treat chronic myelocytic leukemia (CML), as diagnosed according to its symptoms.[208] Working on that premise, scientists at the Institute of Hematology, Chinese Academy of Medical Sciences, first discovered in clinical studies that DLW was indeed effective in patients with CML.[209] Because this herbal formula can contain extracts from 10–11 plant species, plus musk from the Siberian musk deer (*Moschu moschiferus*), simplified combinations of the ingredients were tested to establish the active component(s). Finally, the antileukemia effects were attributed solely to *Indigofera tinctoria*,[210–212] and further study established the alkaloid indirubin as the antileukemic agent, based on both animal and clinical studies.[213–216]

6.9.2 Source Plant and Traditional Uses

Indigofera tinctoria (Family: Fabaceae) is a flowering shrub cultivated worldwide. In addition to its medicinal purposes, the plant is still used as a ground cover to improve soil, and the plant leaves were the previous source of natural indigo dye, a blue coloring agent.

Qing Dai (or Indigo naturalis) is the dark-blue powder extracted from leaves of *Indigofera tinctoria*, *Isatis indigotica*, *Baphicacanthus cusia* or *Polygonum tinctorium*. In addition to Qing Dai, DLW (the multi-herbal formulation mentioned above) contains *Angelica sinensis* root (Dang Gui), *Aloe vera* dried juice (Lu Hui), *Gentiana scabra* root, *Gardenia jasminoides* fruit, *Scutellaria baicalensis* root, *Phellodendron amurense* stem-bark, *Coptis chinensis* rhizome, *Rheum palmatum* root and *Aucklandia lappa* root. DLW has been used for many indications, most commonly strong headache with constipation, restlessness and anxiety/irritability, dark red or yellow urine, dizziness and convulsions.[217,218] In TCM, Qing Dai itself was used to clear heat, cool the blood and relieve toxicity. Topically, it was used to treat oral sores and sore throat.[219,220] Qing Dai was also used alone to treat various skin conditions and is currently being evaluated very successfully in the form of an ointment or oil to treat psoriasis.[221–223]

6.9.3 Chemistry

Chemically, indirubin (also known as isoindigotin or indigo red) is a 3, 2'-bisindole, more specifically, 3-(1,3-dihydro-3-oxo-2*H*-indol-2-ylidene)-1,

3-dihydro-2*H*-indol-2-one. It is a very minor component of Qing Dai.[224,225] In contrast, the major alkaloid indigo (also known as indigotin) is a 2,2 '-indole, more exactly, 2-(1,3-dihydro-3-oxo-2*H*-indol-2-ylidene)-1,2-dihydro-3*H*-indol-3-one. In indigo, both heterocyclic rings are present as pyrrolidin-3-one systems, whereas in indirubin, one ring is a pyrrolidin-2-one system. This structural difference is highly significant, as it is crucial for anticancer activity (indigo is not active as an anticancer agent). Scientists at the Institute of Materia Medica, Chinese Academy of Sciences, Beijing, were able to convert indigo synthetically to indirubin.[226] Another difference in the two compounds is their color; indigo is blue whereas indirubin is red.

Indirubin   **Indigo**

6.9.4 Pharmacology

As mentioned above, indirubin was found to be responsible for the anticancer effects of indigo naturalis. In addition, it and its derivatives have also been linked to anti-inflammatory[227] and neuroprotective[228] effects.

Indirubin and related synthetic derivatives inhibit cancer cell proliferation by inducing cell cycle arrest at the late G1 and G2/M phases. The compounds inhibit the action of cyclin-dependent kinases (CDK1–cyclin B, CDK2–cylin A, CDK2–cyclin E, CDK4–cyclin D1 and CDK5–p35), by competing with ATP for binding to the enzymatic catalytic subunit.[229,230] CDKs play key roles in cell cycle progression and, thus, are molecular targets in anticancer drug discovery.[231] Most indirubins were even more potent inhibitors of glycogen-synthase kinase (GSK-3β), another cell cycle regulator.[232] In addition, indirubin potently inhibited TNF-induced cancer cell invasion, in addition to NF-κB activation induced by various inflammatory agents and carcinogens. More specifically, indirubin blocked the phosphorylation and degradation of IκBα by inhibiting the activation of IκBα kinase and the phosphorylation and nuclear translocation of p65. Indirubin also down-regulated numerous gene products involved in antiapoptosis, proliferation and invasion. Thus, the antiproliferative and anti-inflammatory effects of indirubin may well result from its effects on NF-κB activation pathways.[233]

6.9.5 Development as a Potential Drug

6.9.5.1 Cancer

Early results with indirubin as a potential drug to treat CML were promising, leading to a complete remission in 26% and partial remission in 33% of 314 cases in a clinical trial in China in 1980.[234] Treatment was generally well tolerated, with abdominal pain and diarrhea being the major side effects.[211] Fewer side effects and comparable effectiveness were seen in similar clinical trials with *N*-methylisoindigo (meisoindigo), a synthetic derivative of indirubin.[212] Mechanistic studies with the two compounds indicated that they caused growth inhibition and apoptosis of leukemic cells.[212]

Meisoindigo

In continued studies, both the 5-sulfonate (sodium salt) and the 3′-oxime of indirubrin (IO) exhibited markedly increased inhibition of CDKs compared with indirubin.[229,230,235] The two compounds were bound similarly in crystal structures with CDK2, indicating that effects of substituents at the two modified positions could be additive and offering avenues of further inhibitor development.[229,236] However, the sulfonate was inactive against both LXFL529L (human large cell lung carcinoma) and MCF-7 (human mammary carcinoma cell lines), likely due to its inability to penetrate cell membranes.[230,235] In contrast, the oxime exhibited greater potency than indirubin against the former cell line and comparable potency against the latter cell line.[230] Accordingly, with better solubility than its parent compound, IO has been widely studied and found to exhibit angiogenic activity using a transgenic

Indirubin 5-sulfonate   **Indirubin 3′-monoxime (IO)**

zebrafish model.[237] Recent mechanism of action studies found that IO inhibits the vascular endothelial growth factor-2 (VEGFR-2) signaling pathway, leading to inhibition of angigenesis.[238]

In further structural modifications, hydrophilic moieties were incorporated into the 3′-oxime to increase water solubility and bioavailability, leading to an indirubin analog with the oxime ether of glycerol. In mechanistic studies, this compound inhibited signaling of the signal transducer and activator of transcription (Stat) protein and induced apoptosis in CML, breast and prostate cancer cells.[239,240]

Other indirubin oxime derivatives, particularly IO, 6-bromoindirubin oxime (BIO) and 6-bromoindirubin acetoxime (6-BIA), were also found to block migration of both glioblastoma cells, required for tumor spread, and endothelial cells, required for tumor angiogenesis. The molecular target could be GSK-3, a major contributor to signaling pathways that modulate invasion and proliferation. Thus, indirubin treatment may provide a novel two-pronged chemotherapeutic approach: anti-invasive and antiangiogenic.[241]

6-BIO: R = H
6-BIA: R = Ac

Although all structural modifications performed on indirubin cannot be mentioned in this chapter, a concise description can be found in a mini-review by Karapetyan *et al.*[242] The same paper also gives an extensive review of the syntheses and bioactivities, particularly anticancer activities, of *N*- and *O*-glycoside derivatives of indirubin, indigo, isoindigo and their various heteroanalogs.

6.9.5.2 Neurodegenerative Disorders

The indirubin structure is of interest as a protein kinase-inhibitory scaffold,[243] not only for cancer, but also for neurodegenerative diseases. Leclerc *et al.* found that indirubins not only inhibited GSK3β and CDK5/P35, but also prevented the hyperphosphorylation of tau.[232] Abnormal phosphorylation of this microtubule-binding protein eventually leads to the formation of the neurofibrillary tangles characteristic of Alzheimer's disease. Subsequent studies have been directed at improving solubility and increasing the selectivity of indirubin derivatives towards inhibition of GSK-3β, while decreasing cytotoxicity.[244] Two compounds, 6-BIDECO and 6-BIMYEO, were found to block tau phosphorylation and apoptosis effectively in an *in vitro* model and were identified as potential neuroprotective agents from these studies.[228]

6-BIDECO: R = (acetyl group) $N(CH_2CH_3)_2$

6-BIMYEO: R = ─CH_2CH_2─N(morpholine)O

6.9.5.3 Psoriasis

Studies have shown that indigo naturalis and its component indirubin can decrease the hyperproliferation and help restore normal differentiation of keratinocytes.[245] At a molecular level, the expression of proliferating cell nuclear antigen, a cell cycle regulatory protein that is amplified in psoriasis, was decreased and the expression of involucrin, an early maker of terminal differentiation, was increased.[245] Recruitment for a clinical trial (NCT01445886) in China using indigo naturalis oil extract for treating psoriatic nails began in 2011.[246] Thus, indirubin or indigo naturalis may hold promise as a treatment for psoriasis and other abnormal skin conditions.

6.10 *Huperzia serrata*/Huperzine A

6.10.1 Introduction

Treatment of neurodegenerative disorders, particularly those related to aging, such as Alzheimer's disease (AD), is a growing concern. Several recent reviews are available describing various plant species and their constituent phyto-chemicals as potential treatments for dementia.[247–251] Interestingly, the cholinergic side effects from the medicinal uses of *Huperzia serrata* were the basis of the discovery of its constituent huperazine A (Hup A), which has attracted worldwide attention as a potential therapeutic agent to combat memory deficit disorders, most particularly AD. Thus, the minor effects of a long-used medicinal plant may reveal an entirely new avenue for therapeutic investigation and intervention.

6.10.2   Source Plant and Traditional Uses

H. serrata (Family: Huperziaceae) is a club moss found in deep forests of southeastern Asia, including China and India, and also other countries. It was first recorded as a TCM during the Tang dynasty where it was listed in the Chinese pharmacopeia *Ben Cao Shi Yi* as Shi Song, which was used to treat rheumatism and colds, improve blood circulation and relax muscles.[252] Because Shi Song can also refer to multiple medicinal herbs from the genus *Lycopodium*, *H. serrata* is more specifically known as Qian Ceng Ta, which means thousand-layered pagoda, referring to the plant's shape. In TCM, it is usually employed in the form of a tea to relieve pain and to treat fever, bruises, strains, contusions, swelling and inflammation.[253] More recently, it has been used for organophosphate poisoning, myasthenia gravis, seizures and schizophrenia.[251,254]

Owing to a declining wild population of *H. serrata*, Ma *et al.* surveyed potential natural sources of Hup A.[255] Ma and Gang also established a method for the *in vitro* production of higher levels of Hup A from tissues of *Phlegmariurus squarrosus*.[256] Mass production possibilities of Hup A, and also galantamine (an approved AD drug also from a natural resource, *Galanthus nivalis*), by organic synthesis and *in vitro* culture methods have been described.[257]

6.10.3   Chemistry

The chemical composition of *H. serrata* has been reviewed,[252] but the major research focus is on Hup A, a lycodine alkaloid. Three additional classes of alkaloids (lycopodine, fawcettimine and miscellaneous) and also triterpenes, flavones and phenolic acid are also found.

Chemically, Hup A is 9-amino-13-ethylidene-11-methyl-4-azatricyclo [7.3.1.0(3.8)]tridec-3(8),6,11-trien-5-one. It is a tricyclic molecule, with a tetrahydroquinolinone system bridged by a –CH_2–C(CH_3)=CH– group. Huperzinine is the dimethylamino analog of Hup A, whereas Hup B is tetracyclic, with the pendant primary amino group of Hup A contained in a piperidine ring.

**Structures of Huperzine A and Two Related Alkaloids
found in *H. serrata***

Huperzine A: R = H
Huperzinine: R = CH$_3$

Huperzine B

6.10.4 Pharmacology

6.10.4.1 Anticholinergic Effects

The traditional use of *H. serrata* was often accompanied by cholinergic side effects,[258] which have now been linked to certain alkaloids contained in this plant.[251,252] Among them, Hup A is the most potent acetylcholinesterase (AChE) inhibitor and its action is reversible and highly specific. AChE is the enzyme that hydrolyzes acetylcholine (ACh), a protein that is essential for neurotransmission. Thus, AChE inhibitors can increase neuronal transmission by increasing the quantity of ACh. Hup A shows a marked preferential inhibition for the tetrameric, G4 form of AChE, which is the major form in mammalian brain.[259] In the Hup A–AChE complex, five different sites of interactions are found, accounting for Hup A's improved potency relative to other AChE inhibitors.[260,261] Hup A is also orally bioavailable, readily crosses the blood–brain barrier (BBB) and has a better pharmacokinetic profile than four approved AD drugs with the same mechanism of action, including tacrine (Cognex), donepezil (Aricept), rivastigmine (Exelon) and galantamine (Reminyl).[250,262,263]

6.10.4.2 Neuroprotective Effects

Hup A is also an effective nootropic agent that enhances concentration and memory, while also having neuroprotective effects.[264,265] The latter action may result from activation of muscarinic and nicotinic acetylcholine receptors and antagonism of *N*-methyl-D-aspartate (NMDA) receptors,[251] which can protect neurons against glutamate-induced excitotoxicity.[266] Hup A can also scavenge reactive oxygen species[267] and increase nerve growth factor levels in rats.[268] Importantly, Hup A can enhance the processing of amyloid precursor protein (APP) *in vitro* and *in vivo*,[269] and thus protect against β-amyloid protein (or peptide) toxicity, a hallmark of AD. Interestingly, both enantiomers of Hup A exhibited equal protection against ®-amyloid-induced injury *in vitro*, and the

(–)-isomer was more potent than the (+)-isomer against AChE.[270] Hence the neuroprotective and anticholinergic effects may not be directly related and the therapeutic effects of Hup A may be due to actions on multiple targets.[271] The anticholinergic and neuroprotective effects of Hup A have been well summarized.[263,265,272,273]

6.10.5 Development as a Potential Drug

6.10.5.1 *Neurodegenerative Disorders*

During the 1980s, Chinese researchers targeted certain medicinal plants and their constituent alkaloids, which had effects on the cardiovascular or neuromuscular system or with cholinesterase activity, in a search for new drugs to treat myasthenia gravis.[274] The isolation/characterization of Hup A and the discovery of its AChE activity by Liu and co-workers at the Shanghai Institute of Materia Medica (SIMM)[275,276] were highlights of this research.

Two independent total syntheses of Hup A were first reported in 1989 from groups in China[277] and the USA.[278] Subsequent synthetic efforts have been reviewed.[252,261,271] Hup A was found to be effective in numerous cognition and performance animal models, involving both rats and monkeys.[279,280] In 1996, the research on Hup A successfully led to its approval by the government of China in a tablet form for treating AD in China.[279] It was developed by Zhu and co-workers of SIMM, denoted shuangyiping and marketed as a new drug.[279,280]

Various reviews detailing the numerous clinical studies on Hup A have been published,[261,263,280–282] and an overview of the compound, including synthetic efforts, was published in 2006.[283] In China, Hup A successfully reached Phase IV clinical trials.[254] Overall, Hup A notably improved memory in elderly people with benign senescent forgetfulness and in AD or vascular dementia (VD) patients, without significant side effects or toxicity.[263,284] In 2009, a meta-analysis of Hup A's efficacy and safety in randomized trials to treat AD was published, with the conclusion that Hup A was well tolerated and could notably improve cognitive performance in AD patients based on primary outcome measures [mini-mental state examination (MMSE) and activities of daily living (ADL)].[285] In a Cochrane review of various clinical studies on Hup A for VD, it was noted that Hup A showed a significant beneficial effect on ADL after 6 months of treatment; however, the authors stated that further well-controlled trials were needed for this purpose.[286]

In 1996, synthetic modification studies led to the selection of ZT-1 as a prodrug of Hup A.[282] ZT-1 is a Schiff base between Hup A and 5-chloro-*O*-vanillin. It has been patented internationally and is being co-developed between the SIMM of China and Debiopharm of Switzerland. In 2006, Debiopharm announced that a monthly sustained-release implant of ZT-1 (denoted Debio-9902 SR) was well tolerated and showed no safety issues in a Phase I trial and that a once-daily oral tablet form was also efficacious and safe in a Phase II study. Phase IIb trials (referred to as BRAINz: Better Recollection for Alzheimer's patients with the ImplaNt of ZT-1) began in 2007 in Australia, Canada and the UK. In 2008, a clinical bioequivalence bridging study was begun in the USA for a new oral tablet formulation.[287]

**Structure of ZT-1 (DEBIO 9902),
Prodrug of Huperzine A**

Regarding cognitive clinical trials on Hup A, Shandong Luye Pharmaceutical Co. is currently recruiting patients for a Phase II/III study of Hup A sustained-release tablets in patients with AD to be performed in China (NCT01282619).[288] Biomedisyn Corporation and Yale University are also recruiting for a Phase II study of Hup A for cognitive and functional impairment in schizophrenia (NCT00963846).[288] The study results are eagerly awaited as a step in the continuing development of Hup A/ZT-1 as a new drug from TCM.

In the USA, Hup A (as powdered *H. serrata*) is marketed as a dietary supplement for memory support, but has not yet been approved for treatment of AD or other cognitive disorders.[272] Some trials in the USA have shown beneficial results, such as a completed clinical trial of Hup A co-sponsored by the National Institute on Aging and Neuro-Hitech Inc., which found dose-related improvements at higher dosages in AD patients, with no safety issues (NCT00083590).[289,290] However, other trials have not shown benefits.[291] Thus, the use of US FDA-approved drugs is still encouraged, until further evaluations and regulations are in place.[292]

6.10.5.2 Organophosphate Antidote

Hup A's reversible inhibition of AChE can protect the enzyme from poisoning by the nerve agent soman and other organophosphates (OPs). Because Hup A crosses the BBB, it may be effective when used as a pretreatment drug in preventing seizures and other neuropathology caused by OP agents. Because it is more selective for red cell AChE than plasma butylcholinesterase (BuChE),

Hup A could protect the cerebral AChE, while allowing the plasma BuChE to scavenge OP agents.[293,294] Hup A showed promising activity as a potential protective agent against OP-induced neurotoxicity, neuronal damage, seizures and mortality in several animal experiments.[295–297]

6.11   *Astragalus membranaceus*/PG2

6.11.1   Introduction

TCM can be viewed as a complementary therapy for current Western medicine (WM). Accordingly, the integration of TCM with WM can be either a tactic to increase the efficacy of WM or a strategy to reduce certain unwanted side effects of WM. This combined approach to TCM and WM is more likely to gain the approval of Western regulatory agencies in the application of Chinese herbal medicine or prescriptions. The history of *Astragalus membranaceus* (Huang Qi) and the development of PG2 injection for cancer-related fatigue constitute an excellent illustration of this approach. In addition, this example shows that new drug discovery does not have to be limited to studies on a single bioactive compound, but rather can focus also on herbal extracts and prescriptions.

6.11.2   Source Plant and Traditional Uses

Astragalus membranaceus (Family: Fabaceae) is native to Mongolia and northern China, but is now cultivated worldwide. Its root, called Huang Qi (meaning yellow leader), was cited in *Shen Nong Ben Cao Jing* as a TCM tonic herb that could replenish a person's Wei Qi (the energy force that protects the body). Symptoms of such insufficient energy included weakness, fatigue, apathy, poor appetite, clammy hands and vulnerability to infection,[298] hence *Astragalus* was used to decrease fatigue and promote healing.[299] Even today, Huang Qi is a popular tonic herb, used to improve resistance to disease and increase deficient energy, and is marketed as dried roots, ground roots in capsules or tablets, as a liquid extract and as an ingredient in herbal teas.[300]

In TCM, *Astragalus* is a common ingredient in multicomponent herbal formulas. The *Astragalus* roots are boiled and combined with other herbs to make a decoction or medicinal soup. An example is Buzhong Yiqi Tang (BYT), used for conditions involving lack of strength; it contains *Astragalus* as the major herbal ingredient, along with *Cimicifuga*, *Bupleurum*, ginseng, licorice, *Atractylodes*, ginger, orange peel, jujube and *Angelica*. Another example is Buyong Huanwu decoction (BHD), which is used to restore neurological impairment following cerebral stroke; it contains *Astragalus membranaceus*, *Angelica sinensis*, *Paeonia lactiflora*, *Ligusticum chuanxiong*, *Prunus persica*, *Carthanmus tinctorius* and *Pheretima asperigillum*.[301,302] A third example is Huang Chi Wu Wu Tang, which contains *A. membranaceus*, *Paeonia lactiflora*, cinnamon twig, *Zingiber officinale* and *Ziziphus jujube* and is used as a TCM formula to treat hemiplegia. China Medical University

Hospital, Taiwan, is currently recruiting patients for a Phase IV clinical trial (NCT01553643) of the latter prescription for treating intracranial arterial stenosis (narrowing of arteries in the brain).[303]

6.11.3 Chemistry and Pharmacology

In 1997, Rios and Waterman reviewed the chemical composition of the genus *Astragalus*, particularly regarding the medicinal effects.[304] The main bioactive components include polysaccharides, saponins (particularly astragalosides) and flavonoids.[305] Astragalan is the term for a polysaccharide-rich fraction (molecular weights from 20 000 to 25 000) extracted from *Astragalus*. It has significant effects on the immune system, including increasing the secretion of TNF *in vitro*[306] and activating lymphocytes (B cells) and macrophages.[307] The saponins, particularly astragaloside IV, have been linked to antioxidative, vasodilative and cardioprotective effects.[308] Flavonoids from *A. membranaceus* have been less well studied, but also contribute to immunoregulatory effects.[309]

Both *in vitro* and *in vivo* studies have shown that *Astragalus*-based therapies have potent immune-restorative and immune-stimulating modulating effects.[310,311] The plant can increase the production or action of lymphocytes, natural killer cells and macrophages,[312–314] and can also stimulate macrophages to produce interleukin-6 and TNF.[315]

6.11.4 Clinical Studies

Clinical studies of *Astragalus* in immunotherapy, adjunct cancer therapy, cardiovascular studies and nephritis were reviewed by Sinclair in 1998,[316] and are discussed in more depth below.

6.11.4.1 Immunotherapy and Adjunct Cancer Therapy

Astragalus is a major component in zheng fu therapy, a treatment involving combinations of herbs designed to restore immune function in cancer patients undergoing radiation treatment or chemotherapy.[314,317] In cancer patients treated with a combination of recombinant IL-2 and *Astragalus*, anticancer activity of lymphokine-activated killer (LAK) cells was heightened and deleterious side effects (*e.g.*, acute renal failure, capillary leakage syndrome, myocardial infarction and edema) were reduced.[318] Furthermore, a meta-analysis of randomized trials showed that *Astragalus*-based Chinese herbal medicine may well have effectiveness as an adjuvant treatment to platinum-based cancer chemotherapy for advanced non-small cell lung cancer. The improved response coupled with reduced therapy-related toxicity could be linked to stimulation of macrophages and natural killer cell activity, and also inhibition of T-helper cell type 2 cytokines.[319,320] Among the literature on the use of medicinal herbs as an adjuvant treatment to chemo- or radiotherapy for cancer, two reviews mention *Astragalus* as having positive benefits.[321,322]

6.11.4.2   Nephritis

Studies have confirmed that the herb acts as a diuretic[323] and has a renal protective effect. Although more rigorous clinical trials are needed, *Astragalus* has significant value in treating diabetic nephropathy, as concluded in both a 2009 review[324] and a 2011 meta-analysis.[325]

6.11.4.3   Heart Failure

The single compound astragaloside IV has been found to improve multiple measures of heart function and decrease cerebral infarction size.[326–328] Two groups have recently published three reviews on the use of *Astragalus*, and particularly Huang Qi injection, in the treatment of chronic heart failure, ischemic heart disease and myocardial infarction.[329–331]

6.11.5   Drug Development of PG2

In the early 1990s, the biotechnology company Pharmagenesis wanted to find a natural product-based chemotherapeutic method to boost immunity. Upon scientific consultation with one of the present authors (K.-H. Lee), the company began an investigation of the polysaccharide extract from Huang Qi (*Astragalus membranaceus*).[332] A goal was to find a conjunctive treatment in cancer therapy, where severe weakness and fatigue can be incapacitating obstacles to good quality of life, or to counter cachexia resulting from chronic conditions, in addition to extensive traumatic injury or sepsis. Eventually, this research led to the product PG2, a sterile powder of polysaccharides isolated from extracts of *Astragalus* root (Huang Qi) and administered via intravenous injection. PG2 was approved for use in mainland China and its further development licensed to PhytoHealth Corporation, a Taiwan-based company focusing on botanical-based new drugs.[333] Studies culminated in a randomized double-blind Phase II clinical trial (NCT00523107)[334] of PG2 for cancer-related fatigue, where significant and sustained benefits were seen with no major or irreversible toxicities.[335] Thus, PG2 was validated as an adjunct treatment in cancer chemotherapy and on 20 April 2010 the Taiwan Food and Drug Administration granted PG-2 the "first approval" under the category of Botanical New Drugs in the form of Injectables.[336] PG2 has other possible clinical uses, for instance, for idiopathic thrombocytopenic purpura (ITP), a disorder in which a person has an abnormally low platelet count due to unknown reasons or for hemorrhagic stroke. PhytoHealth completed a Phase II clinical trial (NCT00860600) for ITP, but the results have not yet been published.[337] PhytoHealth is also recruiting patients for a pilot clinical trial (NCT01325233) of PG2 for treating hemorrhagic stroke.[338] In addition, China Medical University Hospital is planning a Phase II trial (NCT01603667) to evaluate the efficacy of PG2 in acute stroke.[339]

6.12 Conclusion

TCM has been used for several thousands of years to treat human illness, which makes it the best source to provide valuable and unique information for modern drug discovery and development. Development of TCM products as adjunct therapies to augment the efficacy and offset the toxicity of Western medicine is an excellent approach for rapid advancement into FDA-approved new drugs. Medicinal chemistry should and will play a very important role in converting TCM products, especially the pure single active principles, through modification and synthesis into clinical trial candidates very efficiently and effectively. The future is bright for research on TCM targeting drug discovery and development.

Acknowledgements

The authors would like to acknowledge support from NIH grant CA017625-32 from the National Cancer Institute awarded to K.-H. Lee and support in part by the Cancer Research Center of Excellence (CRC), Taiwan (DOH-100-TD-C-111-005).

References

1. J. M. Hagel, R. Krizevski, F. Marsolais, E. Lewinsohn and P. J. Facchini, *Trends Plant Sci.*, 2012, **17**, 404.
2. R. C. Hatton, A. G. Winterstein, R. P. McKelvey, J. Shuster and L. Hendeles, *Ann. Pharmacother.*, 2007, **41**, 381.
3. M. R. Lee, *J. R. Coll. Physicians Edinb.*, 2011, **41**, 78.
4. S. R. Mehendale, B. A. Bauer and C. S. Yuang, *Am. J. Chin. Med.*, 2004, **32**, 1.
5. S. Caveney, D. A. Charlet, H. Freitag, M. Maier-Stolte and A. N. Starratt, *Am. J. Bot.*, 2001, **88**, 1199.
6. S. Foster and R. L. Johnson, *Desk Reference to Nature's Medicine*, National Geographic Society, Washington, DC, 2006, p. 146.
7. L. Liu and Z. Liu (eds), *Essentials of Chinese Medicine*, Springer, Dordrecht, 2009.
8. D. Bensky, S. Chaney and E. Stöger, *Chinese Herbal Medicine Materia Medica*, 3rd edn. Eastland Press, Seattle, WA, 2004.
9. L. Zhou, Y. P. Tang, L. Gao, X. S. Fan, C. M. Liu and D. K. Wu, *Molecules*, 2009, **14**, 3942.
10. B. T. Schaneberg, S. Crockett, E. Bedir and I. A. Khan, *Phytochemistry*, 2003, **62**, 911.
11. J. R. Docherty, *Br. J. Pharmacol.*, 2008, **154**, 606.
12. K. J. Broadley, *Pharmacol. Ther.*, 2010, **125**, 363.
13. J. D. Ellis, C. L. German, E. Birdsall, J. E. Hanson, M. A. Crosby, S. D. Rowley, N. A. Sawada, J. N. West, G. R. Hanson and A. E. Fleckenstein, *Synpase*, 2011, **65**, 449.

14. D. Sulzer, *Neuron*, 2011, **69**, 628.

15. L. Reti, *The Alkaloids: Chemistry and Physiology*, Vol. III. ed. R. H. F. Manske and H. L. Holmes, Academic Press, New York, 1953, p. 339.

16. K. Chen and C. F. Schmidt, *Chin. Med. J.*, 1925, **39**, 928.

17. K. K. Chen, *J. Pharmacol.*, 1928, **33**, 237.

18. K. K. Chen and C. F. Schmidt, *Medicine*, 1930, **9**, 1.

19. K. Freudenberg, E. Schoeffel and E. Braun, *J. Am. Chem. Soc.*, 1932, **54**, 234.

20. F. L. Greenway, *Obes. Rev.*, 2011, **2**, 199.

21. C. Haller and N. Benowitz, *N. Engl. J. Med.*, 2000, **343**, 1833.

22. J. Arditti, J. H. Bourdon, M. Spadari, L. de Haro, N. Richard and M. Valli, *Acta Clin. Belg. Suppl.*, 2002, **57**, 34–36.

23. M. H. Pittler, K. Schmidt and E. Ernst, *Obes. Rev.*, 2005, **6**, 93.

24. Food and Drug Administration, *Federal Register Final Rule – FR69 6787 Dietary Supplements Containing Ephedrine Alkaloids Adulterated Because They Present an Unreasonable Risk, Federal Register*, 2004, **69**, 6788. http://www.gpo.gov/fdsys/pkg/FR-2004-02-11/pdf/04-2912.pdf, 2004 (last accessed 25 October 2012).

25. D. Shaw, *Planta Med.*, 2010, **76**, 2012.

26. W. Barker and U. Antia, *Forensic Sci. Int.*, 2007, **166**, 102.

27. International Narcotics Control Board, *List of Precursors and Chemicals Frequently Used in the Illicit Manufacture of Narcotic Drugs and Psychotropic Substances Under International Control*, http://www.incb.org/pdf/e/list/red.pdf, 2007 (last accessed 15 September 2012).

28. J. Chen and T. Chen, *Chinese Medical Herbology and Pharmacology*, Art of Medicine Press, City of Industry, CA, 2001, p. 36.

29. A. Lee, W. D. Ngan Kee and T. Gin, *Anesth. Analg.*, 2002, **94**, 920.

30. K. H. Lee, H. Itokawa and M. Kozuka, in *Asian Functional Foods*, ed. J. Shi, F. Shahidi and C. T. Ho, Marcel Dekker/CRC Press, Boca Raton, FL, 2005, p. 21.

31. K. H. Lee, S. Morris-Natschke, K. Qian, Y. Dong, X. Yang, T. Zhou, E. Belding, S.-F. Wu, K. Wada and T. Akiyama, *J. Trad. Comp. Med.*, 2012, **2**, 6.

32. K. H. Lee and Z. Xiao, *Anticancer Agents from Natural Products*, 2nd edn, ed. G. M. Cragg, D. G. I. Kingston and D. J. Newman, CRC Press, Boca Raton, FL, 2011, p. 95.

33. H.-Y. Hsu, *Oriental Materia Medica, a Concise Guide*, Oriental Healing Arts Institute, Long Beach, CA, 1986.

34. T. F. Imbert, *Biochimie*, 1998, **80**, 207.

35. W. J. Gensler and C. D. Gatsonis, *J. Am. Chem. Soc.*, 1962, **84**, 1748.

36. L. S. King and M. Sullivan, *Science*, 1946, **104**, 244.

37. L. Wilson and M. Friedkin, *Biochemistry*, 1967, **6**, 3126.

38. H. F. Stähelin and A. von Wartburg, *Cancer Res.*, 1991, **51**, 5.

39. K. I. Kaitin, M. Manocchia, M. Seibring and L. Lasagna, *J. Clin. Pharmacol.*, 1994, **34**, 120.

40. M. Gordaliza, M. A. Castro, J. M. Miguel del Corral and A. San Feliciano, *Curr. Pharm. Des.*, 2000, **6**, 1811.
41. H. Xu, M. Lv and X. Tian, *Curr. Med. Chem.*, 2009, **16**, 327.
42. J. D. Loike and S. B. Horwitz, *Biochemistry*, 1976, **15**, 5443.
43. B. H. Long and A. Minocha, *Proc. Am. Assoc. Cancer Res.*, 1983, **24**, 321.
44. W. Ross, T. Rowe, B. Glisson, J. Yalowich and L. Liu, *Cancer Res.*, 1984, **44**, 5857.
45. S. J. Cho, Y. Kashiwada, K. F. Bastow, Y.-C. Cheng and K. H. Lee, *J. Med. Chem.*, 1996, **39**, 1396.
46. Z. Xiao, S. Han, K. F. Bastow and K. H. Lee, *Bioorg. Med. Chem. Lett.*, 2004, **14**, 1581.
47. *Physicians' Desk Reference*, 54th edn, Medical Economics Company, Montvale, NJ, 2000.
48. Y. You, *Curr. Pharm. Des.*, 2005, **11**, 1695.
49. H.-K. Wang, S. L. Morris-Natschke and K. H. Lee, *Med. Res. Rev.*, 1997, **17**, 367.
50. F. Leteurtre, J. Madalengoitia, A. Orr, T. J. Cuzi, E. Lehnert, T. Macdonald and Y. Pommier, *Cancer Res.*, 1992, **52**, 4478.
51. E. Solary, F. Leteurtre, K. D. Paull, D. Scudiero, E. Hamel and Y. Pommier, *Biochem. Pharmacol.*, 1993, **45**, 2449.
52. D. Perrin, B. van Hille, J.-M. Barret, A. Kruczynski, C. Etiévant, T. Imbert and B. T. Hill, *Biochem. Pharmacol.*, 2000, **59**, 807.
53. S. Salerno, F. Da Settimo, S. Taliani, F. Simorini, C. La Motta, G. Fornaciari and A. M. Marini, *Curr. Med. Chem.*, 2010, **17**, 4270.
54. P. Meresse, E. Dechaux, C. Monneret and E. Bertounesque, *Curr. Med. Chem.*, 2004, **11**, 2443.
55. T. L. MacDonald, E. K. Lehnert, J. T. Loper, K.-C. Chow and W. E. Ross, in *DNA Topoisomerases in Cancer*, ed. M. Potmesil and K. W. Kohn, Oxford University Press, New York, 1991, p. 199.
56. C.-C. Wu, T.-K. Li, L. Farh, L.-Y. Lin, T.-S. Lin, Y.-J. Yu, T.-J. Yen, C.-W. Chiang and N.-L. Chan, *Science*, 2011, **333**, 459.
57. W. L. Wu, W. L. Chang and C. F. Chen, *Am. J. Chin. Med.*, 1991, **14**, 207.
58. S. Y. Ryu, C. O. Lee and S. U. Choi, *Planta Med.*, 1997, **63**, 339.
59. B. Q. Wang, *J. Med. Plants Res.*, 2010, **4**, 2813.
60. Y. Dong, S. L. Morris-Natschke and K. H. Lee, *Nat. Prod. Rep.*, 2011, **28**, 529.
61. X. Wang, S. L. Morris-Natschke and K. H. Lee, *Med. Res. Rev.*, 2007, **27**, 33.
62. Q. Shang, H. Xu and L. Huang, *Evid. Based Compl. Alt. Med.*, 2012, **2012**, 716459.
63. S. Gao, Z. Liu, H. Li, P. J. Little, P. Liu and S. Xu, *Atherosclerosis*, 2012, **220**, 3.
64. L. Zhou, Z. Zuo and M. S. S. Chow, *J. Clin. Pharmacol.*, 2005, **45**, 1345.
65. X. Qiu, A. Miles, X. Jiang, X. Sun and N. Yang, *Evid. Based Compl. Alt. Med.*, 2012, **2012**, 715790.

66. P. Chan, I. M. Liu, Y. X. Li, W. J. Yu and J. T. Cheng, *Evid. Based Compl. Alt. Med.*, 2011, **2011**, 392627.

67. R. Yang, A. Liu, X. Ma, L. Li, D. Su and J. Liu, *J. Cardiovasc. Pharmacol.*, 2008, **51**, 396.

68. Z. Y. Fang, R. Lin, B. X. Yuan, Y. Liu and H. Zhang, *Life Sci.*, 2007, **81**, 1339.

69. G. Y. Zhou, B. L. Zhao, J. W. Hou, G. E. Ma and W. J. Xin, *Pharmacol. Res.*, 1999, **40**, 487.

70. B. Jiang, L. Zhang, Y. Wang, M. Li, W. Wu, S. Guan, X. Liu, M. Yang, J. Wang and D. A. Guo, *Food. Chem. Toxicol.*, 2009, **47**, 1538.

71. Y. F. Huang, M. L. Liu, M. Q. Dong, W. C. Yang, B. Zhang, L. L. Luan, H. Y. Dong, M. Xu, Y. X. Wang, L. L. Liu, Y. Q. Gao and Z. C. Li, *J. Ethnopharmacology*, 2009, **125**, 436.

72. K. Takahashi, X. Ouyang, K. Komatsu, N. Nakamura, M. Hattori, A. Baba and J. Azuma, *Biochem. Pharmacol.*, 2002, **64**, 745.

73. X. Tan, J. Li, X. Wang, N. Chen, B. Cai, G. Wang, H. Shan, D. Dong, Y. Liu, X. Li, P. Zhang, X. Li, B. Yang and Y. Liu, *Int. J. Biol. Sci.*, 2011, **7**, 383.

74. H. Weizhe, Z. Dongtao, Z. Ge and L. Tong, *J. Med. Plants Res.*, 2012, **6**, 763.

75. C. Pan, L. Lou, Y. Huo, G. Singh, M. Chen, D. Zhang, A. Wu, M. Zhao, S. Wang and J. Li, *Ther. Adv. Cardiovasc. Dis.*, 2011, **5**, 99.

76. H. Itokawa, S. L. Morris-Natschke, T. Akiyama and K. H. Lee, *J. Nat. Med.*, 2008, **62**, 263.

77. S. H. Won, H. J. Lee, S. J. Jeong, H. J. Lee, E. O. Lee, D. B. Jung, J. M. Shin, T. R. Kwon, S. M. Yun, M. H. Lee, S. H. Choi, J. Lü and S. H. Kim, *Biol. Pharm. Bull.*, 2010, **33**, 1828.

78. Z. Gong, C. Huang, X. Sheng, Y. Zhang, Q. Li, M. W. Wang, L. Peng and Y. Q. Zang, *Endocrinology*, 2009, **150**, 104.

79. Y. Xiao, W. X. Qing, M. S. Lan and C. B. Ying, *Biochem. Pharmacol.*, 2006, **72**, 582.

80. Y. Xu, D. Feng, Y. Wang, S. Lin and L. Xu, *J. Clin. Immunol.*, 2008, **28**, 512.

81. H. B. Kwak, D. Yang, H. Ha, J. H. Lee, H. N. Kim, E. R. Woo, S. Lee, H. H. Kim and Z. H. Lee, *Exp. Mol. Med.*, 2006, **38**, 256.

82. F. L. Li, R. Xu, Q. C. Zeng, X. Li, J. Chen, Y. F. Wang, B. Fan, L. Geng and B. Li, *Evid. Based Compl. Alt. Med.*, 2012, **2012**, 927658.

83. W. Liu, J. Zhou, G. Geng, Q. Shi, F. Sauriol and J. H. Wu, *J. Med. Chem.*, 2012, **55**, 971.

84. Y. F. Bi, H. W. Xu, X. Q. Liu, X. J. Zhang, Z. J. Wang and H. M. Liu, *Bioorg. Med. Chem. Lett.*, 2010, **20**, 4892.

85. X. Wang, K. F. Bastow, C. M. Sun, Y. L. Lin, H. J. Yu, M. J. Don, T. S. Wu, S. Nakamura and K. H. Lee, *J. Med. Chem.*, 2004, **47**, 5816.

86. Y. Dong, Q. Shi, H. C. Pai, C. Y. Peng, S. L. Pan, C. M. Teng, K. Nakagawa-Goto, D. Yu, Y. N. Liu, P. C. Wu, K. F. Bastow, S. L.

Morris-Natschke, A. Brossi, J. Y. Lang, J. L. Hsu, M. C. Hung, E. Y. Lee and K. H. Lee, *J. Med. Chem.*, 2010, **53**, 2299.

87. Y. Dong, Q. Shi, Y. N. Liu, X. Wang, K. F. Bastow and K. H. Lee, *J. Med. Chem.*, 2009, **52**, 3586.

88. Y. Dong, Q. Shi, K. Nakagawa-Goto, P. C. Wu, K. F. Bastow, S. L. Morris-Natschke and K. H. Lee, *Bioorg. Med. Chem. Lett.*, 2009, **19**, 6289.

89. Y. Dong, Q. Shi, K. Nakagawa-Goto, P. C. Wu, S. L. Morris-Natschke, A. Brossi, K. F. Bastow, J. Y. Lang, M. C. Hung and K. H. Lee, *Bioorg. Med. Chem.*, 2010, **18**, 803.

90. Y. Dong, K. Nakagawa-Goto, C. Y. Lai, S. L. Morris-Natschke, K. F. Bastow and K. H. Lee, *Bioorg. Med. Chem. Lett.*, 2010, **20**, 4085.

91. Y. Dong, K. Nakagawa-Goto, C. Y. Lai, S. L. Morris-Natschke, K. F. Bastow and K. H. Lee, *Bioorg. Med. Chem. Lett.*, 2011, **21**, 546.

92. Y. Dong, K. Nakagawa-Goto, C. Y. Lai, S. L. Morris-Natschke, K. F. Bastow and K. H. Lee, *Bioorg. Med. Chem. Lett.*, 2011, **21**, 2341.

93. J. S. Lee, S. Y. Han, M. S. Kim, C. M. Yu, M. H. Kim, S. H. Kim, Y. K. Min and B. T. Kim, *Bioorg. Med. Chem. Lett.*, 2006, **16**, 4733.

94. J. Chang, W. Xi, L. Wang, N. Ma, S. Cheng and J. Xie, *Eur. J. Med. Chem.*, 2006, **41**, 397.

95. J. X. Xie, J. H. Yang, Y. X. Zhao and C. Z. Zhang, *Sci. Sin. B*, 1983, **26**, 931.

96. W. Zheng, L. Wang, L. Meng and J. Liu, *Genetica*, 2008, **132**, 123.

97. J. M. Poupko, S. I. Baskin and E. Moore, *J. Appl. Toxicol.*, 2007, **27**, 116.

98. R. J. Griffin Jr, *Am. Pharm.*, 1979, **19**, 16.

99. P. Qicheng, *J. Ethnopharmacol.*, 1980, **2**, 57.

100. Chinese Academy of Medical Sciences, Institute of Materia Medica, *J. Chin. Med.*, 1975, **55**, 795.

101. J. T. Zhang, *Therapie*, 2002, **57**, 137.

102. H. H. Chen, *Resuscitation*, 1983, **10**, 149.

103. S. H. Zhu, T. Y. Yan, S. Y. Gao and G. Li, *Chin. J. Int. Med.*, 1966, **14**, 19–21.

104. Department of Paediatrics, Peking Friendship Hospital, Department of Pharmacology, Institute of Materia Medica, Chinese Academy of Medical Sciences and First Laboratory of an Institute of Chinese Academy of Medical Sciences. *Chin. Med. J.*, 1973, **53**, 259–263.

105. R. J. Xiu and M. Intaglietta, *Adv. Chin. Med. Mater. Res.*, 1984, **1**, 553.

106. D. R. Varma and T. L. Yue, *Br. J. Pharmacol.*, 1986, **87**, 587.

107. J. Y. Su, *Chin. Med. J.*, 1992, **105**, 976.

108. Q. Li, H. Lei, A. Liu, Y. Yang, D. Su and X. Liu, *Life Sci.*, 2011, **89**, 395.

109. T. Zhao, D. J. Li, C. Liu, D. F. Su and F. M. Shen, *Frontiers Pharmacol.*, 2011, **2**, 1.

110. J. X. Xie, C. X. Liu, X. X. Jia and J. Zhou, *Kexue Tongbao*, 1975, **20**, 197.

111. J. X. Xie, J. Zhou, X. X. Jia, C. X. Liu, H. Q. Su, A. S. Fang, J. Z. Wang and B. Y. Xia, *Acta Pharm. Sin.*, 1980, **15**, 403.

112. J. X. Xie, J. Zhou, C. Z. Zhang, J. H. Yang and X. X. Chen, *Zhongguo Yi Xue Ke Xue Yuan Xue Bao*, 1982, **4**, 92.

113. J. X. Xie, J. H. Yang and C. Z. Zhang, *Acta Pharm. Sin.*, 1981, **16**, 762.

114. J. X. Xie, personal communication.

115. G. Grynkiewicz and M. Gadzikowska, *Pharmacol. Rep.*, 2008, **70**, 439.

116. X. Ying and J. Ruan, *Chin. J. Pharmacol. Toxicol.*, 2001, **15**, 302.

117. B. A. Weissman and L. Raveh, *Toxicology*, 2011, **290**, 149.

118. Q. B. Li, R. Pan, G. F. Wang and S. X. Tang, *J. Nat. Toxins*, 1999, **8**, 327.

119. S. L. Zhang, D. Lax, Y. Li, E. Stejskal, R. V. Lucas Jr and S. Einzig, *J. Ethnopharmacol.*, 1990, **30**, 121.

120. Y. L. Jing, Y. L. Wang, Y. Sun, C. X. Zhao, H. J. Li and X. Y. Kong, *Zhongguo Ying Yong Sheng Li Xue Za Zhi*, 2009, **25**, 557.

121. Z. P. Xu, H. Wang, L. N. Hou, Z. Xia, L. Zhu, H. Z. Chen and Y. Y. Cui, *Int. Immunopharmacol.*, 2011, **11**, 260.

122. A. Lorence and C. L. Nessler, *Phytochemistry*, 2004, **65**, 2735.

123. H.-Y. Hsu, *Oriental Materia Medica, a Concise Guide*, Oriental Healing Arts Institute, Long Beach, CA, 1986.

124. M. E. Wall, M. C. Wani, C. E. Cook, K. H. Palmer, A. T. McPhail and G. A. Sim, *J. Am. Chem. Soc.*, 1966, **88**, 3888.

125. N. J. Rahier, C. J. Thomas and S. M. Hecht, in *Anticancer Agents from Natural Products*, 2nd edn, ed. G. M. Cragg, D. G. I. Kingston and D. J. Newman, CRC Press, Boca Raton, FL, 2011, p. 5.

126. M. E. Wall and M. C. Wani, in *Camptothecins: New Anticancer Agents*, ed. M. Potmesil and H. Pinedo, CRC Press, Boca Raton, FL, 1995, p. 21.

127. Y.-H. Hsiang, R. Hertzberg, S. Hecht and L. F. Liu, *J. Biol. Chem.*, 1985, **260**, 14873.

128. L. F. Liu, in *Camptothecins: New Anticancer Agents*, ed. M. Potmesil and H. Pinedo, CRC Press, Boca Raton, FL, 1995, p. 9.

129. *Physicians' Desk Reference*, 54th edn, Medical Economics Company, Montvale, NJ, 2000.

130. B. L. Staker, K. Hjerrild, M. D. Freese, C. A. Behnke, A. B. Burgin Jr and L. Stewart, *Proc. Natl. Acad. Sci. U. S. A.*, 2002, **99**, 15387.

131. Q.-Y. Li, Y.-G. Zu, R.-Z. Shi and L.-P. Yao, *Curr. Med. Chem.*, 2006, **13**, 2021.

132. R. Wang, J. Kong, D. Wang, L. L. Lien and E. J. Lien, *Chin. Med.*, 2007, **2**, 5.

133. Y. Lu and D. F. Chen, *J. Chromatogr. A*, 2009, **1216**, 1980.

134. K. C. Huang, *The Pharmacology of Chinese Herbs*, 2nd edn, CRC Press, Boca Raton, FL, 1999.

135. J. Batchelor and K. Miyabe, *Trans. Asiatic Soc. Jpn.*, 1893, **51**, 198.

136. S. Foster and R. L. Johnson, *Desk Reference to Nature's Medicine*, National Geographic Society, Washington, DC, 2008, p. 324.

137. J. Gnabre, I. Unlu, T. C. Cheng, P. Lisseck, B. Bourne, R. Scolni, N. E. Jacobsen, R. Bates and R. C. Huang, *J. Chromatogr. B*, 2010, **878**, 2693.

138. H. Hikino, Y. Kiso, H. Taguchi and Y. Ikeya, *Planta Med.*, 1984, **50**, 213.

139. L. Opletal, M. Krenková and P. Havlicková, *Cesk. Slov. Farm.*, 2001, **50**, 173.
140. L. Opletal, H. Sovova and M. Bartlova, *J. Chromatogr. B*, 2004, **812**, 357.
141. L. Q. Yang, X. Y. Wu, Z. Q. Xu, H. R. Hou and H. Z. Hu, *Zhongguo Zhong Yao Za Zhi*, 2005, **30**, 650.
142. C. F. Chien, Y. T. Su and T. H. Tsai, *Biomed. Chromatogr.*, 2011, **25**, 21.
143. A. Panossian, E. Garielian and H. Wagner, *Phytomedicine*, 1999, **6**, 127.
144. A. Panossian, G. Wikman and H. Wagner, *Phytomedicine*, 1999, **6**, 287.
145. A. Panossian and G. Wikman, *J. Ethnopharmacol.*, 2008, **118**, 183.
146. A. Panossian and H. Wagner, *Phytother. Res.*, 2005, **19**, 819.
147. A. Panossian and G. Wikman, *Curr. Clin. Pharmacol.*, 2009, **4**, 198.
148. D. Winston and S. Maimes, *Adaptogens: Herbs for Strength, Stamina and Stress Relief*, Healing Arts Press, Rochester, VT, 2007, p. 129.
149. T. Konishi, *Neurochem. Res.*, 2009, **34**, 711.
150. L. Y. Guo, T. M. Hung, K. H. Bae, E. M. Shin, H. Y. Zhou, Y. N. Hong, S. S. Kang, H. P. Kim and Y. S. Kim, *Eur. J. Pharmacol.*, 2008, **591**, 293.
151. X. J. Li, B. L. Zhao, G. T. Liu and W. J. Xin, *Free Rad. Biol. Med.*, 1990, **9**, 99.
152. N. Chen, P. Y. Chu and K. M Ko, *Biol. Pharm. Bull.*, 2008, **31**, 1387.
153. R. D. Lin, Y. W. Mao, S. J. Leu, C. Y. Huang and M. H. Lee, *Molecules*, 2011, **16**, 4836.
154. X. Ci, R. Ren, K. Xu, H. Li, Q. Yu, Y. Song, D. Wang, R. Li and X. Deng, *Inflammation*, 2010, **33**, 126.
155. S. Y. Oh, Y. H. Kim, K. S. Bae, B. H Um, C. H. Pan, C. Y. Kim, H. J. Lee and J. K Lee, *Biosci. Biotechnol. Biochem.*, 2010, **74**, 285.
156. J. C. Cyong, S. M. Ki, K. Ikjima, T. Kobayahi and M. Furuya, *Am. J. Chin. Med.*, 2000, **28**, 351.
157. H. S. Azzam, C. Goertz, M. Fritts and W. B. Jonas, *Liver Int.*, 2007, **27**, 17.
158. M. Zhu, K. F. Lin, R. Y. Yeung and R. C. Li, *J. Ethnopharmacol.*, 1999, **67**, 61.
159. M. Zhu, R. Y. Yeung, K. F. Lin and R. C. Li, *Planta Med.*, 2000, **66**, 521.
160. G. T. Liu, in *Recent Advances in Chinese Herbal Drugs*, ed. J. H. Zhou and G. Z. Liu, Science Press, Beijing, 1991, p. 100.
161. R. Teraoka, T. Shimada and M. Aburada, *Biol. Pharm. Bull.*, 2012, **35**, 171.
162. G. T. Liu, *Chin. Med. J.*, 1989, **102**, 740.
163. H. Lu and G. T. Liu, *Chem. Biol. Interact.*, 1991, **78**, 77.
164. M. Yasuhiro, S. Tohkan and M. Seiji, *Planta Med.*, 1991, **57**, 11.
165. S. P. Ip, C. Y. Ma, C. T. Che and K. M. Ko, *Biochem. Pharmacol.*, 1997, **54**, 317.
166. Y. Chen, S. P. Ip, K. M. Ko, T. C. Poon, E. W. Ng, P. B. Lai, Q. Q. Mao, Y. F. Xian and C. T. Che, *J. Proteome Res.*, 2011, **10**, 299.
167. T. T. Bao, G. F. Xu, G. T. Liu, R. H. Sun and Z. Y. Song, *Yaoxue Xuebao*, 1979, **14**, 1.

168. Y. Y. Chen, Z. B. Shu and L. N. Li, *Sci. Sin.*, 1976, **2**, 276.
169. T. T. Bao, G. T. Liu, Z. Y. Song, G. F. Xu and R. H. Sun, *Chin. Med. J.*, 1980, **93**, 41.
170. X. Sinclair, *Alt. Med. Rev.*, 1998, **3**, 338.
171. L. N. Li, *Pure Appl. Chem.*, 1998, **70**, 547.
172. J. X. Xie, J. Zhou, C. Z. Zhang, J. H. Yang and J. X. Chen, *Chin. Sci. Bull.*, 1982, **27**, 383.
173. J. X. Xie, J. Zhou, C. Z. Zhang, J. H. Yang and J. X. Chen, *Sci. Sin.*, 1983, **12**, 1291.
174. J. S. Liu, S. D. Fang, M. F. Huang, Y. L. Gao and J. S. Hsu, *Sci. Sin.*, 1978, **21**, 483.
175. J. B. Chang, Q. Wang and Y. F. Li, *Curr. Top. Med. Chem.*, 2009, **9**, 1660.
176. J. X. Xie, J. Zhou, C. Z. Zhang, J. H. Yang and J. X. Chen, *Acta Pharm. Sin.*, 1981, **16**, 306.
177. J. X. Xie and W. S. Feng, *Chin. J. Pharm.*, 1989, **20**, 331.
178. J. X. Xie, personal communication.
179. H. Q. Yu, X. Y. Yang, Y. X. Zhang and J. Z. Shi, *Chin. Med. J. (Engl. Ed.)*, 1987, **100**, 122.
180. R. Huber, B. Hockenjos and H. E. Blum, *Hepatology*, 2004, **39**, 1732.
181. X. K. Kim, Y. J. Cho and Z. G. Gao, *J. Control. Release*, 2001, **70**, 149.
182. Y. K. Ahn and J. H. Kim, *J. Toxicol. Sci.*, 1993, **18**, 185.
183. World Health Organization, *World Malaria Report: 2011*, WHO, Geneva, 2011.
184. L. B. Slater, *War and Disease: Biomedical Research on Malaria in the Twentieth Century*, Rutgers University Press, New Brunswick, NJ, 2009.
185. I. W. Sherman, *Magic Bullets to Conquer Malaria: from Quinine to Qinghaosu*, ASM Press, Washington, DC, 2011.
186. L. H. Miller and X. Su, *Cell*, 2011, **146**, 855.
187. Y. Tu, *Nat. Med.*, 2011, **17**, 1217.
188. Anon, *Chin. Med. J. (Engl. Ed.)*, 1979, **92**, 811.
189. R. K. Haynes, *Curr. Top. Med. Chem.*, 2006, **6**, 509.
190. J. Li and B. Zhou, *Molecules*, 2010, **15**, 1378.
191. P. M. O'Neill and G. H. Posner, *J. Med. Chem.*, 2004, **47**, 2945.
192. N. J. White, *Science*, 2008, **320**, 330.
193. R. T. Eastman and D. A. Fidock, *Nat. Rev. Microbiol.*, 2009, **7**, 864.
194. Y. Li, H. Huang and Y. L. Wu, in *Medicinal Chemistry of Bioactive Natural Products*, ed. X. T. Liang and W. S. Fang, Wiley, Hoboken, NJ, 2006, p. 183.
195. R. K. Haynes, W. C. Chan, C. M. Lung, A. C. Uhlemann, U. Eckstein, D. Taramelli, S. Parapini, D. Monti and S. Krishna, *ChemMedChem*, 2007, **2**, 1480.
196. J. Golenser, J. H. Waknine, M. Krugliak, N. H. Hunt and G. E. Grau, *Int. J. Parasitol.*, 2006, **36**, 1427.
197. W. Li, W. Mo, D. Shen, L. Sun, J. Wang, S. Lu, J. M. Gitschier and B. Zhou, *PLoS Genet.*, 2005, **1**, e36.

198. J. B. Jiang, G. Jacobs, D. S. Liang and M. Aikawa, *Am. J. Trop. Med. Hyg.*, 1985, **34**, 424.
199. J. Wang, L. Huang, J. Li, Q. Fan, Y. Long, Y. Li and B. Zhou, *PLoS ONE*, 2010, **5**, A158.
200. U. Eckstein-Ludwig, R. J. Webb, I. D. A. van Goethem, J. M. East, A. G. Lee, M. Kimura, P. M. O'Neill, P. G. Bray, S. A. Ward and S. Krishna, *Nature*, 2003, **424**, 957.
201. A. Mital, *Curr. Med. Chem.*, 2007, **14**, 759.
202. N. C. Waters, G. S. Dow and M. P. Kozar, *Expert Opin. Ther. Pat.*, 2004, **14**, 1125.
203. C. W. Jefford, J. A. Velarde, G. Bernardinelli, D. H. Bray, D. C. Warhurst and W. K. Milhous, *Helv. Chim. Acta*, 1993, **76**, 2775.
204. Y. Tang, Y. Dong, J. L. Vennerstrom, *Med. Res. Rev.*, 2004, **24**, 425.
205. J. L. Vennerstrom, S. Arbe-Barnes, R. Brun, S. A. Charman, F. C. K. Chiu, J. Chollet, Y. Dong, A. Dorn, D. Hunziker, H. Matile, K. McIntosh, M. Padmanilayam, J. S. Tomas, C. Scheurer, B. Scorneaux, Y. Tang, H. Urwyler, S. Wittlin and W. N. Charman, *Nature*, 2004, **430**, 900.
206. A. Gautam, T. Ahmed, P. Sharma, B. Varshney, M. Kothari, N. Saha, A. Roy, J. J. Moehrle and J. Paliwal, *J. Clin. Pharmacol.*, 2011, **51**, 1519.
207. S. A. Charman, S. Arbe-Barnes, I. C. Bathurst, R. Brun, M. Campbell, W. N. Charman, F. C. K. Chiu, J. Chollet, J. C. Craft, D. J. Creek, Y. Dong, H. Matile, M. Maurer, J. Morizzi, T. Nguyen, P. Papastogiannidis, C. Sheurer, D. M. Shackleford, K. Sriraghavan, L. Stingelin, Y. Tang, H. Urwyler, X. Wang, K. L. White, S. Wittlin, L. Zhou and J. L. Vennerstrom, *Proc. Natl. Acad. Sci. U. S. A.*, 2011, **108**, 4400.
208. *Chinese Pharmacopeia*, Vol. 1, People's Health Publisher, Beijing, 1995.
209. Institute of Hematology, Chinese Academy of Medical Sciences, *Chin. J. Int. Med.*, 1976, **15**, 86.
210. Institute of Hematology, Chinese Academy of Medical Sciences, *Transfusion Hematol.*, 1978, **4**, 44.
211. J. Han, *J. Ethnopharmacol.*, 1988, **24**, 1.
212. Z. Xiao, Y. Hao, B. Liu and L. Qian, *Leukemia Lymphoma*, 2002, **43**, 1763.
213. L. M. Wu, Y. P. Yang and Z. H. Zhu, *Commun. Chin. Herb. Med.*, 1978, **9**, 6.
214. Q. T. Zheng, D. J. Lu and S. L. Yang, *Commun. Chin. Herb. Med.*, 1979, **10**, 35.
215. G. Y. Wu, F. D. Fang, J. Z. Liu, A. Chang and Y. H. Ho, *Chin. Med. J.*, 1980, **60**, 451.
216. P. K Hsiao, in *Natural Products as Medicinal Agents*, ed. J. L. Beal and E. Reinhard, Hippokrates Verlag, Stuttgart, 1981, p. 351.
217. W. Tang and G. Eisenbrand, *Chinese Drugs of Plant Origin. Chemistry, Pharmacology and Use in Traditional and Modern Medicine*, Springer, New York, 1992.
218. R. Han, *Stem Cells*, 1994, **12**, 53.

219. D. Bensky and A. Gamble, *Chinese Herbal Medicine: Materia Medica*, Eastland Press, Seattle, WA, 1993.

220. K. C. Huang, *The Pharmacology of Chinese Herbs*, 2nd edn, CRC Press, Boca Raton, FL, 1999.

221. Y. K. Lin, C. J. Chang, Y. C. Chang, W. R. Wong, S. C. Chang and J. H. S. Pang, *Arch. Dermatol.*, 2008, **144**, 1457.

222. Y. K. Lin, *Arch. Dermatol.*, 2011, **147**, 627.

223. C. Y. Liang, T. Y. Lin and Y. K. Lin, *Pediatric Dermatol.*, 2012, doi: 10.1111/j.1525-1470.2012.01721.x.

224. S. X. Zhang, *Chin. Trad. Herb. Drugs*, 1983, **14**, 247.

225. D. H. Chen and J. X. Xie, *Chin. Trad. Herb. Drugs*, 1984, **14**, 6.

226. D. W. Chen, Y. F. Li and H. P. Ye, *Commun. Chin. Herb. Med.*, 1979, **10**, 7.

227. T. Kunikata, T. Tatefuji, H. Aga, K. Iwaki, M. Ikeda and M. Kurimoto. *Eur. J. Pharmacol.*, 2000, **410**, 93.

228. L. Martin, A. Magnaudeix, C. M. Wilson, C. Yardin and F. Terro. *J. Neurosci. Res.*, 2011, **89**, 1802.

229. R. Hoessel, S. Leclerc, J. A. Endicott, M. E. M. Nobel, A. Lawrie, P. Tunnah, M. Leost, E. Damiens, D. Marie, D. Marko, E. Niederberger, W. Tang, G. Eisenbrand and L. Meijer, *Nat. Cell Biol.*, 1999, **1**, 60.

230. D. Marko, S. Schatzle, A. Friedel, A. Genzlinger, H. Zankl, L. Meijer and G. Eisenbrand, *Br. J. Cancer*, 2001, **84**, 283.

231. J. K. Buolamwini, *Curr. Pharm. Des.*, 2000, **6**, 379.

232. S. Leclerc, M. Garnier, R. Hoessel, D. Marko, J. A. Bibb, G. L. Snyder, P. Greengard, J. Biernati, Y. Z. Wui, E. M. Mandelkowi, G. Eisenbrand and L. Meijer, *J. Biol. Chem.*, 2001, **276**, 251.

233. G. Sethi, K. S. Ahn, S. K. Sandur, X. Lin, M. M. Chaturvedi and B. B. Aggarwal, *J. Biol. Chem.*, 2006, **281**, 23425.

234. Cooperative Group of Clinical Therapy of Indirubin, *Chin. J. Intern. Med.*, 1980, **1**, 132.

235. G. Eisenbrand, F. Hippe, S. Jakobs and S. Muehlbeyer, *J. Cancer Res. Clin. Oncol.*, 2004, **130**, 627.

236. T. G. Davies, P. Tunnah, L. Meijer, D. Marko, G. Eisenbrand, J. A. Endicott and M. E. Noble, *Structure*, 2001, **9**, 389.

237. T. C. Tran, B. Sneed, J. Haider, D. Blavo, A. White, T. Aiyejorun, T. C. Baranowski, A. L. Rubinstein, T. N. Doan, R. Dingledine and E. M. Sandberg, *Cancer Res.*, 2007, **67**, 11386.

238. J. K. Kim, E. K. Shin, Y. H. Kang and J. H. Y. Park, *J. Cell. Biochem.*, 2011, **112**, 1384.

239. S. Nam, R. Buettner, J. Turkson, D. Kim, J. Q. Cheng, S. Muehlbeyer, F. Hippe, S. Vatter, K. Merz, G. Eisenbrand and R. Jove, *Proc. Natl. Acad. Sci. U. S. A.*, 2005, **102**, 5998.

240. S. Nam, A. Scuto, F. Yang, W. Y. Chen, S. Park, H. S. Yoo, H. Konig, R. Bhatia, X. Cheng, K. Merz, G. Eisenbrand, and R. Jove, *Mol. Oncol.*, 2012, **6**, 276.

241. S. P. Williams, M. O. Nowicki, F. Liu, R. Press, J. Godlewski, M. Abedel-Rasoul, B. Kaur, S. A. Fernandez, E. A. Chiocca and S. E. Lawler, *Cancer Res.*, 2011, **71**, 5374.
242. G. Karapetyan, K. Chakrabarty, M. Hein and P. Langer, *ChemMedChem*, 2011, **6**, 25.
243. A. Beauchard, H. Laborie, H. Rouillard, O. Lozach, Y. Ferandin, R. Le Guevel, C. Guguen-Guillouzo, L. Meijer, T. Besson and V. Thiery, *Bioorg. Med. Chem.*, 2009, **17**, 6257.
244. K. Vougogiannopoulou, Y. Ferandin, K. Bettayeb, V. Nyrianthopoulos, O. Lozach, Y. Fan, C. H. Johnson, P. Magiatis, A. L. Skaltsounis, E. Mikros and L. Meijer, *J. Med. Chem.*, 2008, **51**, 6421.
245. Y. K. Lin, Y. L. Leu, S. H. Yang, H. W. Chen, C. T. Wang and J. H. S. Pang, *J. Dermatol. Sci.*, 2009, **54**, 168.
246. Y. K. Lin, *Comparison of Indigo Naturalis Oil Extract and Calcipotriol Solution in Treating Psoriasis*, http://clinicaltrials.gov/ct2/show/NCT 01445886?term=indirubin&rank=1, 2012 (last accessed 15 September 2012).
247. M. J. Howes and E. Perry, *Drugs Aging*, 2011, **28**, 439.
248. E. Perry and M. J. R. Howes, *CNS Neurosci. Ther.*, 2011, **17**, 683.
249. M. D. da Rocha, F. P. D. Viegas, H. C. Campos, P. C. Nicastro, P. C. Fossaluzza, C. A. M. Fraga, E. J. Barreiro and C. Viegas, *CNS Neurol. Disord. Drug Targets*, 2011, **10**, 251.
250. R. Jesky and C. Hailong, *Phytother. Res.*, 2011, **25**, 1105.
251. T. Y. Wu, C. P. Chen and T. R. Jinn, *Taiwan. J. Obstet. Gynecol.*, 2011, **50**, 131.
252. X. Ma, C. Tan, D. Zhu, D. R. Gang and P. Xiao, *J. Ethnopharmacol.*, 2007, **113**, 15.
253. F. Pilotaz and P. Masson, *Ann. Pharm. Fr.*, 1999, **57**, 363.
254. A. Zangara, *Pharmacol. Biochem. Behav.*, 2003, **75**, 675.
255. X. Ma, C. Tan, D. Zhu and D. R. Gang, *J. Ethnopharmacol.*, 2006, **104**, 54.
256. X. Ma and D. R. Gang, *Phytochemistry*, 2008, **69**, 2022.
257. I. E. Orhan, G. Orhan and E. Gurkas, *Mini Rev. Med. Chem.*, 2011, **11**, 836.
258. X. Z. Zhu, *Mem. Inst. Osawaldo Cruz.*, 1991, **86**, 173.
259. Q. Zhao and X. C. Tang, *Eur. J. Pharmacol.*, 2002, **455**, 101.
260. M. L. Raves, M. Harel, Y. P. Pang, I. Silman, A. P. Kozikowski and J. L. Sussman, *Nat. Struct. Biol.*, 1997, **4**, 57.
261. G. T. Ha, R. K. Wong and Y. Zhang, *Chem. Biodiversity*, 2011, **8**, 1189.
262. D. H. Cheng and X. C. Tang, *Pharmacol. Biochem. Behav.*, 1998, **60**, 377.
263. R. Wang, H. Yan and X. Tang, *Acta Pharmacol. Sin.*, 2006, **27**, 1.
264. H. Y. Zhang and X. C. Tang, *Trends Pharmacol. Sci.*, 2006, **27**, 619.
265. R. Wang and X. C. Tang, *Neurosignals*, 2005, **14**, 71.
266. R. K. Gordon, S. V. Nigam, J. A. Weitz, J. R. Dave, B. P. Doctor and H. S. Ved. *J. Appl. Toxicol.*, 2001, **21**, S47.
267. X. Gao and X. C. Tang, *J. Neurosci. Res.*, 2006, **83**, 1048.
268. L. Tang, R. Wang and X. Tang, *Acta Pharmacol. Sin.*, 2005, **26**, 673.

269. H. Y. Zhang, H. Yan and X. C. Tang, *Neurosci. Lett.*, 2004, **360**, 21.
270. H. Y. Zhang, Y. Q. Liang, X. C. Tang, X. C. He and D. L. Bai, *Neurosci. Lett.*, 2002, **317**, 143.
271. D. Bai, *Pure Appl. Chem.*, 2007, **79**, 469.
272. J. T. Little, S. Walsh and P. S. Aisen, *Expert Opin. Investig. Drugs*, 2008, **17**, 209.
273. H. Y. Zhang, C. Y. Zheng, H. Yan, Z. F. Wang, L. L. Tang, X. Gao and X. C. Tang, *Chem. Biol. Interact.*, 2008, **175**, 396.
274. Y. S. Cheng, C. Z. Lu, Z. L. Ying, W. Y. Ni, C. L. Zhang and G. W. Sang, *Chin. J. New Drugs Clin. Remedies*, 1986, **5**, 197.
275. J. S. Liu, C. M. Yu, Y. Z. Zhou, Y. Y. Han, F. W. Wu, B. F. Qi and Y. L. Zhu, *Acta Chim. Sin.*, 1986, **44**, 1035.
276. J. S. Liu, Y. L. Zhu, C. M. Yu, Y. Z. Zhou, Y. Y. Han, F. W. Wu and B. F. Qi, *Can. J. Chem.*, 1986, **64**, 837.
277. L. Qian and R. Ji, *Tetrahedron Lett.*, 1989, **30**, 2089.
278. Y. Xia and A. P. Kozikowski, *J. Am. Chem. Soc.*, 1989, **111**, 416.
279. X. C. Tang, *Zhongguo Yao Li Xue Bao*, 1996, **17**, 481.
280. X. Z. Zhu, X. Y. Li and J. Liu, *Eur. J. Pharmacol.*, 2004, **500**, 221.
281. H. Jiang, X. Luo and D. Bai, *Curr. Med. Chem.*, 2003, **10**, 2231.
282. X. Ma and D. R. Gang, *Nat. Prod. Rep.* 2004, **21**, 752.
283. D. Y. Zhu, C. H. Tan and Y. M. Li, in *Medicinal Chemistry of Bioactive Natural Products*, ed. X. T. Liang and W. S. Fang, WileyHoboken, NJ, 2006, p. 143.
284. M. J. R. Howes and P. J. Houghton, *Int. J. Biomed. Pharm. Sci.*, 2009, **3**, 67.
285. B. S. Wang, H. Wang, Z. H. Wei, Y. Y. Song, L. Zhang and H. Z. Chen, *J. Neural Transm.*, 2009, **116**, 457.
286. Z. Hao, M. Liu, Z. Liu and D. Lu, *Cochrane Database Syst. Rev.*, 2009, CD007365.
287. Debiopharm, *Debio 9902-ZT-1*, http://www.debiopharm.com/media/press-releases/36-debio-9902-zt-1.html, 2012 (last accessed 15 September 2012).
288. National Institutes of Health, *Huperazine A*, http://clinicaltrials.gov/ct2/results?term=huperzine+a, 2012 (last accessed 15 September 2012).
289. A. Mazurek, *Alt. Ther.*, 1999, **5**, 97.
290. A. A. Mazurek, *N. Engl. J. Med.*, 2000, **342**, 821.
291. M. S. Rafii, S. Walsh, J. T. Little, K. Behan, B. Reynolds, C. Ward, S. Jin, R. Thomas and P. S. Aisen, *Neurology*, 2011, **76**, 1389.
292. A. R. Desilets, J. J. Gickas and K. C. Dunican, *Ann. Pharmacother.*, 2009, **43**, 514.
293. G. Lallement, V. Baille, D. Baubichon. P. Carpentier, J. M. Collombet, P. Filliat, A. Foquin, E. Four, C. Masqueliez, T. Testylier, L. Tonduli and F. Dorandeu, *Neurotoxicology*, 2002, **23**, 1.
294. G. Lallement, J. P. Demoncheaux, A. Foquin, D. Baubichon, M. Galonnier, D. Clarençon and F. Dorandeu, *Drug Chem. Toxicol.*, 2002, **25**, 309.

295. J. Grunwald, L. Raveh, B. Doctor and Y. Ashani, *Life Sci.*, 1994, **54**, 991.
296. L. S. Tonduli, G. Testylier, C. Masqueliez, G. Lallement and P. Monmaur, *Neurotoxicology*, 2001, **22**, 29.
297. J. Bajgar, J. Fusek, J. Kassa, K. Kuca and D. Jun, *Curr. Med. Chem.*, 2009, **16**, 2977.
298. H. Beinfeield and E. Korngold, *Between Heaven and Earth*, Ballantine, New York, 1991, p. 265.
299. P. A. Batch, *Prescription for Nutritional Healing: a Practical A-to-Z Reference to Drug-Free Remedies Using Vitamins, Minerals, Herbs and Food Supplements*, 4th edn, Avery Penguin Putnam, New York, 2010.
300. S. Foster and R. L. Johnson, *Desk Reference of Nature's Medicine*, National Geographic Society, Washington, DC, 2008, p .28.
301. Q. R. Wang, *Yilin Gaicuo*, People's Medical Publishing House, Beijing, 2005.
302. H. W. Wang, K. T. Liou, Y. H. Wang, C. K. Lu, Y. L. Lin, I. J. Lee, S. T. Huang, Y. H. Tsai, Y. C. Cheng, H. J. Lin and C. Shen, *J. Ethnopharmacol.*, 2011, **138**, 22.
303. China Medical University Hospital, *A Randomized Double Blind Placebo Control Study of Huang-Chi-Wu-Wu-Tang in Patients With Intracranial Arterial Stenosis*, http://clinicaltrials.gov/ct2/show/NCT01553643, 2012 (last accessed 15 September 2012).
304. J. L. Rios and P. G. Waterman, *Phytother. Res.*, 1997, **11**, 411.
305. J. Zhang, X. Xie, C. Li and P. Fu, *J. Ethnopharmacol.*, 2009, **126**, 189.
306. K. W. Zhao and H. Y. Kong, *Chung Kuo His I Chieh Ho Tsa Chih*, 1993, **13**, 263.
307. B. M. Shao, W. Xu, H. Dai, P. Tu, Z. Li and X. M. Gao, *Biochem. Biophys. Res. Commun.*, 2004, **320**, 1103.
308. J. Purmova and L. Opletal, *Cesk. Slov. Farm.*, 1995, **44**, 246.
309. Y. H. Kuo, W. J. Tsai, S. H. Loke, T. S. Wu and W. F. Chiou, *J. Ethnopharmacol.*, 2009, **122**, 28.
310. W. C. Cho and K. N. Leung, *J. Ethnopharmacol.*, 2007, **113**, 132.
311. J. Wang, X. Tong, P. Li, H. Cao and W. Su, *J. Ethnopharmacol.*, 2012, **139**, 788.
312. Anon, *Alt. Med. Rev.*, 2003, **8**, 72.
313. Y. Sun, E. M Hersh, S. L. Lee, M. McLaughlin, T. L. Loo and G. M. Mavligit, *J. Biol. Resp. Modif.*, 1983, **2**, 227.
314. M. Smith and H. S. Boon, *Patient Educ. Couns.*, 1999, **38**, 109.
315. Y. Yoshida, M. Q. Wang, J. N. Liu, G. E. Shang and U. Yamashita, *Int. J. Immunopharmacol.*, 1997, **19**, 359.
316. S. Sinclair, *Alt. Med. Rev.*, 1998, **3**, 338.
317. S. Foster and Y. Chongxi, *Herbal Emissaries, Bringing Chinese Herbs to the West*, Healing Arts Press, Rochester, VT, 1992.
318. Y. Wang, X. J. Wang, H. R. Hadley and G. H. Lau, *Mol. Biother.*, 1992, **4**, 143.

319. M. McCulloch, C. See, X. J. Shu, M. Broffman, A. Kramer, W. Y. Fan, J. Gao, W. Lieb, K. Shieh and J. M. Colford Jr, *J. Clin. Oncol.*, 2006, **24**, 419.

320. V. B. Konkimalla and T. Efferth, *J. Ethnopharmacol.*, 2008, **116**, 207.

321. K. I. Block and M. N. Mead, *Integr. Cancer Ther.*, 2003, **2**, 247.

322. F. Qi, A. Li, Y. Inagaki, J. Gao, J. Li, N. Kokudo, X. K. Li and W. Tang, *BioSci. Trends*, 2010, **4**, 297.

323. S. Z. Li, *Compendium of Materia Medica*, People's Public Health Press, Beijing, 2004.

324. J. Zhang, X. Xie, C. Li and P. Fu, *J. Ethnopharmacol.* 2009, **126**, 189.

325. M. Li, W. Wang, J. Xue, Y. Gu and S. Lin, *J. Ethnopharmacol.*, 2011, **133**, 412.

326. Y. Luo, Z. Qin, Z. Hong, X. Zhang, D. Ding, J. H. Fu, W. D. Zhang and J. Chen, *Neurosci. Lett.*, 2004, **363**, 218.

327. W. D. Zhang, H. Chen, C. Zhang, R. H. Liu, H. L. Li and H. Z. Chen, *Planta Med.*, 2006, **72**, 4.

328. Z. Zhao, W. Wang, F. Wang, K. Zhao, Y. Han, W. Xu and L. Tang, *Chin. Med.*, 2009, **4**, 6.

329. S. Lu, K. J. Chen, Q. Y. Yang and H. R. Sun, *Chin. J. Integr. Med.*, 2011, **17**, 473.

330. S. Fu, J. Zhang, F. Menniti-Ippolito, X. Gao, F. Galeotti, M. Massari, L. Hu, B. Zhang, R. Ferrelli, A. Fauci, F. Firenzuoli, H. Shang, R. Guerra and R. Raschetti, *PLoS ONE*, 2011, **6**, e19604.

331. Q. Y. Yang, K. J. Chen, S. Lu and H. R. Sun, *Chin. J. Integr. Med.*, 2012, **18**, 236.

332. K. H. Lee, T. W. Tao-Wiedman and H. Wuh, *Pharmagenesis, Advisory Board Meeting*, 1993.

333. Pharmagenesis, *Hematopoiesis Product: PG2*, www.pharmagenesis.net/PG2.htm, 2012 (last accessed 15 September 2012).

334. PhytoHealth Corporation, *PG2 Treatment for Improving Fatigue Among Advanced Cancer Patients Under Standard Palliative Care*, http://clinicaltrials.gov/ct2/show/NCT00523107, 2009 (last accessed 15 September 2012).

335. W. W. Chen, I. H. Lin, Y. J. Chen, K. H. Chang, M. H. Wu, W. H. Su, G. C. Huang and Y. L. Lai, *Clin. Invest. Med.*, 2012, **35**, E1.

336. PhytoHealth Corporation, *Asian Leader*, Global Niche, http://www.phytohealth.com.tw/en/, 2012 (last accessed 15 September 2012)

337. PhytoHealth Corporation, *Safety and Efficacy Study of PG2 to Treat Idiopathic Thrombocytopenic Purpura (ITP) Patients*, http://clinicaltrials.gov/ct2/show/NCT00860600, 2012 (last accessed 15 September 2012).

338. PhytoHealth Corporation, *The Pilot Clinical Study of PG2 Injection on Hemorrhagic Stroke*, http://clinicaltrials.gov/ct2/show/NCT01325233, 2012 (last accessed on 25 October 2012).

339. China Medical University Hospital, *PG2 Injection 500 mg in Acute Stroke Study (Pass)*, http://clinicaltrials.gov/ct2/show/NCT01603667, 2012 (last accessed 15 September 2012).

CHAPTER 7

PI3K–AKT Signaling in Cell Growth and Metabolism

YANG LI[†][a], ANKETSE KASSA[†][a] AND
BANGYAN L. STILES[*][a,b]

[a] Pharmacology and Pharmaceutical Sciences, School of Pharmacy, University of Southern California, Los Angeles, CA 90089, USA; [b] Department of Pathology, Keck School of Medicine, University of Southern California, Los Angeles, CA 90033, USA
*E-mail: bstiles@usc.edu

7.1 Introduction

Phosphatidylinositol 3-kinases (PI3Ks) are lipid kinases that regulate a variety of cellular processes including growth, proliferation, differentiation, survival and motility, in addition to cellular glucose and lipid metabolism. Hyperactivation of the PI3K signaling cascade is one of the most frequent events associated with human cancers. Most of the PI3K hyperactivation cases identified in tumors resulted from mutations of the PI3K gene or other genes downstream of PI3K, such as AKT kinases.[1] In addition, mutation, silencing, aberrant transcripts and allelic imbalance of its negative regulator PTEN (phosphatase and tensin homolog deleted on chromosome 10) are common occurrences.[1] Studies over the last two decades have suggested PI3K and its downstream targets to be promising drug targets for cancer therapy. However, the current inhibitors of PI3K signals inevitably lead to hyperglycemia due to the role of PI3K signaling in glucose metabolism.[2] In this chapter, we describe the PI3K signals in regulation of both growth and metabolism. We also

RSC Drug Discovery Series No. 31
Traditional Chinese Medicine: Scientific Basis for Its Use
Edited by James David Adams, Jr. and Eric J. Lien
© The Royal Society of Chemistry 2013
Published by the Royal Society of Chemistry, www.rsc.org

summarize the functional consequences of PI3K signals in various diseases that arise from dysregulation of this signaling pathway. Traditional Chinese Medicine applications of PI3K signaling are discussed in chapters 8 and 11.

7.2   The Basics of PI3K Signaling

7.2.1   Classifications of PI3K and Its Activation

PI3K functions to catalyze the phosphorylation of Ptdlns (4,5) P2 to make Ptdlns (3,4,5) P3. It achieves this task by phosphorylating the hydroxyl group of the third position on the inositol ring of phosphatidylinositols (Ptdlns).[3] PI3Ks are heterodimeric enzymes that are composed of a catalytic subunit and a regulatory subunit. PI3Ks are classified into three classes, Class I, Class II and Class III, based on structure, substrate preference and sequence homology.[4] Class I is divided into class 1A and 1B based on its regulatory subunit. The regulatory subunit for class IA PI3Ks contains binding sites for phosphotyrosine motifs, indicating its potential importance in tyrosine kinase signaling. The regulatory subunit of class 1B PI3K does not possess the ability to bind phosphotyrosine. It is, however, involved in signaling through the G$\beta\gamma$ subunits in the G protein-coupled receptor pathway.[5] Class II PI3K is poorly understood but is believed to be involved in clathrin-mediated membrane trafficking.[6] Class III PI3K contains catalytic vacuolar protein-sorting defective 34 (Vps34) and regulatory Vps15 subunits that are required for trafficking of proteins and vesicles from Golgi to endosomes in vacuolar protein sorting.[7] For this chapter, we will focus on the signals and the biological functions mediated by class IA PI3K due to its role in both growth/survival and metabolic regulation. So far, there are five variants of the regulatory subunit: p85α, p55α, p50α, p85β or p55γ, and three variants of the catalytic (p110) subunit, p110α, -β or -δ, identified for class IA PI3Ks. P85α is the most highly expressed regulatory subunit among all the variants. P110α and -β are globally expressed whereas p110δ is primarily enriched in leukocytes.[8] The *PIK*3r1 gene is responsible for encoding p85α, p55α and p50α. *PIK*3r2 encodes p85β and *PIK*3r3 encodes p55γ. Three genes, *PIK*3ca, *PIK*3cb and *PIK*3cd, respectively encode catalytic subunits p110α, -β and -δ, which pair with one of the five regulatory subunits.[9]

PI3K is activated in response to growth factor binding to their receptors (Figure 7.1). Such binding induces autophosphorylation of the receptors and also phosphorylation of downstream scaffolding proteins due to the activation of receptor tyrosine kinases (RTKs). The phosphotyrosine residues on the receptors or scaffolding proteins, for example, insulin receptor substrate-1 (IRS-1), then serve as docking sites to recruit and activate PI3K. The regulatory subunit of PI3K, *e.g.*, p85, contains an src homology (SH2) domain that allows the PI3K to bind the phosphotyrosine motif pYXXM on RTK or IRS.[10] In the absence of stimulation, PI3K remains in a resting status in the cytoplasm as the inactive p85–p110 complex. In response to signals such as

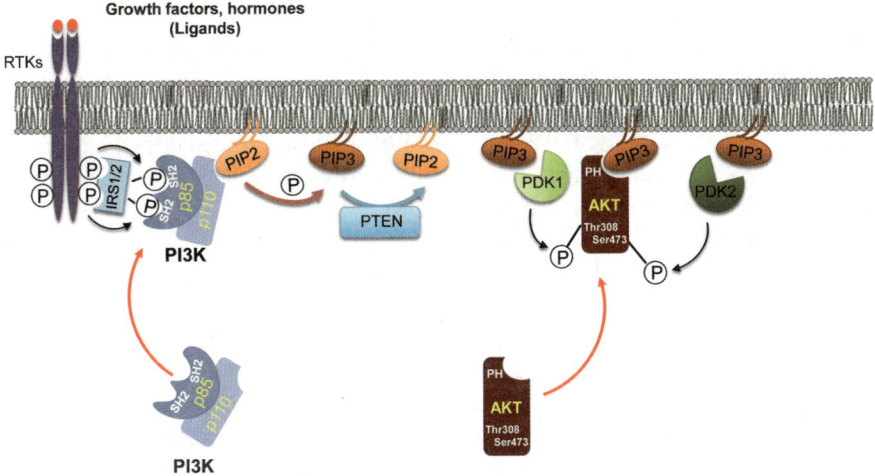

Figure 7.1   Basics of PI3K–AKT signaling. Binding of growth factors/hormones to receptor tyrosine kinases (RTKs) leads to recruitment of insulin receptor substrate (IRS) 1/2, which serve as adaptor proteins and provide docking sites for the SH2 domain on the p85 subunit of PI3K. Interaction between PI3K's SH2 domain and IRS recruits PI3K to the plasma membrane where PI3K is in close proximity to its substrate, PtdIns (4,5) P2 (PIP$_2$). The SH2–IRS interaction also relieves the inhibitory action of the p85 subunit on the p110 catalytic subunit. PtdIns (3,4,5) P3 (PIP$_3$) produced by PI3K causes AKT translocation to the plasma membrane *via* interaction with the pleckstrin-homology (PH) domain on AKT. PIP$_3$ also recruits PDK 1/2, which phosphorylate Thr308 and Ser473 on AKT. Phosphorylation on these two sites fully activates AKT. By dephosphorylating PIP$_3$, PTEN antagonizes PI3K–AKT signaling.

ligand-mediated RTK activation, the SH2 domain of p85 interacts with consensus phosphotyrosine residues on RTK or IRS. This interaction brings PI3K to the cell membrane where the lipid substrates are located. Moreover, it is believed that the interaction between p85 and RTK or IRS releases the inhibitory effect of p85 on p110, leading to the induction of PI3K's catalytic activity.[11]

7.2.2   AKT as a Major Target of PI3K Activation

The action of PI3K leads to the activation of a number of molecules that contain the pleckstrin homology (PH) domain. The PH domain is a lipid-binding domain involved in recruiting proteins to the inner surface of plasma membrane.[12] One primary effect of PI3K action is activation of AKT, a serine/threonine kinase also known as protein kinase B (PKB). AKT harbors a PH domain at the amino terminal.[13] Following PI3K activation, accumulation of PIP$_3$ allows recruitment of AKT via direct interaction with its PH domain.[1] Meanwhile, another PH domain-containing kinase, 3-phosphoinositide-dependent protein kinase 1 (PDK1), also interacts with PIP$_3$ and translocates to the cell membrane, where it phosphorylates AKT at Thr308 (Figure 7.1).[14] Phosphorylation on Ser473 is important for initial activation of AKT whereas Thr308 phosphorylation by PDK1 is required for maximal AKT activation.[15] It has been shown that AKT is further phosphorylated by mTOR2 at Ser473.[16,17] Activated AKT has a plethora of downstream functions including the regulation of kinases such as GSK3β,[18] IκB kinases (IKKα and IKKβ);[19] apoptotic factors such as BAD,[20] MDM2, a ubiquitin ligase for p53,[21] GTPases such as Rac and Rho,[22] cell cycle inhibitors p21 and p27[23] and transcription factors such as forkhead transcription family (FoxO) members (Figure 7.2).[24–26]

There are three known isoforms of AKT: AKT1 (PKBα), AKT2 (PKBβ) and AKT3 (PKBγ). Since these isoforms share a conserved kinase domain, they do not differ significantly in terms of substrate specificity and molecular action *in vitro*.[13] However, recent studies have identified potential structural bases for substrate specificities among AKT isoforms.[27] By ablating the different isoforms alone or in combination, we now have a better understanding of the different functions associated with each isoform.[28–30]

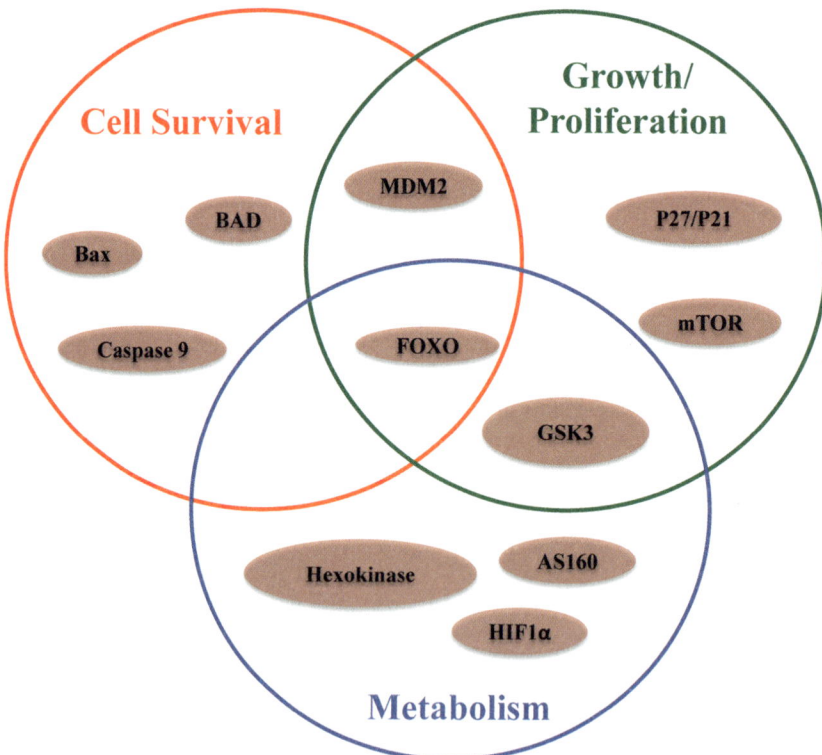

Figure 7.2 Major biological functions controlled by PI3K–AKT substrates. Phosphorylated AKT regulates cell proliferation, growth, survival and metabolism *via* phosphorylation of a number of substrates.

AKT1 is ubiquitously expressed in most tissues and primarily involved in cell survival pathways by inducing pro-survival genes while concomitantly down-regulating pro-apoptotic genes, thereby inhibiting programmed cell death.[31] In addition, AKT1 is also able to promote protein synthesis through the activation of mammalian target of rapamycin (mTOR).[32] Hyperactivation of AKT1 is implicated in various types of cancer.[3,33] Constitutive activation of AKT1 in pancreatic β-cells in transgenic mice induced a significant increase in islet size and mass.[34] Similar increases in cell size are also observed in the heart upon overexpression of AKT1.[35,36] Mice lacking AKT1 exhibit growth retardation, reduced body size and perinatal lethality.[30] Together, these animal model studies strongly suggest that AKT1 actively promotes cell growth and proliferation.

AKT2 is mainly found in the skeletal muscle and the pancreas. It has been found to be involved in insulin signaling and glucose metabolism. In humans, a dominant negative AKT2 mutation has been found to associate with severe hyperinsulinemia and a spike in blood glucose.[37] This result is further confirmed by the phenotypes of glucose intolerance and impaired glucose uptake observed in muscle and adipose tissue upon global deletion of AKT2.[38] Consistent with the above results, mutation in the catalytic domain of AKT2 causes insulin resistance and a diabetic phenotype.[39]

AKT3 is abundant in cortical progenitor cells and is important for brain development and neurogenesis. Somatic mutations of AKT3 cause complex neurogenetic diseases.[40] Global AKT3 deletion only affects brain size, highlighting the isoform-specific mechanism for brain size regulation. The reduction in cell size of AKT3 knockout mice is believed to be due to inhibition of the mTOR pathway.[41] Deletion of the AKT3 gene in hyperlipidemic ApoE-null mice led to increased lipoprotein uptake and macrophage cholesteryl ester accumulation, leading to atherosclerosis.[42] Furthermore, mutations of AKT3 are found in melanomas and AKT3 inhibition has also been attempted in melanoma prevention.[43]

Although the knockout studies of AKT isoforms as stated above have shown different physiological phenotypes, the three AKTs exhibit overlapping functions. Mice with a germline deletion of AKT1 are protected from high fat diet-induced obesity and insulin resistance, indicating an unexpected role of AKT1 in regulating energy metabolism.[44] Moreover, mice lacking AKT1 showed improved insulin sensitivity, lower blood glucose and higher serum glucagon concentrations, revealing the role of AKT1 in metabolic regulation and glucose homeostasis.[45] The role of AKT2 in glucose and lipid metabolism has been intensively investigated.[29,38,46] However, AKT2 is also robustly induced in several types of cancers, including breast cancer,[47] lung cancer[48] and colorectal cancer.[49] We showed recently that AKT2 (but not AKT1) is critically required for the tumorigenic effects of PTEN loss in liver cancer development.[50] Whether this effect is due only to AKT2's metabolic functions or is compounded with its function as a survival kinase is unclear.

7.2.3 PTEN: the Negative Regulator of PI3K Signaling

PTEN is the negative regulator of the RTK–PI3K–AKT signaling pathway. It is encoded on chromosome 10q23, a region where loss of heterozygosity frequently occurs. The protein encoded by the PTEN gene is a 403 amino acid protein whose amino-terminal region shares sequence homology with tensin and auxilin.[51,52] Crystal structure analysis for PTEN has revealed a C2 domain that retains affinity for phospholipids on the membrane and a phosphatase domain that displays dual specific activity towards both protein and lipid substrates *in vitro*.[53] As the PIP$_3$ phosphatase, PTEN antagonizes the function of PI3K by removing the phosphate from the third position on PIP$_3$, leading to reduced PIP$_3$ production and the down-regulation of signals downstream of PI3K (Figure 7.1).[1] Whereas the lipid substrate is well characterized to be PIP$_3$, the identities of the protein substrates for PTEN *in vivo* have been elusive.[17] However, *in vitro* studies have revealed that PTEN is able to regulate cell migration by dephosphorylating itself, providing insight for investigating potential protein substrates for PTEN.[54] Later structural analysis found that PTEN also harbors a PDZ domain and two PEST sequences in the C-terminal region.[55] The PDZ domain is thought to regulate PTEN's subcellular localization whereas the two PEST sequences regulate its protein stability.[56]

PTEN was cloned from differential display of tumor and non-tumor samples.[57] It was found to be frequently lost or mutated in various forms of cancer and thus identified as a tumor suppressor. Germline mutations of PTEN are associated with several familial syndromes, including Cowden syndrome, Bannayan–Zonana syndrome and Lhermitte–Duclos disease.[51] These syndromes are characterized by multiple hyperplastic lesions known as harmatomas within multiple organs that are susceptible to malignant transformation. Genetic studies using transgenic and knockout animals targeted at PTEN supported the role of PTEN as tumor suppressor.[51]

Similarly to PTEN, which facilitates the removal of phosphate from the third position on PIP$_3$, SHIP (src homology 2-containing inositol 5'-phosphatase) also converts PIP$_3$ to PIP$_2$ by removing the phosphate on the fifth position of PIP$_3$.[1] Studies using SHIP knockout mice showed that SHIP-deficient mice have hyperproliferative myeloid cells, increased cell survival and AKT activation, indicating that SHIP also regulates the mitogenic activity *via* AKT.[58] Moreover, SHIP has also been implicated in the regulation of B lymphoid development and antigen responsiveness.[59]

7.3 The Biological Functions of PI3K Signal

The PI3K–AKT pathway mediates signals from growth factor and plays a mitogenic role in regulating cell metabolism, growth and proliferation (Figure 7.2). In this section, we summarize the biological effects of PI3K–AKT signaling.

7.3.1 Cell Survival

Apoptosis or programmed cell death is an endogenous process that functions to eliminate damaged cells in order to maintain the proper cell population and tissue homeostasis. The apoptosis process can be regulated either by an extrinsic cell surface receptor-mediated pathway or an intrinsic mitochondria-mediated pathway.[60] The extrinsic apoptotic pathway relies on the binding of "death ligands," such as Fas (apoptosis antigen 1), tumor necrosis factor 1 (TNF-1) and TRAIL (TNF-related apoptosis-inducing ligand) to their receptors on the cell surface. Whereas intrinsic apoptosis depends on mitochondrial cytochrome c release stimulated by an internal "death signal" such as cellular damage or stress. Both death ligand and the internal "death signals" are targets of AKT regulation. The internal signals are initiated at the mitochondrial membrane where association of pro-apoptotic factors Bax and BAD with the anti-apoptotic factors Bcl-X$_L$ and Bcl-2 controls the release of cytochrome c from the mitochondria.[61] BAD induces cell death by forming a heterodimer with Bcl-X$_L$ and diminishing Bcl-X$_L$'s anti-apoptotic function. Phosphorylation of BAD at Ser9 by AKT interrupts this interaction and allows Bcl-X$_L$ and Bcl-2 to associate and prevent pore formation on the mitochondrial membrane.[20]

Both death receptor–ligand interaction and cytochrome c release initiate a signaling cascade that leads to the activation of caspases, a family of cysteine-dependent aspartate-specific proteases. Caspases are critical players in the regulation and execution of the apoptosis process.[62] Upon cleavage by apoptosome formed by apaf-1 and cytochrome c, caspases become active and cease essential cellular events by breaking down key components of the cells such as poly(ADP–ribose) polymerase,[63] nuclear lamins[64] and DNA-dependent protein kinase.[65] AKT is found to phosphorylate directly two members of the caspase family, caspase 3, an executor caspase, and caspase 9, an initiator caspase, leading to their inactivation and cell survival.[66,67]

Studies from previous decades have demonstrated that the PI3K–AKT axis also functions as an anti-apoptotic pathway by phosphorylating several Bcl-2 family members that are involved in mediating cell apoptosis.[1] In response to an apoptotic signal, Bax translocates onto the mitochondrial outer membrane and forms pores in order to allow the mitochondria to release cytochrome c.[68] Studies using molecular manipulations of PI3K–AKT signaling confirmed that Bax is directly phosphorylated by AKT. This phosphorylation inhibits Bax's function by preventing its translocation on to the mitochondria and sequestering Bax in the cytoplasm where it forms a dimer with anti-apoptotic Bcl-2 members, Bcl-X$_L$ and Bcl-2.[69] AKT also phosphorylates BAD on Ser136 and promotes its interaction with 14-3-3 proteins. Therefore, BAD's binding partners Bcl-2 and Bcl-X$_L$ could be released and transported on to mitochondria in order to execute the anti-apoptotic function.[20]

In addition to regulating directly the activity of apoptotic proteins, AKT also controls apoptosis regulators *via* phosphorylating key transcription factors. Members of the forkhead transcription family (FoxO), *e.g.*, FoxO1 and FoxO3, promote apoptosis *via* inducing expression of apoptosis activators such as Fas ligand (FasL), TRAIL and Bim.[70] Both FoxO1 and FoxO3 are phosphorylated by AKT on three conserved threonine/serine sites, leading to their nuclear exclusion and activity inhibition.[71]

Another pathway targeted by AKT for suppressing apoptosis is MDM2–p53 interaction. By binding to promoters, p53 induces the expression of Puma and Noxa, both of which encode pro-apoptotic Bcl-2 family members that are necessary for apoptosis initiation.[72] Phosphorylation of MDM2 on Ser166 and Ser186 by AKT promotes its translocation to the nucleus and accelerates p53 degradation.[73]

7.3.2   Cell Proliferation and Growth

In addition to negatively regulating cell apoptosis as a pro-survival kinase, AKT also drives cell proliferation and growth by phosphorylating target proteins, including those that directly regulate cell cycle progression. Cyclins, the partners for cell cycle accelerator cyclin-dependent kinases (CDKs), and inhibitors of CDK are both regulated by AKT signaling. p27, one of the CDK inhibitors, is directly phosphorylated by AKT, leading to its cytosolic sequestration *via* binding with 14-3-3 protein and thus its inhibitory effect on the cell cycle is released.[23] Another CDK inhibitor, p21, was also reported to be inhibited by AKT phosphorylation.[74] In addition, AKT may also regulate the expression of p21 by controlling the stability of p53, the transcriptional factor regulating p21 expression. Phosphorylation of MDM2, the ubiquitin ligase of p53, by AKT induces its activity, thus leading to reduced p53 and downregulation of p21.[73] Cyclin D1, the cyclin regulating the G1 cell cycle, is regulated by AKT both transcriptionally and post-translationally. Glycogen synthase kinase-3β (GSK3β), a substrate inhibited by AKT, phosphorylates cyclin D1 and targets it to degradation by the proteasomal system.[75] Moreover, the unphosphorylated active form of FoxO3 is thought to repress cyclin D1's transcription.[76] AKT phosphorylation of FoxO3 leads to its nuclear exclusion. Thus, the transcription of cyclin D1 is blocked.[77] Similarly, negative regulation of FoxO1 by AKT also contributes to enhanced cell survival and cell cycle progression.[78]

In addition to directly regulating the progression of the cell cycle, PI3K–AKT signals also control the mammalian target of rapamycin (mTOR) complex. mTOR is a serine/threonine kinase regulating nutrient sensing that couples nutrient availability to cell growth. In response to growth signal and nutrient changes, activated mTOR phosphorylates S6 kinase 1 (S6K1) and eIF4E-binding protein 1 (4EBP1), two molecules involved in protein translation.[32] The upstream regulatory machinery of mTOR involves tuberous sclerosis1/2 (TSC1/2) complex and RHEB, a small GTPase Ras homolog enriched in the brain. GTP-loaded RHEB plays an important role in stimulating the kinase activity of mTOR. TSC1/2 complex functions as a GTPase-activating protein (GAP) towards RHEB, resulting in the generation of GDP-bound RHEB, which is unable to bind to and activate mTOR.[70,79] PI3K–AKT signals phosphorylate and inhibit the GAP activity of TSC complexes and allow activation of mTOR.[79] In addition, GSK3β may also phosphorylate TSC and contribute to mTOR activation.[80] Direct phosphorylation of mTOR by AKT has also been reported but the functional consequence of such phosphorylation is unclear.[32,81] These results support a role of the PI3K–AKT signal axis in stimulating cell growth *via* mTOR activation, a process that is fulfilled by synthesis of macromolecules such as protein, nucleic acids and lipids.

7.3.3 Metabolism

A major signal transmitted through the PI3K–AKT pathway is that of insulin binding to its receptor. Binding of insulin to insulin receptor (IR) results in recruitment and activation of IRS. Treatment with the PI3K-specific inhibitor wortmannin can effectively inhibit glucose transport, suggesting a role for PI3K in glucose metabolism. Moreover, overexpression of a dominant negative form of the p85 subunit inhibits insulin-induced PI3K activation in the liver whereas overexpression of constitutively active PI3K led to increased insulin responsiveness.[82,83] These studies indicate that PI3K mediates insulin action.

A number of AKT substrates relevant to insulin signaling have been identified. As mentioned previously, GSK3β is phosphorylated and inhibited by AKT. This action blocks the phosphorylation of glycogen synthase by GSK3β and relieves the inhibition of glycogen synthesis. Therefore, insulin signals through AKT led to increased glycogen synthesis through GSK3β.[84] The active form of FoxO1 promotes expression of several genes involved in gluconeogenesis.[76] Inhibition of FoxO1 in the liver was shown to inhibit gluconeogenesis and lead to hypoglycemia in mice.[85] Thus, AKT activation directly controls the glucose metabolic process through GSK3β and FoxO1.

PI3K–AKT signaling also controls the translocation of glucose transporters, particularly in the muscle and adipose tissue where glucose uptake depends on insulin-sensitive translocation of glucose transporter 4 (Glut4). The shuttling of Glut4 to the cell membrane is negatively regulated by AS160, a protein that shares homology with the Rab GTPase activating protein (GAP) family. When stimulated by insulin binding, AKT phosphorylates AS160 on five sites. Mutagenesis studies showed that phosphorylation of two (S588 and T642) or more of these sites is necessary for insulin-stimulated Glut4 translocation to the membrane. Hence it is proposed that phosphorylation at S588 and T642 by AKT is required for inactivating the GAP activity of AS160 and the subsequent stimulation of Glut4 translocation.[86] In non-insulin-sensitive organs, glucose transporter 1 (Glut1) is the major transporter for glucose uptake. Studies have suggested that the PI3K–AKT–mTOR cascade promotes Glut1's mRNA translation through the inhibition of eukaryotic initiation factor 4E-binding protein-1 (4E-BP1), a translational suppressor.[87] Moreover, the PI3K–AKT–mTOR pathway increases the protein level of hypoxia-inducible factor 1α (HIF1α), which also induces Glut1's gene expression.[88] The HIF1α induction stimulated by AKT signaling not only affects Glut1's expression but also promotes the expression of other glycolytic enzymes including hexokinase I/II, lactate dehydrogenase A (LDHA) and phosphofructokinase L (PFKL),[89] leading to a metabolic shift towards glycolysis. Hyperactivation of the PI3K–AKT signal and the subsequent HIF1α accumulation are proposed to underlie the dramatically elevated glucose utilization observed in tumors.

Recent discoveries suggest that AKT may also interact directly with mitochondrial metabolism. Hexokinase II converts glucose into the active form glucose 6-phosphate (G-6-P).[90] This action traps glucose inside the cells and is the first rate-limiting step of glycolysis. Activated AKT promotes the binding of hexokinase II to mitochondrial voltage-dependent anion channel (VDAC). This association between hexokinase II and VDAC stabilizes hexokinase II and primes it to phosphorylate glucose upon ATP release from mitochondria, allowing glycolysis to be carried out with high efficiency. This evidence provides major support for the role of AKT activation in the induction of glycolytic metabolism in the unique cancer metabolic signature known as "Warburg effects."

Another change induced by insulin signals is lipid metabolism within cells. Sterol regulatory element-binding proteins (SREBPs) are a family of transcription factors that control cholesterol and lipid metabolism by up-regulating lipogenic genes including fatty acid synthase (FASN), glycerol-3-phosphate (GPAT), acetyl-CoA carboxylase (ACC) and stearoyl-CoA desaturase 1(SCD1).[91] Insulin treatment leads to rapid SREBP nuclear accumulation and the expression of lipogenic genes.[92] Recent studies suggest that the AKT2 but not other PH domain-containing proteins such as atypical PKC is responsible for this action of insulin.[93,46,94] Studies found that GSK3β phosphorylates T426 and S430 on SREBP.[95] These two phosphorylation sites promote the association between SREBP and SCF ubiquitin ligase and target SREBP for proteasomal degradation.[95] Therefore, by phosphorylating and inhibiting GSK3β, PI3K–AKT signals may stabilize SREBP and thus activate lipid production. mTOR is also involved in the regulation of lipid homeostasis.[92]

Furthermore, the growth and survival effect of PI3K signaling has been shown to protect pancreatic β-cells from apoptosis and promote their growth.[96] Genetic studies have shown that this signaling pathway is needed to maintain the mass of pancreatic islets and activation of the pathway leads to improved tolerance to hyperglycemic induction due to the ability of pancreatic β-cells to compensate for such effects.[97,98] Constitutive expression of AKT or components of insulin–IGF-1 receptor (IR) or substrates (IRS) or PI3K leads to higher islet mass.[99–101] On the other hand, deletion of various components, such as AKT, GSK-3β, S6K, IR, IRS, mTOR or FoxO, has been shown to result in smaller islets and reduced tolerance to glycemic challenge.[29,102–107] We showed that deleting PTEN, the negative regulator of the pathway, leads to increased islet mass particularly when animals age.[108] These genetic manipulations together suggest that enhancing the PI3K signaling pathway may be needed to maintain the homeostasis of some tissues and inhibiting this pathway as a cancer target may require targeted delivery to achieve the needed efficacy.

7.4 Controversies in Targeting PI3K Signaling and Therapies

The PI3K–AKT axis plays a pivotal role in cell growth, proliferation and survival and its dysregulation is commonly linked with cancers. Studies over the past decades have indicated that the constitutive activation of the PI3K–AKT cascade is associated with a wide spectrum of human cancers, including glioblastoma, ovarian cancer, breast cancer, hepatocellular carcinoma and lung cancer. Most of the PI3K hyperactivation cases identified in tumors resulted from different types of PTEN alterations such as mutation, silencing, aberrant transcripts and allelic imbalance.[1] In addition, PI3K alterations such as p110α amplification, p85α mutation and p85–EPH fusion frequently occurred in ovarian, breast and lymphoid cancers.[109–111] Other PI3K activations are due to AKT overexpression and amplification.[112] Therefore, both PTEN alterations and PI3K mutations are associated with tumorigenesis. Specific inhibitors targeting PI3K–AKT signaling components have been developed and provided promising opportunities in the field of translational cancer research. Wortmannin and LY294002 are first-generation compounds that have been used as PI3K inhibitors *in vitro* for many years. Administration of both LY294002 and wortmannin have shown efficacy in xenograft cancer models.[1] However, neither LY294002 nor wortmannin has been developed for clinical application due to excessive toxicity and unfavorable pharmacokinetic properties. Recently, the second generation of PI3K inhibitors has been developed and tested in human clinical Phase I/II trials, such as XL-147 (Exelixis, in combination with chemotherapy), PX866 (Oncothyreon), BKM120 (Norvatis) and CAL101 (Calistoga Pharma). Overall, the PI3K inhibitors can be divided into two classes. The first class is pan-PI3K inhibitors that target all classes of IA PI3K isoforms in the tumor. These inhibitors (*e.g.,* XL-147) are commonly tested in association with chemotherapeutic agents. However, complications such as hyperglycemia and immunosuppression are commonly observed with many PI3K inhibitors. For example, PX-886 (Oncothyreon)-treated animals experienced hyperglycemia and glucose intolerance.[113] PI-103 (Genentech and Calibiochem), a PI3K inhibitor used in melanoma therapy, promotes immunosuppression in a xenograph mouse model.[114] In order to reduce toxicity, the second class of PI3K inhibitors targeting specific PI3K isoforms has also been developed. For example, CAL101 (Calistoga Pharma), selectively targeting the PI3Kδ isoform, showed limited impact on the immune system and promising clinical efficacy in patients with chronic lymphocytic leukemia.[115] In addition, recent studies have shown a predominant role of p110β, but not p110α in some PTEN-null cancers. Hence p110β-specific inhibitors might be more effective than general or other isoform-targeting PI3K inhibitors in this subtype of cancer.[116,117] These isoform-specific PI3K inhibitors might provide a therapeutic advantage in certain cancer types.

As a prominent effector in PI3K signaling, AKT is also targeted by both ATP-non-competitive (allosteric) and ATP-competitive inhibitors.[118] Allosteric AKT inhibitors, including Perifosine (Keryx Phase I/II) and MK-2206 (Merck Phase I), block AKT translocation to the membrane and the subsequent AKT phosphorylation by disrupting the binding of the PH domain to phosphoinositides.[119] In contrast, ATP-competitive inhibitors such as GSK690693 (GSK, Phase I) and VQD002 (Vioquest, Phase I) might not block AKT phosphorylation and are commonly non-selective against AKT isoforms. They are also poorly selective against related kinases due to a conserved ATP binding domain across the kinome.[120]

The conflict between the anti-tumor and hyperglycemia/immunosuppression effects of the inhibitors is reminiscent of the functions of PI3K–AKT as a pro-growth and pro-survival signal. Blocking the action of this signaling pathway suppresses the growth and survival of tumor cells. At the same time, activation of immune cells depends on the same growth/survival signaling to induce the immune system. Thus, pan-PI3K or AKT inhibitors are expected to suppress the growth of tumor cells and block the activation of the immune system. Similarly, activation of the PI3K–AKT pathway is needed for the clearance of glucose. When this pathway is inhibited, insulin is incapable of inducing the transport of glucose from the bloodstream to peripheral tissues. The gluconeogenesis process that is shut down due to insulin–PI3K–AKT signaling in the liver also remains active. These metabolic changes due to inhibition of PI3K–AKT underlie the mechanisms for the hyperglycemic effects of the class I PI3K inhibitors and some of the AKT inhibitors. The ability to specifically target PI3K–AKT signaling without inducing immunosuppression and hyperglycemia is necessary for moving these compounds to the clinic. Targeting specific isoforms of the enzymes may prove to be promising in this effort.

References

1. I. Vivanco and C. L. Sawyers, *Nat. Rev. Cancer*, 2002, **2**, 489.
2. K. D. Courtney, R. B. Corcoran and J. A. Engelman, *J. Clin. Oncol.*, 2010, **28**, 1075.
3. T. L. Yuan and L. C. Cantley, *Oncogene*, 2008, **27**, 5497.
4. J. A. Engelman, J. Luo and L. C. Cantley, *Nat. Rev. Genet.*, 2006, **7**, 606.
5. M. P. Wymann, K. Bjorklof, R. Calvez, P. Finan, M. Thomast, A. Trifilieff, M. Barbier, F. Altruda, E. Hirsch and M. Laffargue, *Biochem. Soc. Trans.*, 2003, **31**, 275.

6. I. Gaidarov, M. E. Smith, J. Domin and J. H. Keen, *Mol. Cell*, 2001, **7**, 443.
7. G. Odorizzi, M. Babst and S. D. Emr, *Trends Biochem. Sci.*, 2000, **25**, 229.
8. B. Vanhaesebroeck, K. Ali, A. Bilancio, B. Geering and L. C. Foukas, *Trends Biochem. Sci.*, 2005, **30**, 194.
9. S. G. Ward, J. Westwick and S. Harris, *Immunol. Lett.*, 2011, **138**, 15.
10. K. Yonezawa, H. Ueda, K. Hara, K. Nishida, A. Ando, A. Chavanieu, H. Matsuba, K. Shii, K. Yokono, Y. Fukui, *et al.*, *J. Biol. Chem.*, 1992, **267**, 25958.
11. J. Yu, Y. Zhang, J. McIlroy, T. Rordorf-Nikolic, G. A. Orr and J. M. Backer, *Mol. Cell. Biol.*, 1998, **18**, 1379.
12. J. P. DiNitto, T. C. Cronin and D. G. Lambright, *Science's STKE*, 2003, **2003** (213), re16.
13. R. W. Matheny Jr and M. L. Adamo, *Exp. Biol. Med.*, 2009, **234**, 1264.
14. B. Vanhaesebroeck and D. R. Alessi, *Biochem. J.*, 2000, **346** (Pt 3), 561.
15. D. R. Alessi, S. R. James, C. P. Downes, A. B. Holmes, P. R. Gaffney, C. B. Reese and P. Cohen, *Curr. Biol.*, 1997, **7**, 261.
16. R. C. Hresko and M. Mueckler, *J. Biol. Chem.*, 2005, **280**, 40406.
17. M. A. Knowles, F. M. Platt, R. L. Ross and C. D. Hurst, *Cancer Metastasis Rev.*, 2009, **28**, 305.
18. M. Pap and G. M. Cooper, *J. Biol. Chem.*, 1998, **273**, 19929.
19. J. A. Gustin, O. N. Ozes, H. Akca, R. Pincheira, L. D. Mayo, Q. Li, J. R. Guzman, C. K. Korgaonkar and D. B. Donner, *J. Biol. Chem.*, 2004, **279**, 1615.
20. S. R. Datta, H. Dudek, X. Tao, S. Masters, H. Fu, Y. Gotoh and M. E. Greenberg, *Cell*, 1997, **91**, 231.
21. M. Ashcroft, R. L. Ludwig, D. B. Woods, T. D. Copeland, H. O. Weber, E. J. MacRae and K. H. Vousden, *Oncogene*, 2002, **21**, 1955.
22. J. Liliental, S. Y. Moon, R. Lesche, R. Mamillapalli, D. Li, Y. Zheng, H. Sun and H. Wu, *Curr. Biol.*, 2000, **10**, 401.
23. M. Collado, R. H. Medema, I. Garcia-Cao, M. L. Dubuisson, M. Barradas, J. Glassford, C. Rivas, B. M. Burgering, M. Serrano and E. W. Lam, *J. Biol. Chem.*, 2000, **275**, 21960.
24. W. H. Biggs, 3rd, J. Meisenhelder, T. Hunter, W. K. Cavenee and K. C. Arden, *Proc. Natl. Acad. Sci. U. S. A.*, 1999, **96**, 421.
25. A. Brunet, A. Bonni, M. J. Zigmond, M. Z. Lin, P. Juo, L. S. Hu, M. J. Anderson, K. C. Arden, J. Blenis and M. E. Greenberg, *Cell*, 1999, **96**, 857.
26. S. Guo, G. Rena, S. Cichy, X. He, P. Cohen and T. Unterman, *J. Biol. Chem.*, 1999, **274**, 17184.
27. A. Toker, *Adv. Enzyme Regul.*, 2011, 2 October, PMID 21986444.
28. S. S. Bae, H. Cho, J. Mu and M. J. Birnbaum, *J. Biol. Chem.*, 2003, **278**, 49530.

29. H. Cho, J. Mu, J. K. Kim, J. L. Thorvaldsen, Q. Chu, E. B. Crenshaw III, K. H. Kaestner, M. S. Bartolomei, G. I. Shulman and M. J. Birnbaum, *Science*, 2001, **292**, 1728.

30. H. Cho, J. L. Thorvaldsen, Q. Chu, F. Feng and M. J. Birnbaum, *J. Biol. Chem.*, 2001, **276**, 38349.

31. B. Stiles, V. Gilman, N. Khanzenzon, R. Lesche, A. Li, R. Qiao, X. Liu and H. Wu, *Mol. Cell. Biol.*, 2002, **22**, 3842.

32. B. T. Nave, M. Ouwens, D. J. Withers, D. R. Alessi and P. R. Shepherd, *Biochem. J.*, 1999, **344** (Pt 2), 427.

33. D. Hanahan and R. A. Weinberg, *Cell*, 2011, **144**, 646.

34. R. L. Tuttle, N. S. Gill, W. Pugh, J. P. Lee, B. Koeberlein, E. E. Furth, K. S. Polonsky, A. Naji and M. J. Birnbaum, *Nat. Med.*, 2001, **7**, 1133.

35. T. Matsui, L. Li, J. C. Wu, S. A. Cook, T. Nagoshi, M. H. Picard, R. Liao and A. Rosenzweig, *J. Biol. Chem.*, 2002, **277**, 22896.

36. Z. Z. Yang, O. Tschopp, M. Hemmings-Mieszczak, J. Feng, D. Brodbeck, E. Perentes and B. A. Hemmings, *J. Biol. Chem.*, 2003, **278**, 32124.

37. I. Hers, E. E. Vincent and J. M. Tavare, *Cell. Signal.*, 2011, **23**, 1515.

38. R. S. Garofalo, S. J. Orena, K. Rafidi, A. J. Torchia, J. L. Stock, A. L. Hildebrandt, T. Coskran, S. C. Black, D. J. Brees, J. R. Wicks, J. D. McNeish and K. G. Coleman, *J. Clin. Invest.*, 2003, **112**, 197.

39. S. George, J. J. Rochford, C. Wolfrum, S. L. Gray, S. Schinner, J. C. Wilson, M. A. Soos, P. R. Murgatroyd, R. M. Williams, C. L. Acerini, D. B. Dunger, D. Barford, A. M. Umpleby, N. J. Wareham, H. A. Davies, A. J. Schafer, M. Stoffel, S. O'Rahilly and I. Barroso, *Science*, 2004, **304**, 1325.

40. A. Poduri, G. D. Evrony, X. Cai, P. C. Elhosary, R. Beroukhim, M. K. Lehtinen, L. B. Hills, E. L. Heinzen, A. Hill, R. S. Hill, B. J. Barry, B. F. Bourgeois, J. J. Riviello, A. J. Barkovich, P. M. Black, K. L. Ligon and C. A. Walsh, *Neuron*, 2012, **74**, 41.

41. R. M. Easton, H. Cho, K. Roovers, D. W. Shineman, M. Mizrahi, M. S. Forman, V. M. Lee, M. Szabolcs, R. de Jong, T. Oltersdorf, T. Ludwig, A. Efstratiadis and M. J. Birnbaum, *Mol. Cell. Biol.*, 2005, **25**, 1869.

42. L. Ding, S. Biswas, R. E. Morton, J. D. Smith, N. Hay, T. V. Byzova, M. Febbraio and E. A. Podrez, *Cell Metab.*, 2012, **15**, 861.

43. S. V. Madhunapantula, P. J. Mosca and G. P. Robertson, Cancer Biol. Ther., 2011, **12**, 1.

44. M. Wan, R. M. Easton, C. E. Gleason, B. R. Monks, K. Ueki, C. R. Kahn and M. J. Birnbaum, *Mol. Cell. Biol.*, 2012, **32**, 96.

45. F. Buzzi, L. Xu, R. A. Zuellig, S. B. Boller, G. A. Spinas, D. Hynx, Z. Chang, Z. Yang, B. A. Hemmings, O. Tschopp and M. Niessen, *Mol. Cell. Biol.*, 2010, **30**, 601.

46. K. F. Leavens, R. M. Easton, G. I. Shulman, S. F. Previs and M. J. Birnbaum, *Cell Metab.*, 2009, **10**, 405.

47. M. Sun, J. E. Paciga, R. I. Feldman, Z. Yuan, D. Coppola, Y. Y. Lu, S. A. Shelley, S. V. Nicosia and J. Q. Cheng, *Cancer Res.*, 2001, **61**, 5985.

48. H. Sasaki, Y. Hikosaka, O. Kawano, S. Moriyama, M. Yano and Y. Fujii, *J. Thorac. Oncol.*, 2010, **5**, 597.

49. P. G. Rychahou, J. Kang, P. Gulhati, H. Q. Doan, L. A. Chen, S. Y. Xiao, D. H. Chung and B. M. Evers, *Proc. Natl. Acad. Sci. U.S.A.*, 2008, **105**, 20315.

50. V. A. Galicia, L. He, H. Dang, G. Kanel, C. Vendryes, B. A. French, N. Zeng, J. A. Bayan, W. Ding, K. S. Wang, S. French, M. J. Birnbaum, C. B. Rountree and B. L. Stiles, *Gastroenterology*, 2010, **139**, 2170.

51. B. Stiles, M. Groszer, S. Wang, J. Jiao and H. Wu, Dev. *Biol.*, 2004, 273, 175.

52. D. M. Li and H. Sun, *Cancer Res.*, 1997, **57**, 2124.

53. J. O. Lee, H. Yang, M. M. Georgescu, A. Di Cristofano, T. Maehama, Y. Shi, J. E. Dixon, P. Pandolfi and N. P. Pavletich, *Cell*, 1999, **99**, 323.

54. M. Raftopoulou, S. Etienne-Manneville, A. Self, S. Nicholls and A. Hall, *Science*, 2004, **303**, 1179.

55. F. B. Furnari, H. J. Huang and W. K. Cavenee, *Cancer Res.*, 1998, **58**, 5002.

56. X. Jiang, S. Chen, J. M. Asara and S. P. Balk, *J. Biol. Chem.*, 2010, **285**, 14980.

57. L. Salmena, A. Carracedo and P. P. Pandolfi, *Cell*, 2008, **133**, 403.

58. Q. Liu, T. Sasaki, I. Kozieradzki, A. Wakeham, A. Itie, D. J. Dumont and J. M. Penninger, *Genes Dev.*, 1999, **13**, 786.

59. C. D. Helgason, C. P. Kalberer, J. E. Damen, S. M. Chappel, N. Pineault, G. Krystal and R. K. Humphries, *J. Exp. Med.*, 2000, **191**, 781.

60. I. Chowdhury, B. Tharakan and G. K. Bhat, Cell. Mol. *Biol. Lett.*, 2006, **11**, 506.

61. A. Burlacu, *J. Cell. Mol. Med.*, 2003, **7**, 249.

62. G. S. Salvesen and V. M. Dixit, *Cell*, 1997, **91**, 443.

63. M. Tewari, L. T. Quan, K. O'Rourke, S. Desnoyers, Z. Zeng, D. R. Beidler, G. G. Poirier, G. S. Salvesen and V. M. Dixit, *Cell*, 1995, **81**, 801.

64. Y. A. Lazebnik, A. Takahashi, R. D. Moir, R. D. Goldman, G. G. Poirier, S. H. Kaufmann and W. C. Earnshaw, *Proc. Natl. Acad. Sci. U.S.A.*, 1995, **92**, 9042.

65. Q. Song, S. P. Lees-Miller, S. Kumar, Z. Zhang, D. W. Chan, G. C. Smith, S. P. Jackson, E. S. Alnemri, G. Litwack, K. K. Khanna and M. F. Lavin, *EMBO J.*, 1996, **15**, 3238.

66. P. Kermer, N. Klocker, M. Labes and M. Bahr, *J. Neurosci.*, 2000, **20**, 2.

67. M. H. Cardone, N. Roy, H. R. Stennicke, G. S. Salvesen, T. F. Franke, E. Stanbridge, S. Frisch and J. C. Reed, *Science*, 1998, **282**, 1318.

68. M. S. Ola, M. Nawaz and H. Ahsan, *Mol. Cell. Biochem.*, 2011, **351**, 41.

69. S. J. Gardai, D. A. Hildeman, S. K. Frankel, B. B. Whitlock, S. C. Frasch, N. Borregaard, P. Marrack, D. L. Bratton and P. M. Henson, *J. Biol. Chem.*, 2004, **279**, 21085.

70. B. D. Manning and L. C. Cantley, *Cell*, 2007, **129**, 1261.
71. K. U. Birkenkamp and P. J. Coffer, *Biochem. Soc. Trans.*, 2003, **31**, 292.
72. A. Villunger, E. M. Michalak, L. Coultas, F. Mullauer, G. Bock, M. J. Ausserlechner, J. M. Adams and A. Strasser, *Science*, 2003, **302**, 1036.
73. L. D. Mayo and D. B. Donner, *Proc. Natl. Acad. Sci. U.S.A.*, 2001, **98**, 11598.
74. B. P. Zhou, Y. Liao, W. Xia, B. Spohn, M. H. Lee and M. C. Hung, *Nat. Cell Biol.*, 2001, **3**, 245.
75. F. Takahashi-Yanaga and T. Sasaguri, *Cell. Signal.*, 2008, **20**, 581.
76. D. Accili and K. C. Arden, *Cell*, 2004, **117**, 421.
77. R. H. Medema, G. J. Kops, J. L. Bos and B. M. Burgering, *Nature*, 2000, **404**, 782.
78. E. D. Tang, G. Nunez, F. G. Barr and K. L. Guan, *J. Biol. Chem.*, 1999, **274**, 16741.
79. R. Zoncu, A. Efeyan and D. M. Sabatini, Nat. Rev. *Mol. Cell Biol.*, 2011, **12**, 21.
80. K. Inoki, H. Ouyang, T. Zhu, C. Lindvall, Y. Wang, X. Zhang, Q. Yang, C. Bennett, Y. Harada, K. Stankunas, C. Y. Wang, X. He, O. A. MacDougald, M. You, B. O. Williams and K. L. Guan, *Cell*, 2006, **126**, 955.
81. G. J. Brunn, J. Williams, C. Sabers, G. Wiederrecht, J. C. Lawrence Jr and R. T. Abraham, *EMBO J.*, 1996, **15**, 5256.
82. K. Miyake, W. Ogawa, M. Matsumoto, T. Nakamura, H. Sakaue and M. Kasuga, *J. Clin. Invest.*, 2002, **110**, 1483.
83. T. Asano, A. Kanda, H. Katagiri, M. Nawano, T. Ogihara, K. Inukai, M. Anai, Y. Fukushima, Y. Yazaki, M. Kikuchi, R. Hooshmand-Rad, C. H. Heldin, Y. Oka and M. Funaki, *J. Biol. Chem.*, 2000, **275**, 17671.
84. D. A. Cross, P. W. Watt, M. Shaw, J. van der Kaay, C. P. Downes, J. C. Holder and P. Cohen, *FEBS Lett.*, 1997, **406**, 211.
85. K. Zhang, L. Li, Y. Qi, X. Zhu, B. Gan, R. A. Depinho, T. Averitt and S. Guo, *Endocrinology*, 2012, **153**, 631.
86. H. Sano, S. Kane, E. Sano, C. P. Miinea, J. M. Asara, W. S. Lane, C. W. Garner and G. E. Lienhard, *J. Biol. Chem.*, 2003, **278**, 14599.
87. C. Taha, Z. Liu, J. Jin, H. Al-Hasani, N. Sonenberg and A. Klip, *J. Biol. Chem.*, 1999, **274**, 33085.
88. N. Pore, Z. Jiang, H. K. Shu, E. Bernhard, G. D. Kao and A. Maity, Mol. *Cancer Res.*, 2006, **4**, 471.
89. G. L. Semenza, *Nat. Rev. Cancer*, 2003, **3**, 721.
90. R. B. Robey and N. Hay, *Oncogene*, 2006, **25**, 4683.
91. Z. Cheng, Y. Tseng and M. F. White, *Trends Endocrinol. Metab.*, 2010, **21**, 589.
92. M. Laplante and D. M. Sabatini, *Curr. Biol.*, 2009, **19**, R1046.
93. L. He, X. Hou, G. Kanel, N. Zeng, V. Galicia, Y. Wang, J. Yang, H. Wu, M. J. Birnbaum and B. L. Stiles, *Am. J. Pathol.*, 2010, **176**, 2302.

94. E. L. Whiteman, H. Cho and M. J. Birnbaum, *Trends Endocrinol. Metab.*, 2002, **13**, 444.
95. A. Sundqvist, M. T. Bengoechea-Alonso, X. Ye, V. Lukiyanchuk, J. Jin, J. W. Harper and J. Ericsson, *Cell Metab.*, 2005, **1**, 379.
96. N. Zeng, J. Bayan, L. He and B. Stiles, *Open Endocrinol. J.*, 2010, **4**, 23.
97. K. Ueki, T. Okada, J. Hu, C. W. Liew, A. Assmann, G. M. Dahlgren, J. L. Peters, J. G. Shackman, M. Zhang, I. Artner, L. S. Satin, R. Stein, M. Holzenberger, R. T. Kennedy, C. R. Kahn and R. N. Kulkarni, *Nat. Genet.*, 2006, **38**, 583.
98. R. N. Kulkarni, J. N. Winnay, M. Daniels, J. C. Bruning, S. N. Flier, D. Hanahan and C. R. Kahn, *J. Clin. Invest.*, 1999, **104**, R69.
99. E. Bernal-Mizrachi, W. Wen, S. Stahlhut, C. M. Welling and M. A. Permutt, *J. Clin. Invest.*, 2001, **108**, 1631.
100. S. Mohanty, G. A. Spinas, K. Maedler, R. A. Zuellig, R. Lehmann, M. Y. Donath, T. Trub and M. Niessen, *Exp. Cell Res.*, 2005, **303**, 68.
101. J. Petrik, J. M. Pell, E. Arany, T. J. McDonald, W. L. Dean, W. Reik and D. J. Hill, *Endocrinology*, 1999, **140**, 2353.
102. M. Pende, S. C. Kozma, M. Jaquet, V. Oorschot, R. Burcelin, Y. Le Marchand-Brustel, J. Klumperman, B. Thorens and G. Thomas, *Nature*, 2000, **408**, 994.
103. J. A. Kushner, L. Simpson, L. M. Wartschow, S. Guo, M. M. Rankin, R. Parsons and M. F. White, *J. Biol. Chem.*, 2005, **280**, 39388.
104. H. Mori, K. Inoki, D. Opland, H. Munzberg, E. C. Villanueva, M. Faouzi, T. Ikenoue, D. J. Kwiatkowski, O. A. Macdougald, M. G. Myers Jr and K. L. Guan, *Am. J. Physiol. Endocrinol. Metab.*, 2009, **297**, E1013.
105. C. L. Buller, R. D. Loberg, M. H. Fan, Q. Zhu, J. L. Park, E. Vesely, K. Inoki, K. L. Guan and F. C. Brosius IIIO, *Am. J. Physiol. Cell Physiol.*, 2008, **295**, C836.
106. R. N. Kulkarni, J. C. Bruning, J. N. Winnay, C. Postic, M. A. Magnuson and C. R. Kahn, *Cell*, 1999, **96**, 329.
107. T. Kitamura, J. Nakae, Y. Kitamura, Y. Kido, W. H. Biggs III, C. V. Wright, M. F. White, K. C. Arden and D. Accili, *J. Clin. Invest.*, 2002, **110**, 1839.
108. B. L. Stiles, C. Kuralwalla-Martinez, W. Guo, C. Gregorian, Y. Wang, J. Tian, M. A. Magnuson and H. Wu, *Mol. Cell. Biol.*, 2006, **26**, 2772.
109. L. Shayesteh, Y. Lu, W. L. Kuo, R. Baldocchi, T. Godfrey, C. Collins, D. Pinkel, B. Powell, G. B. Mills and J. W. Gray, *Nat. Genet.*, 1999, **21**, 99.
110. A. J. Philp, I. G. Campbell, C. Leet, E. Vincan, S. P. Rockman, R. H. Whitehead, R. J. Thomas and W. A. Phillips, *Cancer Res.*, 2001, **61**, 7426.
111. C. Jimenez, D. R. Jones, P. Rodriguez-Viciana, A. Gonzalez-Garcia, E. Leonardo, S. Wennstrom, C. von Kobbe, J. L. Toran, R. B. L. V. Calvo, S. G. Copin, J. P. Albar, M. L. Gaspar, E. Diez, M. A. Marcos, J. Downward, A. C. Martinez, I. Merida and A. C. Carrera, *EMBO J.*, 1998, **17**, 743.

112. M. D. Ringel, N. Hayre, J. Saito, B. Saunier, F. Schuppert, H. Burch, V. Bernet, K. D. Burman, L. D. Kohn and M. Saji, *Cancer Res.*, 2001, **61**, 6105.

113. N. T. Ihle, G. Paine-Murrieta, M. I. Berggren, A. Baker, W. R. Tate, P. Wipf, R. T. Abraham, D. L. Kirkpatrick and G. Powis, *Mol. Cancer Ther.*, 2005, **4**, 1349.

114. M. Lopez-Fauqued, R. Gil, J. Grueso, J. Hernandez-Losa, A. Pujol, T. Moline and J. A. Recio, *Int. J. Cancer*, 2010, **126**, 1549.

115. L. Willems, J. Tamburini, N. Chapuis, C. Lacombe, P. Mayeux and D. Bouscary, *Curr. Oncol. Rep.*, 2012, **14**, 129.

116. S. Jia, Z. Liu, S. Zhang, P. Liu, L. Zhang, S. H. Lee, J. Zhang, S. Signoretti, M. Loda, T. M. Roberts and J. J. Zhao, *Nature*, 2008, **454**, 776.

117. S. Wee, D. Wiederschain, S. M. Maira, A. Loo, C. Miller, R. deBeaumont, F. Stegmeier, Y. M. Yao and C. Lengauer, *Proc. Natl. Acad. Sci. U.S.A.*, 2008, **105**, 13057.

118. J. A. Engelman, *Nat. Rev. Cancer*, 2009, **9**, 550.

119. P. Liu, H. Cheng, T. M. Roberts and J. J. Zhao, *Nat. Rev. Drug Discov.*, 2009, **8**, 627.

120. C. W. Lindsley, *Curr. Top. Med. Chem.*, 2010, **10**, 458.

CHAPTER 8

The Scientific Evidence for Using Astragalus in Human Diseases

WILLIAM C. S. CHO

Department of Clinical Oncology, Queen Elizabeth Hospital, Kowloon, Hong Kong
E-mail: williamcscho@gmail.com

8.1 Introduction

Radix Astragali prepared from the roots of *Astragalus membranaceus* (Fisch.) Bge., also known as astragalus or Huangqi, is one of the most commonly used and valuable traditional Chinese medicines (TCMs). Historically, Hunyuan County of Shanxi Province in China is the geo-authentic producing area of astragalus. According to tradition, geo-authentic TCMs define both authenticity and quality. There are reports indicating that the content of active components in astragalus depends on the interaction of genotype and environment. At the varietal level, genetic properties appear to be more important than environmental factors for pharmaceutical quality. At the intraspecific level, environmental factors may be more important than genetic properties.[1]

It has also been reported that the application of different proportions of nitrogen, phosphorus and potassium increased the content of astragalus polysaccharides (APS) and astragalosides, but had no distinct effect on the content of total flavonoids. The effect on the content of polysaccharides decreased in the order potassium > phosphorus > nitrogen, but the effect on the content of astragaloside was nitrogen > potassium > phosphorus. As a rule of thumb, nitrogen and potassium fertilizer application had more

RSC Drug Discovery Series No. 31
Traditional Chinese Medicine: Scientific Basis for Its Use
Edited by James David Adams, Jr. and Eric J. Lien
© The Royal Society of Chemistry 2013
Published by the Royal Society of Chemistry, www.rsc.org

important effects on growth, yield and the contents of polysaccharides and astragaloside in astragalus. During the medicinal plant cultivation process, attention should be paid to the application of nitrogen and potassium fertilizers and to making a balanced application of nitrogen, phosphorus and potassium fertilizers.[2]

Traditionally, high-performance liquid chromatography (HPLC) coupled with electrospray ionization time-of-flight mass spectrometry has been developed for the qualitative and quantitative analysis of isoflavonoids and saponins, and also for the quality control of astragalus and its preparations. The selectivity, reproducibility and sensitivity are comparable to those of HPLC with diode-array detection and evaporative light scattering detection (ELSD). Extracted ion chromatograms using a narrow mass window for quantification ensure that the chromatographic peaks are free from background or co-eluting interferences and the great enhancement of selectivity and sensitivity allows the identification and quantification of low levels of constituents in complex astragalus matrices.[3]

As expected, owing to the variability of growing conditions, the qualities of astragalus vary, which can give rise to differences in clinical therapy. Detecting adulteration is a routine requirement in pharmaceutical practice. Fourier transform infrared spectrometry combined with a chemometric method has been developed and demonstrated to be a useful tool to discriminate the geographical origin and adulteration of astragalus.[4]

Nowadays, DNA barcoding provides distinction between astragalus and its adulterants. DNA barcoding is widely applicable in species identification, biodiversity studies, forensic analyses and authentication of medicinal plants. Astragalus is commonly used in various Asian countries, including China, Japan and Korea. However, in addition to the two species recorded in the Chinese Pharmacopeia, there are 23 species from different genera including astragalus, oxytropis, hedysarum and glycyrrhiza, which have been used as adulterants not only in trading markets but also by the herbal medicine industry. Moreover, two indels detected in the matK sequence are useful for polymerase chain reaction (PCR) studies for distinguishing astragalus from its adulterants. It has been suggested that the combined barcoding regions of internal transcribed spacer and matK are superior barcodes for astragalus.[5]

8.2 Applications of Astragalus in the Treatment of Various Diseases

Astragalus is often used in China for the treatment of heart disease, diabetes, chronic hepatitis and nephritis and as an adjunctive therapy in cancer. In Oriental countries, it is widely used to prevent and treat common colds and upper respiratory tract infections. In Western countries, the most common use of astragalus is as an immune support supplement to the daily diet and as an immunostimulant to counteract the immune suppression associated with cancer chemotherapy and radiotherapy. It has gained popularity since the

1980s.[6] Traditionally, astragalus is a tonic well known for its vital energy tonifying, abscess draining, skin reinforcing, tissue generating and diuretic actions. It is generally used in combination with other Chinese herbs. In TCM, astragalus is often combined with other herbs, such as angelica and ginseng, in various complex prescription formulas.[7] Such herbal formulas have been used for centuries in Asia to treat cancers, diabetes, kidney infections, strokes and many other diseases.[8,9]

8.3 Immunomodulatory Effects of Astragalus

Astragalus is a TCM herb that has immunoregulatory and immunorestorative effects in many diseases.[10] Zhao *et al.*[11] studied the effect of APS on RAW264.7 macrophages and reported that APS increases the level of cytokines including tumor necrosis factor alpha (TNF-α), granulocyte macrophage-colony stimulating factor and the production of nitric oxide (NO). Nuclear factor kappa B (NF-κB) protein levels are increased in response to APS. Blocking NF-κB with a specific inhibitor resulted in decreased levels of NO and TNF-α. These findings suggest that APS possess potent immunomodulatory activity by stimulating macrophages and can be used as an immunotherapeutic adjuvant.

On the other hand, Xu *et al.*[12] investigated the effects of APS and astragalosides on the phagocytosis of *Mycobacterium tuberculosis* by macrophages. Their study provided evidence that APS and astragalosides have strong promoting effects on the phagocytosis of *M. tuberculosis* by macrophages and the secretion of interleukin (IL)-1β, IL-6 and TNF-α by activated macrophages.

Concerning the effects and mechanism of astragalus extract in macrophage migration and immune response mediator release, Qin *et al.*[13] determined the activity of heparanase by a heparin-degrading enzyme assay. It was suggested that astragalus extract may increase the release of immune response mediators and cell migration *via* heparanase to activate the immune response in macrophages.

The potential adjuvant effect of APS on humoral and cellular immune responses to hepatitis B subunit vaccine was also investigated. Coadministration of APS with recombinant hepatitis B surface antigen significantly increased antigen-specific antibody production, T cell proliferation and cytotoxic T lymphocyte activity. Production of interferon (IFN)-γ, IL-2 and IL-4 in CD4$^+$ T cells and of IFN-γ in CD8$^+$ T cells was dramatically increased. Furthermore, expression of the genes PFP, GraB, FasL and Fas was upregulated. Interestingly, expression of transforming growth factor beta (TGF)-β and the frequency of CD4$^+$ CD25$^+$ Foxp3$^+$ regulatory T cells (Treg cells) were downregulated. Expression of Toll-like receptor 4 (TLR4) was significantly increased by administration of APS. Together, these results suggest that APS is a potent adjuvant for the hepatitis B subunit vaccine and it can enhance both humoral and cellular immune responses *via* activating the TLR4 signaling pathway and inhibiting the expression of TGF-β and the frequency of Treg cells.[14]

Among the humoral responses, Wang *et al.*[15] showed that lysozyme activity significantly increased after feeding with superfine powder of astragalus supplemented diet for 20, 40 or 60 days. Furthermore, the lectin titer showed significant enhancement after 20 and 60 days of feeding with APS-supplemented diet. These results indicated that dietary intake containing astragalus or its polysaccharides could enhance the immune responses and improve resistance to infection.

8.4 Anti-Inflammatory Effects of Astragalus

APS has also been reported to possess anti-inflammatory activities. Mechanistic studies indicated that APS strongly suppresses NF-κB activation and downregulates the phosphorylation of ERK and JNK, which are important signaling pathways involved in the production of TNF-α and IL-1β. It has been demonstrated that APS can suppress the production of TNF-α and IL-1β in lipopolysaccharide-stimulated macrophages by inhibiting NF-κB activation and ERK/JNK phosphorylation.[16]

Foot ulceration, if not treated properly, will eventually result in amputation. Inflammation may impede the wound healing process if not properly controlled. In TCM, astragalus has been used for hundreds of years for many kinds of ulcerated wounds. It is one of the Chinese herbs commonly found in Chinese herbal formulas used for treating foot ulcers. It has been found that one of the active ingredients, formononetin, could significantly inhibit NO production.[17]

Apart from formononetin, new isoflavonoid glycosides and related constituents from astragalus were also found to exhibit inhibitory activity on NO production. Twenty-four secondary metabolites, including 16 isoflavonoids, seven astragalasides and one benzoquinone, have been isolated from astragalus. Among these isolated isoflavonoids, (–)-methylinissolin 3-*O*-β-D-(6'-acetyl)glucoside, (–)-methylinissolin 3-*O*-β-D-{6'-[(*E*)-but-2-enoyl]} glucoside and calycosin 7-*O*-β-D-(6''-acetyl)glucoside have been identified as new compounds on the basis of spectroscopic analysis, in which (–)-methylinissolin 3-*O*-β-D-glucoside was isolated from natural products for the first time. The NO production inhibitory activity of the major compounds has been assessed in lipopolysaccharide-stimulated RAW 264.7 cells. To identify astragalus, a fingerprint method was developed using the HPLC–ELSD method. Furthermore, characteristic peaks for the 11 major compounds in the chromatogram were unambiguously confirmed.[18]

Astragaloside IV is the major active constituent of astragalus, and has been widely used for the treatment of diseases in China for its antioxidant properties. Xiong *et al.*[19] showed that astragaloside IV significantly reduced auditory brainstem response deficits, and also decreased the expression of reactive oxygen species and active caspase-3. They suggested that the beneficial effect of astragaloside IV on impulse noise-induced hearing loss might be due to its ability to inhibit reactive oxygen species and prevent apoptosis.

Astragaloside IV has also been widely used to treat cardiovascular disease. Recent studies have shown that astragaloside IV could potentially protect the heart from myocardial ischemic injury; the protective effects of astragaloside IV in ischemia-reperfusion injury might be related to the upregulation of several ATP-sensitive potassium (KATP) channel subunits and facilitation of KATP currents.[20] On the other hand, another study indicated that astragaloside IV was a novel regulator of hypoxia-inducible factor-1α and angiogenesis through the PI3K/Akt pathway in human umbilical vein endothelial cells that were exposed to hypoxia.[21]

Re-epithelialization is a crucial step towards wound healing. Recent research has indicated that astragaloside IV could also mediate mouse keratinocyte proliferation and migration *via* regulation of the Wnt signaling pathway. Downregulating β-catenin to increase keratinocyte migration and proliferation was one mechanism by which astragaloside IV could promote ulcerated wound healing.[22]

Inhibition of inflammatory responses, acceleration of basal cell growth and balanced synthesis of the extracellular matrix are important in the healing of open cutaneous wounds. On evaluating the wound-healing effects of astragalus, the astragalus extracts were found to accelerate cutaneous wound healing significantly by suppressing inflammation and stimulating basal cell growth in the wound area compared with epidermal growth factor as a positive control. Promotion of basal cell proliferation and angiogenesis by the astragalus extracts was remarkable in the early stages of wound healing, resulting in a significant reduction in the duration of the wound-healing process. These findings suggested that astragalus extracts could be useful in enhancing cutaneous wound healing.[23]

Astragalus is often used in formulas for treating deficiency of vital energy characterized by limb weakness, pale face and dizziness. Using a hemorrhagic shock rat model to examine the effect of astragalus on intestinal mucosa injury induced by ischemia–reperfusion, it was found that treatment with 20 g of crude drugs per kilogram produced antioxidative effects in the intestinal mucosa of rats after ischemia–reperfusion. Astragalus could partly attenuate intestinal mucosa ischemia–reperfusion injury. These results suggested that astragalus reduced intestinal mucosa injury induced by ischemia–reperfusion in rats, at least in part, through its antioxidative effects.[24]

You *et al.*[25] reported that TNF-α strongly increased the expression of VCAM-1 and ICAM-1 accompanied by increased expression of phospho-IκB and NF-κB proteins. However, the expression levels of VCAM-1 and ICAM-1 were reduced by astragalus extract in dose- and time-dependent manners, with the strongest effect at a dose of 120 μg mL^{-1} incubated for 4 h. This was accompanied by significantly decreased expression of phospho-IκB and inhibited activation of NF-κB. Immunofluorescence analysis also revealed that oral administration of astragalus extract resulted in downregulation of adhesion molecules and decreased expression of macrophages in the aortic endothelium of apoE$^{-/-}$ mice. Astragalus extract could suppress the inflammatory reaction and inhibit the progression of atherosclerotic lesions in apoE$^{-/-}$ mice. Their study demonstrated that astragalus extract possesses an anti-atherosclerotic function and might be an effective anti-inflammatory agent for the treatment of atherosclerosis, possibly acting *via* the decreased expression of adhesion molecules.

Furthermore, astragalus extract 75 µg mL^{-1} significantly increased the proliferation, migration and tube formation on human umbilical vein endothelial cells, and a study showed that astragalus extract exerted cardiac protective and angiogenic effects in the ischemic injured heart. The activation of AKT/GSK3β and AKT/mTOR pathways and elevated expression of vascular endothelial growth factor (VEGF) may contribute to the promoted neovascularization by astragalus extract.[26]

Sepsis remains the leading cause of death in intensive care units. Uncontrolled systemic inflammation and an impaired protein C pathway are two important contributors to sepsis pathophysiology. Based on the beneficial effects of the saponin fraction from astragalus against inflammation, liver dysfunction and endothelium injury, the potential protective roles and underlying mechanisms of the saponin fraction from astragalus on poly-microbial sepsis induced by cecal ligation and puncture were investigated in mice. The findings suggested that the saponin fraction from astragalus was able to restore the impaired protein C pathway and accounted for the protective effects on polymicrobial sepsis in mice. The mechanisms of action involved anti-inflammation and upregulation of the protein C pathway.[27]

8.5   Anticancer Effects of Astragalus

It has been well reported that astragalus has promising anti-tumor effects.[28] APS is associated with a variety of immunomodulatory activities. The increase in numbers of tumor-associated Treg cells may play a role in modulation of the immune response against hepatocellular carcinoma (HCC). Li *et al.*[29] demonstrated that APS could restore the cytokine balance in the tumor microenvironment and suppress the expression of FOXp3 mRNA that inhibits the immune suppressive effects of Treg cells. The application of APS in the tumor microenvironment might act to enhance the antitumor effects of the immunotherapy-based methods and consequently to increase the survival rate in HCC.

Plant lectins have attracted great interest owing to their various biological activities such as anticancer, antifungal and antiviral activities. It was reported that astragalus lectin could exert antiproliferation effects on human cancer cells. It could induce apoptosis in a caspase-dependent manner in chronic myeloid leukemia cells. The cytotoxicity and apoptosis of these cells were completely abolished in the presence of lactose or galactose.[30]

On the other hand, astragaloside IV was found to reduce the protein expressions of bax and caspase-3 and increase the protein expressions of bcl-2, which improved the antiapoptosis mechanism of astragaloside IV on human adenovirus type 3 (HAdV-3) in adenocarcinomic human alveolar basal epithelial cells. These findings suggested that astragaloside IV possessed anti-HAdV-3 capabilities and the underlying mechanisms might involve inhibition of HAdV-3 replication and HAdV-3-induced apoptosis.[31]

Similarly, astragalus saponins (ASTs) could induce apoptosis in human gastric adenocarcinoma (AGS) cells by activating caspase-3 with subsequent cleavage of poly(ADP–ribose) polymerase. Furthermore, cell cycle arrest at the G_2/M phase was observed in AST-treated cells, leading to substantial growth inhibition. The antiproliferative effect of AST was associated with the regulation of cyclin B1, p21 and c-myc. Laboratory results indicated that the number of AGS cells that invaded through the Matrigel membrane was significantly reduced upon AST treatment, with concomitant downregulation of the proangiogenic protein VEGF and the metastatic proteins metalloproteinase (MMP)-2 and MMP-9. ASTs could modulate the invasiveness and angiogenesis of AGS cells in addition to its proapoptotic and antiproliferative activities. These findings suggest that AST has the potential to be further developed into an effective chemotherapeutic agent in treating advanced and metastatic gastric cancer.[32]

Astragalus has been used to ameliorate the side effects of antineoplastic drugs because of its immunomodulating nature. ASTs not only possessed anticarcinogenic and proapoptotic properties in human colon cancer cells and tumor xenograft, but also Auyeung *et al.*[33] indicated that NSAID-activated gene NAG-1 is a potential molecular target of AST in its antitumorigenic and proapoptotic actions, which would have additive effects when used along with PI3K–Akt inhibitors. This could facilitate the future development of a novel target-specific chemotherapeutic agent with a known molecular pathway.

It is known that astragalus is capable of reducing the adverse effects of conventional chemotherapy. A study investigated the sperm count and motility in mice treated with cyclophosphamide (CP). CP treatment has detrimental effects on the expression of cAMP-responsive element modulator (CREM), a transcription factor that is highly expressed in male germ cells and is crucial to post-meiotic germ cell differentiation. It was found that astragalus can restore CREM at both the mRNA and protein levels. Astragalus could ameliorate relative testes weight, sperm parameters and CREM expression against CP-induced reproductive toxicity.[34]

8.6 Randomized, Placebo-Controlled Clinical Trials and Meta-Analyses

PG2 is a partially purified extract of astragalus used as a complementary and palliative medicine for managing cancer-related fatigue. There was a Phase II double-blind, randomized placebo-controlled study on this novel infusible botanically derived drug. The results indicated that in the first 4 week cycle, PG2 administration resulted in a greater fatigue improvement response rate than controls treated with normal saline alone. No major or irreversible toxicities were observed with the PG2 treatment. Hence PG2 may be an effective and safe treatment for relieving cancer-related fatigue among cancer patients.[35]

On the other hand, APS was reported to increase tumor response, stabilize and improve performance status and reduce chemotherapy toxicity. A trial was designed to determine whether APS injection integrated with vinorelbine and cisplatin (VC) offered an improved quality of life (QoL) over VC for patients with advanced non-small cell lung cancer (NSCLC). After three cycles of treatment, there were significant differences in the overall patient QoL, physical function, fatigue, nausea and vomiting, pain and loss of appetite between the two study groups. These results indicated that the treatment with APS integrated with VC significantly improved the QoL in patients with advanced NSCLC compared with VC alone.[36]

Several recent meta-analyses supported the use of astragalus-based Chinese herbal therapy combined with chemotherapy for advanced lung cancer, nasopharyngeal carcinoma and HCC. A meta-analysis evaluated the evidence from randomized trials to assess whether astragalus-based Chinese herbal medicine combined with platinum-based chemotherapy could improve survival, increase tumor response, improve performance status and reduce chemotherapy toxicity in NSCLC. It was found that astragalus-based Chinese herbal medicine might increase the effectiveness of platinum-based chemotherapy compared with platinum-based chemotherapy alone.[37]

Another systematic review was conducted to assess the clinical efficacy of astragalus-based TCM as a concomitant therapy for nasopharyngeal carcinoma patients. The meta-analysis demonstrated a significant increase in the number of patients with prolonged survival, enhanced tumor response, improved Karnofsky performance status, reduced risk of adverse effects and significant immunostimulation in those treated by astragalus-based TCM combined with conventional cancer therapy *versus* patients using conventional cancer therapy alone.[38]

A meta-analysis was also performed for clinical trials comparing astragalus-based Chinese herbal therapy *versus* no Chinese herbal therapy given to HCC patients receiving transcatheter arterial chemoembolization. The results showed that astragalus-based Chinese herbal therapy was associated with a significant rise in the number of patients with prolonged survival, improved treatment response and non-deterioration performance status. This evidence supported the use of astragalus-based Chinese herbal therapy to enhance the efficacy of transcatheter arterial chemoembolization in HCC patients.[39]

8.7 Prospective

This chapter aims to provide scientific evidence for the application of astragalus in human diseases. Owing to its immunomodulatory, anti-inflammatory and anticancer effects, this herbal medicine is often administered to treat cancer, heart disease, diabetes, chronic hepatitis and nephritis. Astragalus is also used as an immune support supplement to daily diet and as an immunostimulant to counteract the immune suppression associated with cancer chemotherapy and radiotherapy. With increasing large-scale and stringent clinical trials, it is expected that there will be more applications of astragalus in human diseases, together with its integration with conventional Western treatments.

References

1. J. Liu, H. B. Chen, B. L. Guo, Z. Z. Zhao, Z. T. Liang and T. Yi, *Biol. Pharm. Bull.*, 2011, **34** (9), 1404.
2. W. L. Wang, Z. Wang and F. L. Xu, *Zhongguo Zhong Yao Za Zhi*, 2008, **33** (15), 1802.
3. L. W. Qi, P. Li, M. T. Ren, Q. T. Yu, X. D. Wen and Y. X. Wang, *J. Chromatogr. A*, 2009, **1216** (11), 2087.
4. L. Zhang and L. Nie L, *Phytochem. Anal.*, 2010, **21** (6), 609.
5. H. Guo, W. Wang, N. Yang, B. Guo, S. Zhang, R. Yang, Y. Yuan, J. Yu, S. Hu, Q. Sun and J. Yu, *Sci. China Life Sci.*, 2010, **53** (8), 992.
6. W. C. Cho, in *Recent Progress in Medicinal Plants, Vol. 28, Drug Plants II*, ed. A. S. Awaad, V. K. Singh and J. N. Govil, Studium Press, Houston, TX, 2010, p. 31.
7. Q. T. Gao, J. K. Cheung, J. Li, G. K. Chu, R. Duan, A. W. Cheung, K. J. Zhao, T. T. Dong and K. W. Tsim, *Planta Med.*, 2006, **72** (13), 1227.
8. Y. W. Zhang, D. Xie, Y. X. Chen, H. Y. Zhang and Z. X. Xia, *Exp. Clin. Endocrinol. Diabetes*, 2006, **114** (10), 563.
9. S. Li, Y. Zhang and J. Zhao, *Int. Immunopharmacol.*, 2007, **7** (1), 23.
10. W. C. Cho and K. N. Leung, *J. Ethnopharmacol.*, 2007, **113** (1), 132.
11. L. H. Zhao, Z. X. Ma, J. Zhu, X. H. Yu and D. P. Weng, *Cell. Immunol.*, 2011, **271** (2), 329.
12. H. D. Xu, C. G. You, R. L. Zhang, P. Gao and Z. R. Wang, *J. Int. Med. Res.*, 2007, **35** (1), 84.
13. Q. Qin, J. Niu, Z. Wang, W. Xu, Z. Qiao and Y. Gu, *Molecules*, 2012, **17** (6), 7232.
14. X. Du, X. Chen, B. Zhao, Y. Li, H. Zhang, H. Liu, Z. Chen, Y. Chen and X. Zeng, *FEMS Immunol. Med. Microbiol.*, 2011, **63** (2), 228.
15. T. Wang, Y. Sun, L. Jin, Y. Xu, L. Wang, T. Ren and K. Wang, *Fish Shellfish Immunol.*, 2009, **27** (6), 757.
16. X. He, J. Shu, L. Xu, C. Lu and A. Lu, *Molecules*, 2012, **17** (3), 3155.
17. P. K. Lai, J. Y. Chan, L. Cheng, C. P. Lau, S. Q. Han, P. C. Leung, K. P. Fung and C. B. Lau, *Phytother. Res.*, 2012, doi: 10.1002/ptr.4759.
18. L. J. Zhang, H. K. Liu, P. C. Hsiao, L. M. Kuo, I. J. Lee, T. S. Wu, W. F. Chiou and Y. H. Kuo, *J. Agric. Food Chem.*, 2011, **59** (4), 1131.
19. M. Xiong, Q. He, H. Lai and J. Wang, *Acta Otolaryngol.*, 2012, **132** (5), 467.

20. X. H. Han, P. Liu, Y. Y. Zhang, N. Zhang, F. R. Chen and J. F. Cai, *J. Tradit. Chin. Med.*, 2011, **31** (4), 321.
21. L. Zhang, Q. Liu, L. Lu, X. Zhao, X. Gao and Y. Wang, *J. Pharmacol. Exp. Ther.*, 2011, **338** (2), 485.
22. F. L. Li, X. Li, Y. F. Wang, X. L. Xiao, R. Xu, J. Chen, B. Fan, W. B. Xu, L. Geng and B. Li, *Evid. Based Comp. Alt. Med.*, 2012, **20**, 9561.
23. D. O. Han, H. J. Lee and D. H. Hahm, *Methods Fundam. Exp. Clin. Pharmacol.*, 2009, **31** (2), 95.
24. R. Chen, H. Shao, S. Lin, J. J. Zhang and K. Q. Xu, *Am. J. Chin. Med.*, 2011, **39** (5), 879.
25. Y. You, Y. Duan, S. W. Liu, X. L. Zhang, X. L. Zhang, J. T. Feng, C. H. Yan and Y. L. Han, *BMC Comp. Alt. Med.*, 2012, **12** (1), 54.
26. L. Zhang, Y. Yang, Y. Wang and X. Gao, *Pharmazie*, 2011, **66** (2), 144.
27. X. H. Gao, X. X. Xu, R. Pan, Y. Li, Y. B. Luo, Y. F. Xia, K. Murata, H. Matsuda and Y. Dai, *J. Nat. Med.*, 2009, **63** (4), 421.
28. W. C. Cho and K. N. Leung, *Cancer Lett.*, 2007, **252** (1), 43.
29. Q. Li, J. M. Bao, X. L. Li, T. Zhang and X. H. Shen, *Chin. Med. J. (Engl.)*, 2012, **125** (5), 786.
30. L. H. Huang, Q. J. Yan, N. K. Kopparapu, Z. Q. Jiang and Y. Sun, *Cell. Prolif.*, 2012, **45** (1), 15.
31. L. Shang, Z. Qu, L. Sun, Y. Wang, F. Liu, S. Wang, H. Gao and F. Jiang, *J. Pharm. Pharmacol.*, 2011, **63** (5), 688.
32. K. K. Auyeung, P. K. Woo, P. C. Law and J. K. Ko, *J. Ethnopharmacol.*, 2012, **141** (2), 635.
33. K. K. Auyeung, C. H. Cho and J. K. Ko, *Int. J. Cancer*, 2009, **125** (5), 1082.
34. W. Kim, S. H. Kim, S. K. Park and M. S. Chang, *Phytother. Res.*, 2012, **26** (9), 1418.
35. H. W. Chen, I. H. Lin, Y. J. Chen, K. H. Chang, M. H. Wu, W. H. Su, G. C. Huang and Y. L. Lai, *Clin. Invest. Med.*, 2012, **35** (1), E1.
36. L. Guo, S. P. Bai, L. Zhao and X. H. Wang, *Med. Oncol.*, 2011, **29** (3), 1656.
37. M. McCulloch, C. See, X. J. Shu, M. Broffman, A. Kramer, W. Y. Fan, J. Gao, W. Lieb, K. Shieh and M. Colford, *J. Clin. Oncol.*, 2006, **24** (3), 419.
38. W. C. Cho and H. Y. Chen, *Cancer Invest.*, 2009, **27** (3), 334.
39. W. C. Cho and H. Y. Chen, *Expert Opin. Investig. Drugs*, 2009, **18** (5), 617.

CHAPTER 9

Traditional Chinese Medicine: Anti-Inflammation for Cancer Prevention

CONSTANCE L. L. SAW*[a], CHEE L. SAW[b] AND LITA S. T. CHEW[c,d]

[a] Center for Cancer Prevention Research, Department of Pharmaceutics, Ernest Mario School of Pharmacy, Rutgers, The State University of New Jersey, Piscataway, NJ 08854, USA; [b] Suchen Clinic, Winnipeg, Manitoba R3G 1X6, Canada; [c] Department of Pharmacy, Faculty of Science, National University of Singapore, Singapore 117543; [d] Department of Pharmacy, National Cancer Centre Singapore, Singapore 169610
*E-mail: constancesaw@gmail.com

9.1 Introduction

Non-communicable diseases (NCDs), mainly cardiovascular diseases, cancers, chronic respiratory diseases and diabetes, are the world's biggest killers and require an urgent global response.[1] The burden of death and disability attributable to these diseases has major adverse social, economic and health effects. In 2008, of the total 57 million deaths globally, 63% (36 million) were due to NCDs and one-third of these NCD deaths were in people younger than 60 years old.[2]

The main risk factors for NCDs include tobacco use, foods high in saturated and *trans* fats, salt and sugar (especially in sweetened drinks), physical inactivity and alcohol consumption.[3] The current burden of NCDs reflects

RSC Drug Discovery Series No. 31
Traditional Chinese Medicine: Scientific Basis for Its Use
Edited by James David Adams, Jr. and Eric J. Lien
© The Royal Society of Chemistry 2013
Published by the Royal Society of Chemistry, www.rsc.org

past exposure to these risk factors and the future burden is determined by current exposure. The WHO mentions that unless efforts are made, the disease burden and mortality from these health problems will continue to increase.[4] In the light of the importance of addressing this disease burden, this chapter is focused on the potential of traditional Chinese medicine (TCM) in preventing cancer. Owing to the complexity of the nature of cancer, anti-inflammatory activities of TCM will be the focus. Before we consider the pathogenesis and mechanism of the disease and treatment, below we review cancer epidemiology in brief.

9.1.1 Cancer Epidemiology

9.1.1.1 Global Trends

The burden of cancer, one of the NCDs, is a growing health concern worldwide. It is already the leading cause of death in many high-income countries and will become a major cause of morbidity and mortality in the coming decades in every region of the world.[5,6] Based on the GLOBOCAN 2008 estimates, about 12.7 million cancer cases and 7.6 million cancer deaths are estimated to have occurred in 2008, with 56% of the cases and 64% of the deaths in the economically developing world.[7] Breast cancer is the most common cancer in females, accounting for 23% of the total cancer cases and 14% of the total cancer deaths. Lung cancer is the leading cancer in males, accounting for 17% of the total cancer cases and 23% of the total cancer deaths.

9.1.1.2 Incidence and Mortality

In a recent population-based study on global cancer transitions according to the Human Development Index (HDI), a marker of socioeconomic development, the authors predicted an increase in the incidence of all-cancer cases from 12.7 million new cases in 2008 to 22.2 million by 2030.[8] In regions with either high or very high levels on the HDI, breast, lung, colorectal and prostate cancers accounted for half the overall cancer incidence burden. In medium HDI regions, lung cancer is the most common cancer, ahead of stomach, liver and breast cancers. In low HDI regions, cervical cancer was most common. In men, cancers of the prostate, lung and liver were the most common cancers. Breast and cervical cancers were the most common in women.

Cancer survival tends to be poorer in developing countries compared with developed countries. A recent population-based study based on patients diagnosed in the 1990s indicated substantially lower survival rates in parts of Africa, India and the Philippines than for those diagnosed in Singapore, South Korea and parts of China.[9] The observed difference is most likely because of a combination of later stage at diagnosis and limited access to timely and standard treatment.

The global burden of cancer will continue to increase because of the aging population and adoption of cancer-causing behaviors, particularly smoking. It is, therefore, important for countries to initiate and capitalize on preventive strategies such as preventive programs (tobacco control), vaccination (for liver and cervical cancer), screening/early detection and treatment, in addition to public health campaigns promoting healthy diets and physical activity.[10]

9.1.2   Overview of Cancer Management

Cancer is a group of more than 100 different diseases that are characterized by uncontrolled cellular growth, local tissue invasion and distant metastases. There have been dramatic advances in cellular and molecular biology in the last two decades that have led to a greater understanding of the nature of malignant transformation. Progress made in early detection, diagnosis and treatment of cancer has also positively affected cancer survival rates.

The goals of cancer treatment include cure, prolongation of life and relief of symptoms. The selection of cancer treatment is guided by many factors, including type and stage of disease, patient performance status, treatment availability and treatment goals. Four primary modalities are employed in the approach to cancer treatment – surgery, radiation, chemotherapy and biological therapy.[11,12] Surgery and radiation therapy provide the best chance of cure for patients with localized cancers, but systemic treatment methods are required for disseminated cancers.

9.1.2.1   Surgical Therapy

Surgery is the oldest treatment for cancer. It remains the treatment of choice for most early stage solid tumors. The role of surgery in the treatment of cancer patients includes the following: (1) definitive resection of the primary cancer; (2) reducing the bulk of the disease (*e.g.*, ovarian cancer); (3) resection of metastatic disease with curative intent (*e.g.*, pulmonary metastases in sarcoma patients, hepatic metastases in colorectal cancer); (4) treatment of oncological emergencies; (5) palliation; and (6) reconstruction and rehabilitation. In each of the treatment intents, integration with other treatment modalities can be essential for a successful outcome.

9.1.2.2   Radiation Therapy

Radiation therapy is the use of ionizing radiation to treat cancer. The aim of radiation therapy is to deliver a precisely measured dose of radiation to a defined tumor volume with minimal damage to the surrounding healthy tissue, resulting in eradication of the tumor, maintaining a high quality of life and prolongation of survival.

In addition to curative efforts, radiation therapy can be used for palliative treatment and plays a major role in supportive care such as pain management.

Better radiation therapy techniques, such as three-dimensional conformal radiation therapy, stereotactic radiosurgery and brachytherapy (radioactive seeds), which are designed to deliver high doses of radiation to tumors while minimizing the doses delivered to nearby healthy tissue, are now widely available. These advances have resulted in positive benefits to patients, such as greater tissue, organ and limb preservation.

Recent research on the use of targeted agents in combination with radiation may offer new opportunities.[13,14] Enthusiasm for such strategies is fueled by the first randomized controlled trial demonstrating a significant clinical benefit from combining radiation with a molecular targeted agent, cetuximab, a recombinant mouse/human chimeric monoclonal antibody against the epidermal growth factor (EGF) receptor, in the treatment of head and neck cancer.[15] However, dosing and timing of these agents in relation to radiation appear to be important in maximizing the benefit of such combinations, in addition to appropriate selection of the tumor stage and histology.

9.1.2.3   Chemotherapy

Chemotherapy, including hormonal therapy, is the use of drugs to treat cancer. Adjuvant therapy is defined as the use of systemic agents to eradicate micro-metastatic disease following localized modalities such as surgery or radiation, or both. Thus, it is prescribed to patients with potentially curable malignancies who have no clinically detectable disease. The goal of systemic therapy in an adjuvant setting is to improve treatment outcome and for curative intent. Drug therapy may also be given in the neoadjuvant or preoperative setting, in an attempt to shrink large tumors and to make them more amenable to later surgical resection and possibly sparing critical organs and improving treatment outcome.

9.1.2.4   Biological Therapy

Biological therapies refer to the use of immunotherapy or "targeted therapies" to treat cancer. Immunotherapy usually involves stimulating the host's immune system to fight the cancer. The agents used in immunotherapy are usually naturally occurring cytokines, which have been produced with recombinant DNA technology. Examples of agents used in immunotherapy include interferons and interleukins (ILs). Targeted therapies include monoclonal antibodies, tyrosine kinase inhibitors, proteosome inhibitors and others. As described in the following, many TCM approaches have been tested for anti-inflammatory activities targeting ILs, which hold promising potential for the scientific understanding of the working mechanism of TCM in cancer prevention.

9.2   Methods

A full scientific literature search up to July 2012 was conducted in PubMed using keywords such as "Cortex Moutan anti-inflammatory or anti-cancer" with the herb name as the changing term for the use of TCM. The Latin names for herbs such as *Poenia rubra*, *Rhizoma corydalis*, *Faeces trogopterorum*, *Rhizoma zedoariae* and *Cortex phellodendri* were searched along with *Cortex moutan* because these herbs are among those indicated for inflammation and masses. Terms such as "anti-inflammatory," "anti-inflammation," "anti-cancer" and "cancer" were also employed. For the TCM herbs used for tumor treatment, the canons were used as references and the relevant formulas are listed. Owing to the paucity of information on TCM, translation of Chinese terms was done to the best knowledge of the authors and the translated names may not be as standardized as done in other publications. Readers are referred to original Chinese texts for all queries and interest.

9.3 Current Scientific Understanding of Cancer and Inflammation: Pathogenesis of Cancer and Inflammation

Carcinogenesis is a process involving multiple steps, including transition of normal cells to pre-initiated cells and ultimately invasive carcinoma, thus providing plentiful opportunities for cancer prevention using cancer chemo-preventive compounds, including TCM.[16,17] In general, tumor development follows three distinct yet closely interrelated phases, namely initiation, promotion and progression.[18,19] When cells are exposed to oxidative stress, DNA may suffer oxidative damage[20] coupled with persisting aberrant inflammation[21] in addition to DNA adduct formation, leading to increased genomic instability, enhanced neoplastic transformation and, if not stopped, cancer may form. Recently, cancer-related inflammation (inflammatory microenvironment) has been included as the seventh hallmark of cancer,[22,23] which is an additional hallmark to the identified six hallmarks of cancer summarized by Hanahan and Weinberg[24] in 2000, namely (i) self-sufficiency in growth signals, (ii) insensitivity to anti-growth signals, (iii) evading apoptosis, (iv) sustained angiogenesis, (v) limitless replicative potential and (vi) tissue invasion and metastasis. In 2011, Hanahan and Weinberg updated the hallmarks of cancer in their latest review.[25]

Significantly, many preclinical studies have been conducted to study the anti-inflammatory activities of TCM herbs by investigating the downstream key transcriptional factor in inflammatory responses, nuclear factor kappa B (NF-κB) (some references are listed in Table 9.1). Among those TCM herbs, some have been found to have anti-inflammatory and anticancer properties (Table 9.2), which suggests that the potential cancer prevention of TCM herbs by anti-inflammatory activities is an important mechanism. From the TCM perspective, the selection of the herbs in Tables 9.1 and 9.2 were based on TCM use for the treatment of cancer as discussed in Section 9.2 (Methods) and the following sections.

Table 9.1 Preclinical studies of TCM with anti-inflammatory activities.

Test models	Test compounds and methods	Outcome effects	Ref.
Lipopolysaccharide (LPS)-induced acute lung injury (ALI) model in Sprague–Dawley rats	Paeonol, an active component of *Moutan cortex radicis*, a commonly used TCM analgesic, antipyretic and anti-inflammatory agent	Decreased histopathological scores, myeloperoxidase-reactive cells and inducible nitric oxide synthase expression in lung tissue; decreased proinflammatory cytokine tumor necrosis factor alpha (TNF-α), IL-1β, IL-6 and plasminogen-activated inhibition factor-1 in bronchoalveolar lavage fluid	32
Type II collagen-induced arthritis (CIA) mouse model	Herbal extracts from *Trachelospermi caulis* (TC) and *Moutan cortex radicis* (MC)	Inhibition of inflammation: decreased clinical arthritis index, histological deformation of joints, serum levels of rheumatoid arthritis biomarkers and Th1-related responses. Suppression of the activation of the transcription NF-κB and activator protein (AP)-1	33
LPS-induced inflammation in rat primary microglia and 6-hydroxydopamine-induced oxidative damage in cortical neurons	Paeonol	In LPS-treated microglia: inhibition of inducible nitric oxide synthase (iNOS) and cyclooxygenase 2 (COX-2), decreased nitric oxide and prostaglandin E2 (PGE 2) production. In 6-hydroxydopamine-treated cortical neurons: decreased reactive oxygen species production	34
Ovalbumin-sensitized BALB/c mice, airway inflammation and hyper-responsiveness in a mouse model of allergic asthma	Paeonol	Decreased total inflammatory cell and eosinophil count in bronchoalveolar lavage fluid, IL-4 and IL-13 levels in bronchoalveolar lavage fluid and total immunoglobulin E levels in serum, eosinophilic inflammation and mucus-producing goblet cells in the airways	35
The atherosclerotic model developed in rabbits fed a high-fat diet for 12 weeks	Paeonol	Significant improvement in atherosclerotic plaque. Decreased blood levels of TNF-α, IL-1β and the translocation of NF-κB to the nucleus and lipid peroxidation	36

Table 9.1 (*Continued*)

Test models	Test compounds and methods	Outcome effects	Ref.
IL-1β-stimulated rat synoviocytes	Bioactivities of ethyl acetate and ethanol extracts of *Moutan cortex* fractions and pure compounds paeonol, paeoniflorin and pentagalloylglucose	Inhibition of TNF-α synthesis and IL-6 production by the purified compounds	37
Rat peritoneal mast cells (RPMCs) and cytokine production from HMC-1	PentaHerbs formula (PHF) containing *Cortex moutan, Cortex phellodendri, Flos lonicerae, Herba menthae* and *Rhizoma atractylodis*	PHF, *Cortex moutan* and *Herba menthae* significantly attenuated histamine release and prostaglandin D(2) synthesis from RPMC activated by anti-IgE and compound 48/80; *Cortex moutan* was the only herb that affected cytokine production in HMC-1	38
LPS-induced IL-8 production in the human monocytic cell line THP-1	Tetrahydropalmatine, found in *Corydalis rhizoma*	Inhibition of LPS-induced IL-8 production in a dose-dependent manner; extracellular signal-regulated kinase and p38 MAPK phosphorylation	39
LPS-stimulated mouse macrophage RAW 264.7 cells	Curdione from the rhizome of *Curcuma zedoaria*	Inhibition of the production of PGE 2 and COX-2 mRNA	40
12-*O*-Tetradecanoylphorbol-13-acetate (TAP)-induced inflammation of mouse ears	11 anti-inflammatory sesquiterpenes from rhizome of *Curcuma zedoaria*	Suppression of the TPA-induced inflammation of mouse ears by 75 and 53% by furanodiene and furanodienone, respectively, at a dose of 1.0 μmol; activities are comparable to that of a standard anti-inflammatory agent, indomethacin	41
Human hepatic myofibroblast cells (hMF)	Water extracts of *Zedoariae rhizoma*	Inhibition of hMF growth (IC$_{50}$ = 8.5 μg mL^{-1}), through early COX-2-dependent release of PGE 2 and cAMP and delayed COX-2 induction	42
LPS-activated macrophages	1,7-Bis(4-hydroxyphenyl)-1,4,6-heptatrien-3-one from *Curcuma zedoaria*	Inhibition of NO synthesis and the expression of iNOS	43

Table 9.1 (Continued)

Test models	Test compounds and methods	Outcome effects	Ref.
LPS-activated RAW 264.7 cells	Xanthorrhizol, a sesquiterpenoid, from the rhizome of *Curcuma xanthorrhiza* Roxb. (Zingiberaceae), sesquiterpenoids β-turmerone and ar-turmerone from the rhizome of *Curcuma zedoaria* Roscoe (Zingiberaceae)	All inhibited activities of COX-2 and iNOS	44
LPS-activated macrophages	1,7-Bis(4-hydroxyphenyl)-1,4,6-heptatrien-3-one and procurcumenol from *Curcuma zedoaria*	Inhibition of TNF-α production	45
LPS-activated BV2 cells (a mouse microglia cell line) and primary mouse microglia	Methanol extract of *Cortex phellodendri*	Inhibition of LPS-stimulated TNF-α, IL-1β and NO in both cells; mechanism involved inhibition of LPS-stimulated phosphorylation of ERK and activation of NF-κB	46
TPA-induced inflammation of mouse ears	Ethanol extracts of *Cortex phellodendri chinensis* (ECPC) and *Cortex phellodendri amurensis* (ECPA)	Both ECPC and ECPA significantly decreased the ear thickness, myeloperoxidase activity and the reactive oxygen species level, also inhibition of the protein and mRNA levels of TNF-α, IL-1β, IL-6 and COX-2	47
Carrageenan-induced mice paw edema	2000 years of Chinese recipe Huang-Lian-Jie-Du decoction (HLJDD) to treat inflammation: *Rhizoma coptidis, Radix scutellariae, Cortex phellodendri* and *Fructus gardeniae*	Exhibited anti-inflammatory effect	48

Table 9.2 Preclinical studies of TCM with anti-inflammatory and anticancer activities.

Test models	Test compounds and methods	Outcome effects	Ref.
Human hepatocellular carcinoma (HCC) cell lines BEL-7404, SMMC-7721 and MHCC97-H *in vitro*	Cytotoxicity of paeonol was compared with that of 5-fluorouracil (5-FU)	Dose-dependent inhibition of the proliferation of all three human HCC cell lines; these compounds led to the upregulation of the anti-oncongene PTEN and the downregulation of the oncogene AKT; paeonol induced DNA fragmentation in BEL-7404	30
Murine gastric cancer cell line mouse forestomach carcinoma (MFC) and human gastric cancer cell line SGC-7901	Paeonol	*In vitro*, dose-dependent inhibition of cell proliferation and induced apoptosis, reduced the expression of Bcl-2 and increased the expression of Bax. Treatment to MFC tumor-bearing mice caused significant tumor regression	49
Human umbilical vein endothelial cells (HUVECs) and HT-1080 fibrosarcoma cell line	Paeonol oxime, a paeonol derivative	Inhibition of basic fibroblast growth factor (bFGF)-mediated angiogenesis in HUVECs and pro-survival activity in HT-1080 fibrosarcoma cell line: decreased the levels of PI3K, phospho-AKT and VEGF	50
Human hepatoma (HepG2) cells	Gallic acid (GA), active compound from *Radix paeoniae rubra*; active compound from *Radix astragali*, designated RA-C	Combination of GA and RA-C significantly induced G2/M cell cycle arrest and attenuated the oncogenic transformation: the focus formation and anchorage-independent growth	51
Human breast cancer cells (MDA-MB-231)	Combination of ezhu (*Rhizoma curcumae*) and yanhusuo (*Rhizoma corydalis*) in a ratio of 3:2 (ezhu to yanhusuo; referred to as E3Y2)	E3Y2 markedly reduced cell invasion ability, induced cytochrome *c* release and significantly suppressed the level of p-ERK	52
Orthotopic implantation of Lewis lung carcinoma (LLC)	Berberine in *Phellodendri cortex*	Marked inhibition of tumor growth and lymphatic metastasis by the combination treatment of berberine and an anticancer drug, CPT-11, as compared with either single treatment alone. Anti-activator protein-1 (anti-AP-1) transcriptional activity of berberine inhibited LLC cells through the repression of expression of urokinase-type plasminogen activator (u-PA)	31

9.4 TCM Perspective of Cancer

9.4.1 Classical View of Masses, Growth or Inflamed Symptoms

The observation of tumors and the pathogenesis of tumors have been mentioned in various classical canons of TCM. Excessive growth is one of the characteristics of tumors. Tumors as known in the modern era of medicine can be categorized as benign or malignant types as opposed to tumors that were defined in TCM – which were differentiated by their motility and texture. TCM does not specify how a benign or malignant tumor should take shape as they were never defined in this context; nevertheless, it is understood that the pathogenesis of tumors may involve various mechanisms according to TCM differentiation, *e.g.*, loss of coordination between the organs, *Qi* 氣 and blood 血 stagnation and blood stasis. Blood stagnation is a situation where the blood does not circulate optimally whereas blood stasis refers to pathological hematoma or congealing blood that usually causes pain. While *Qi* stagnation causes the formation of *Jia* 瘕 (soft mobile tumor), blood stasis would be the basis of the formation of *Zheng* 癥 (hard non-mobile tumor). Other pathological definitions, including *Ji* 積 (aggregation) and *Ju* 聚 (accumulation), explain the different etiologies of tumors. However the natures of *Ji* and *Ju* tend towards not being inflammatory; *Ji* has a swelling form causing stationary pain and *Ju* has no swelling form and causes inconsistent sensations. Furthermore, *Ju* syndrome is mainly caused by *Qi* stagnation and is clinically characterized by *Qi* gathering in the abdomen with wandering and distending pain whereas *Ji* syndrome results from blood stasis and appears clinically as fixed abdominal masses. These masses are normally located in the liver, spleen, stomach and intestine.

Yong 癰 (pruritic tumor) and *Zhong* 腫 (swollen tumor), two further different types of tumors, could have closer etymological backgrounds that explain a growth or tumor that is pruritic and inflamed.

Detailed expositions on tumors can be found in numerous classical canons. For example, the *Synopsis of the Golden Chamber* (*Jin Kui Yao Lue* 金匱要略) devoted a chapter to the description of pathogenesis, pathology and treatment for pruritic swellings. The *Yellow Emperor's Classic of Internal Medicine* (*Huang Di Nei Jing* 黃帝內經) detailed systematically the characteristics of *Ji*, *Ju*, *Yong* and *Zhong* in Chapter 81. The *Detailed Analysis of Epidemic Warm Diseases* (*Wen Bing Tiao Bian* 溫病條辨) listed diagnoses and treatments for febrile diseases with concomitant masses.

Following the general pathway, every tumor growth has a starting point where cold accumulates and blood becomes stagnated. Once the congealing mass has formed the tumor, it is difficult to revert. The inflammation process begins with heat generated from blood accumulation. Over a long time where heat lingers, the growth process would erode flesh (myofibril tissue) into pus. Pus is a secretion of intense inflammation that could erode tendons which are in the proximity and eventually bone that attaches to the tendon. Bone that is affected by the inflammation will have impaired marrow capacity. When marrow is reduced to lower capacity, the blood supply cannot be replenished at an optimal rate and thus results in a weak constitution. With a weaker constitution, the organs will be driven into compensatory conditions in order to supply or strengthen their constitution to reach homeostasis and eventually the compensated organ will overwork, resulting in insufficiency and impaired function that could be fatal.

Most early stages of tumor pathogenesis may appear somewhat painful and not inflamed. These stages are said to have characteristics of *Ji* or *Ju*. In later stages when the tumors advance, they may become painful inflammatory tumors which will have characteristics of *Yong* or *Zhong*. However, it should be clarified that certain *Yong* symptoms in the lung and intestine could take the forms of acute appendicitis, colitis or bronchitis; they do not necessarily become cancerous, but are nevertheless fatal.

Tumor growth, from the TCM perspective, is further classified by its patterns and affected organs. The often reported but not exhausted list of tumor growth is as follows: tumor growth on the neck, brain, shoulder, throat, axilla, breast, lung, abdomen, lower abdomen, anus, knee, popliteal area, ankle, dorsal area of the foot, plantar area of the foot, sides of the foot and toes. However, we restrict our review of classical canons, formulas and herbs to those only for internal medicine, specifically in the region of the abdomen and pelvis.

9.4.2 Inflammation as the Clinical Sign

Employing the examination techniques of TCM practice, inflammation can usually be palpated to feel for the temperature regardless of cancerous growth. If the temperature is warmer than the superficial skin, the growth can be pruritic. If the temperature is latent and non-superficial, the growth should not be pruritic. For growth that is located underneath deep tissue, other physical symptoms determined through tongue examination, pulse examination or self-reported information can be used to understand inflammation of the growth in the body. Physical examination skills are not within the scope of this chapter; interested readers are referred to *The Foundations of Chinese Medicine*.[26]

9.4.3 Contemporary Understanding of Masses (Zheng Jia Ji Ju)

Masses may take specific shapes on palpation, and can be large or small, soft or hard. Patients demonstrate clinical symptoms such as nausea, vomiting, weakness, anorexia and emaciation, and pain usually precedes the formation of masses.[27,28]

Two types of masses, *Ji* 積 and *Ju* 聚, are usually considered together owing to their similar manifestation. In general, their etiology is a result of (i) irregular emotions, (ii) improper diet, (iii) external contraction of cold–dampness or (iv) secondary to other diseases.[27,28] *Ju*, as mentioned earlier, is a *Qi* disorder thus is thought to be a less severe symptom than the *Ji*, a syndrome that has advanced into blood stasis and is associated with internal organs that are painful and distended. The common syndrome patterns of *Ju* syndrome are described in Table 9.3. On the other hand, according to Liu and Liu[27] and Peng,[28] abdominal masses in terms of *Ji* can be categorized into three stages according to the syndrome patterns (Table 9.4).

Another two types of masses, *Zheng* 癥 and *Jia* 瘕, are specific mass growths in or around the uterus. The gynecological syndrome patterns can be divided into blood stasis due to *Qi* stagnation, syndrome of accumulation of phlegm–dampness and syndrome of stagnation of damp–heat. The understanding of *Zheng Jia* is that *Jia* 瘕 (soft mobile tumor) is a result of *Qi*-related illnesses and has the characteristic of a mobile mass with unfixed pain, while *Zheng* 癥 (hard non-mobile tumor) is a result of blood stasis and has the characteristic of a hard mass with fixed pain. The therapeutic principles are similar to those of *Ji Ju*; using herbs that promote circulation, break blood stasis, dissipate mass, cool heat and regulate *Qi* were the considerations. Table 9.5 describes the common pathologies of *Zheng Jia* and the respective treatments.

An abdominal mass is a typical internal medicine disorder that would be considered a tumor based on common practical knowledge in TCM. Xiao Yao San (Carefree Powder), Liu Mo Tang (Decoction of Six Ground Substances), Jin Ling Zi San (Melia Powder), Shi Xiao San (Stasis-Relieving Powder), Ge Xia Zhu Yu Tang (Decoction for Relieving Stasis Below the Diaphragm), Ba Zhen Tang (Eight Treasures Decoction), Hua Ji Wan (Mass-Dissipating Pill), Xiang Leng Wan (Cyperus and Sparganium Pills), Cang Fu Dao Tan Wan (Atractylodes-Poria Phlegm-Dissipating Decoction) and Dai Huang Mu Dan Tang (Rhubarb and Moutan Bark Decoction) are among the important formulas that are used for treating masses (*Zheng Jia Ji Ju*). The multitude herbs that are indicated for swelling, painful tumors have many actions with selected properties corresponding to symptom patterns. Herbs that regulate *Qi* stagnation (for treating symptoms of aching pain, *i.e.*, *Rhizoma corydali*s and *Rhizoma zedoariae*), stimulate blood circulation and remove blood stasis (symptoms of sharp pain, *i.e.*, *Faeces trogopterorum*, *Flos carthami*), dissipate dampness (phlegm dampness, *i.e.*, Poria) and cool heat accumulation (*i.e.*, *Poenia rubra* and *Cortex moutan*) are among the chief herbs. Herbs that are able to warm the interior, drain water, relieve food retention and stabilize and restore form are assistant and envoy herbs. From the perspective of masses, herbs specific to acting on the symptoms distension and pain were selected in this chapter. The next section is dedicated to the associated formulas and herbs that are indicated to treat inflammation.

In general TCM understanding, a few herbs are indicated for dissolving masses; for example, Bie Jia Carapax Trionycis, Shui Zhi Hirudo Nipponica, Mang Chong Tanabus and Shan Jia Squama Manitis are among the recognized herbs in the traditional canons, *i.e.*, *Treatise on Febrile Disease*,

Table 9.3 TCM pathology and therapeutic approach for abdominal mass *Ju* (no swelling, inconsistent sensation).

Differentiation diagnosis	Main symptoms	Therapeutic principle	Treatment	Ref.
Liver-*Qi* stagnation	Intermittent soft abdominal masses with wandering pain. Masses change with emotional fluctuations	Soothe liver, relieve stagnation and dissipate mass	Xiao Yao San (Carefree Powder) (prescriptions from the *Taiping Benevolent Pharmaceutical Bureau/Tai Ping Hui Min He Ji Ju Fang*, 太平惠民和劑局方) is indicated. *Chai Hu Radix Bupleuri* 9 g, *Bai Zhu Radix Atratylodis Macrocephalae* 9 g, *Bai Shao Yao Radix Paeoniae Alba* 9 g, *Dang Gui Radix Angelica Sinensis* 9 g, *Fu Ling Poria* 9 g, *Zhi Gan Cao Radix Glycyrrhizae Praeparata* 5 g, *Bo He Herba Menthae* a small quantity, *Wei Sheng Jiang Roast Rhizoma Zingiberie Recens* a small quantity	53
Food retention and phlegm accumulation	Abdominal distension or pain which can be aggravated upon pressure, poor appetite, anorexia	Regulate *Qi*, promote digestion, resolve phlegm and remove masses	Liu Mo Tang (Decoction of Six Ground Substances) (*Standards of Diagnosis and Treatment/Zheng Zhi Zhun Sheng*, 証治准繩) is indicated. *Wu Yao Radix Lindera* 6 g, *Chen Xiang Lignum Aquilariae Resinatum* 4.5 g, *Bing Lang Semen Arecae* 6 g, *Zhi Shi Fructus Citri Immaturus* 6 g, *Dai Huang Radix et Rhizoma Rhei* 6 g	54

Table 9.4 TCM pathology and therapeutic approach for abdominal masses *Ji* (swelling form causing stationary pain).

Differentiation diagnosis and severity	Main symptoms	Therapeutic principle	Treatment	Ref.
Qi stagnation and blood stasis in an early stage	Soft abdominal mass accompanied by distension and pain	Regulate *Qi*, mobilize blood, unblock channels and dissipate mass	Jin Ling Zi San (Melia Powder) (*Liu Wansu's Exposition on Pathogenesis in Nei Jing/Su Wen Bing Ji Qi Yi Bao Ming Ji*, 素問病機氣宜保命集) and Shi Xiao San (Carefree Powder) (prescriptions from the Taiping Benevolent Pharmaceutical Bureau/Tai Ping Hui Min He Ji Ju Fang, 太平惠民和劑局方) are indicated. Melia powder and Shi Xiao San: *Chuan Lian Zi Fructus Meliae Toosendan* 12 g, *Yan Hu Suo Rhizoma Corydalis* 10 g, *Wu Ling Zhi Faeces Trogopterorum* 6 g, *Pu Huang Pollen Typhae* 6 g.	53,55
Static blood gelling in the interior in a middle stage	Mass becoming larger and harder, the growth and pain are fixed in location, anorexia, weakness, cyclical chills and fever, dull complexion and emaciation may be observed	Remove blood stasis, weaken hard masses and regulate the spleen and the stomach	Ge Xia Zhu Yu Tang (Decoction for Relieving Stasis Below the Diaphragm) (*Corrections of Medical Errors/ Yi Lin Gai Cuo*, 醫林改錯) is indicated. Ge Xia Zhu Yu Tang: *Wu Ling Zhi Faeces Trogopterorum* 9 g, *Dang Gui Radix Angelica Sinensis* 9 g, *Chuan Xiong Rhizoma Ligustici Chuangxiong* 6 g, *Tao Ren Semen Persicae* 9 g, *Dan Pi Cortex Moutan Radicis* 6 g, *Chi Shao Yao Radix Paeoniae Rubra* 6 g, *Wu Yao Radix Lindera* 6–12 g, *Yan Hu Suo Rhizoma Corydalis* 3–15 g, *Gan Cao Radix Glycyrrhizae Praeparata* 9 g, *Xiang Fu Rhizoma Cyperi* 4.5 g, *Hong Hua Flos Carthami* 9 g, *Zhi Qiao Fructus Aurantii* 4.5 g	56

Table 9.4 *(Continued)*

Differentiation diagnosis and severity	Main symptoms	Therapeutic principle	Treatment	Ref.
Genuine-*Qi* depletion in a late stage	Hard mass present with worsening pain; anorexia, sallow complexion and severe emaciation are observed	Strengthen *Qi*, nourish blood, break blood stasis and dissolve masses	Combined Ba Zhen Tang (Eight Treasures Decoction) (*Classification and Treatment of Traumatic Injuries/ Zheng Ti Lei Yao*, 正體類要) and Hua Ji Wan (Mass-Dissipating Pill) (*Different Kinds of Diseases/Lei Zheng Zhi Cai*, 類証治裁) are indicated. Hua Ji Wan is composed of *San Leng Radix Sparganii* 6 g, *E Zhu Rhizoma Zedoariae* 6 g, *Su Mu Lignum Sappan* 6 g, *A Wei Resina Ferulae*, *Wu Ling Zhi Faeces Trogopterorum* 10 g, *Xiang Fu Rhizoma Cyperi* 10 g, *Bin Lang Semen Arecae* 6 g, *Hai Fu Shi Pumice* 6 g, *Wa Leng Zi Concha Arcae* 6 g, *Xiong Huang Realgar* 3 g, and Ba Zhen Tang is composed of *Ren Shen Radix Ginseng* 9 g, *Shu Di Huang Radix Rehmanniae Praeparata* 9 g, *Bai Zhu Rhizoma Atractylodis Macrocephalae* 9 g, *Fu Ling Poria* 9 g, *Dang Gui Radix Angelica Sinensis* 9 g, *Bai Shao Yao Radix Paeoniae Alba* 9 g, *Chuan Xiong Rhizoma Ligustici Chuangxiong* 9 g, *Gan Cao Radix Glycyrrhizae Praeparata* 5 g	57,58

Table 9.5 TCM pathology and therapeutic approach for abdominal masses *Zheng Jia.*

Differentiation diagnosis	Main symptoms	Therapeutic principle	Treatment	Ref.
Syndrome of blood stasis due to *Qi* stagnation	Palpable lumps in the uterus with a distinct shape, tender or not tender when touched, irregular menstruation, heavy flow of menstrual blood containing clots, dark complexion, depression	Cool heat, resolve stasis, eliminate dampness	Xiang Leng Wan (Cyperus and Sparganiu Pills) (*Life-Saving Prescription*/*Ji Sheng Fang* 濟生方) is indicated. *Mu Xiang Radix Aucklandiae* 15 g, *Ding Xiang Flos Caryophylli* 15 g, *San Leng Rhizoma Sparganii* 30 g, *Zhi Ke Fructus Aurantii* 30 g, *E Zhu Rhizoma Zedoariae* 30 g, *Qing Pi Pericarpium Citri Reticulatae Viridae* 30 g, *Chuan Lian Zi Fructus Melia Toosendan* 30 g, *Xiao Hui Xiang Fructus Foeniculi* 30 g	59
Accumulation of phlegm–damp	Stagnated lumps, excessive menstrual flow, increased leucorrhea between periods, oppression in the chest, pain in the back and abdomen	Dissolve phlegm, eliminate dampness, promote blood flow	Cang Fu Dao Tan Wan (Atractylodes-Poria Phlegm-Dissipating Decoction) (*Secrets of Diagnosis and Treatment on Gynecology by Ye Tianshi*/*Ye Tianshi Nu Ke Zheng Zhi Mi Fang* 葉天士女科診治秘方) with Dang Gui and Chuan Xiong are indicated. *Fu Ling Poria* 25 g, *Ban Xia Rhizoma Pinelliae* 10 g, *Chen Pi Pericarpium Citri Reticulatae* 6 g, *Gan Cao Radix Glycyrrhizae Praeparata* 6 g, *Cang Zhu Rhizoma Atractylodis* 10 g, *Xiang Fu Rhizoma Cyperi* 12 g, *Nan Xing Rhizoma Arisaematis* 10 g, *Zhi Ke Fructus Aurantii* 10 g, *Sheng Jiang Rhizoma Zingiberis Recens* 6 g, *Shen Qu Massa Medicara Fermentata* 15 g, *Dang Gui Radix Angelica Sinensis* 12 g, *Chuang Xiong Rhizoma Ligustici Chuangxiong* 10 g	60

Table 9.5 (*Continued*)

Differentiation diagnosis	Main symptoms	Therapeutic principle	Treatment	Ref.
Stagnation of damp heat	Masses in the uterus, burning pain which is aggravated by pressing, excessive menstrual flow, low-grade fever, yellowish leucorrhea, restlessness, dry stool	Clear heat, break stasis and eliminate dampness	Dai Huang Mu Dan Tang (Rhubarb and Moutan Bark Decoction) (*Synopsis of Golden Chamber/Jin Kui Yao Lue* 金匱要略) are indicated. *Dai Huang Radix et Rhizoma Rhei* 12 g, *Dan Pi Cortex Moutan* 9 g, *Mang Xiao Natrii Sulfas* 9 g, *Tao Ren Semen Persicae* 12 g, *Dong Gua Zi Semen Benincasae* 30 g	61

Synopsis of the Golden Chamber and *Detailed Analysis of Epidemic Warm Diseases*. We performed searches on PubMed with these herbal names and found no work in relation to the investigation of their anti-inflammatory or anticancer properties. Further scientific research in this respect is warranted.

9.4.4 Contemporary Understanding of Pelvic Inflammation

Inflammation is the manifestation of heat generated from dampness from the perspective of TCM. Inflammation in special sensory organs is not discussed here because it rarely forms cancers. Only the internal organs in the pelvic region will be discussed since pelvic inflammation is a sign of cancer. As described by Liu and Liu (Chapter 34, p. 297),[27] patients with chronic pelvic inflammation suffer from distension and pain in the lower abdomen. Vaginal discharge, low-grade fever, fatigue, weakness and lassitude are observed. A tender uterus that could be accompanied by masses is noticeable. The leukocyte count in the blood is often elevated. Fluid in the posterior fornix and thickening of the fallopian tubes with fluid accumulation can be seen through ultrasound scanning. Pelvic inflammation and adhesion can be confirmed by laparoscopy or culdoscopy. According to TCM, the pathology is the result of the accumulation of dampness and heat. The symptoms, therapeutic principle and treatment are described in Table 9.6.

9.5 Selected TCM Herbs That Are Indicated for Inflammation and Masses

A recent study on about 190 000 TCM prescriptions for 30 types of cancers found that each prescription consists of an average of two formulas and four herbs.[29] Furthermore, based on the effects of formulas and the natures of herbs that have been frequently prescribed for cancers, it has been proposed that cancers are a "warm and stagnant" syndrome in TCM, suggesting that anti-inflammatory regimens could be indicated for better prevention and treatment of cancers. In the previous two subsections, the combination of the key herbs that were indicated in treating both masses and inflammation were examined. As a result, a number of herbs that were indicated were compiled from the formulas. We screened herbs that were categorized as "heat-relieving herbs," "circulation-promoting herbs" and "blood stasis-removing herbs" exclusively. We narrowed down the herbs to *Cortex moutan*, *Poenia rubra*, *Rhizoma corydalis*, *Faeces trogopterorum*, *Rhizoma zedoariae* and *Cortex phellodendri* because these herbs are frequently indicated for inflammation and masses. Searches in PubMed were performed as described in the Section 9.2 and the resultant publications were analyzed. The studies found demonstrated that *Cortex moutan*, *Poenia rubra*, *Rhizoma corydalis*, *Rhizoma zedoariae* and *Cortex phellodendri* have anti-inflammatory activities that involve major IL mechanisms, *i.e.*, through the NF-κB pathway (Table 9.1) and some of them have demonstrated anticancer activities (Table 9.2). Scientific tests of the use of TCM herbs in comparison to

Table 9.6 TCM pathology of and therapeutic approach in chronic pelvic inflammation.

Differentiation diagnosis	Main symptoms	Therapeutic principle	Treatment	Ref.
Dampness and heat	Pain with guarding in the lower abdomen, often with a burning quality, possible masses, ongoing low fever, yellowish odorous vaginal discharge and dark oliguria	Cool heat, break blood stasis, dispel dampness and stop pain	Zhi Dai Tang (Discharge-Stopping Decoction) (*Shibuzhai's Miraculous Prescriptions/Shi Bu Zhai Bu Xie Fang* 世補齋不謝方) with *Yan Hu Suo Rhizoma Corydalis* 12 g and *E Zhu Rhizoma Zedoariae* 10 g. Zhi Dai Tang is composed of *Zhu Ling Polyporus Umbellatus* 20 g, *Fu Ling Poria* 30 g, *Che Qian Zi Semen Plantaginis* 15 g, *Ze Xie Rhizoma Alismatis* 15 g, *Yin Chen Herba Artemisiae Scopariae* 15 g, *Chi Shao Yao Radix Paeoniae Rubra* 12 g, *Dan Pi Cortex Moutan Radicis* 10 g, *Huang Bai Cortex Phellodendri* 12 g, *Zhi Zi Fructus Gardeniae* 10 g, *Niu Xi Radix Achyranthis Bidentatae* 10 g	62

or together with conventional chemotherapy[30,31] have been conducted and the results are promising. However, no report on the anti-inflammatory activity of *Faeces trogopterorum* was found.

9.6   Conclusion

It is challenging to provide scientific bases for the long and historical use of TCM for the treatment of cancer, as the concept of cancer according to the classical canons of TCM is rather different from Western ideas. As such, at this stage, it is even more challenging to provide scientific bases for the cancer preventive properties of TCM. However, we attempted to explore the scientific bases of TCM in cancer prevention as we believe that aberrant inflammation, the seventh hallmark of cancer, is one of the underlying etiological causes by definition of both the modern scientific understanding of the disease itself and the historical TCM perspective. This chapter has reviewed the modern treatment modalities for cancer and it is hoped that the TCM used for cancer according to the definition of TCM canons will be tested more in a modern approach and form the basis for cancer prevention using TCM as many of the TCM products do have anti-inflammatory activities, which is critical in cancer prevention as gathered from our latest understanding of carcinogenesis.

References

1. R. Horton, *Lancet*, 2007, **370**, 1881.
2. World Health Organization, *Global Status Report on Non-Communicable Diseases*, WHO, Geneva, 2010.
3. R. Beaglehole, R. Bonita, R. Horton, C. Adams, G. Alleyne, P. Asaria, V. Baugh, H. Bekedam, N. Billo, S. Casswell, M. Cecchini, R. Colagiuri, S. Colagiuri, T. Collins, S. Ebrahim, M. Engelgau, G. Galea, T. Gaziano, R. Geneau, A. Haines, J. Hospedales, P. Jha, A. Keeling, S. Leeder, P. Lincoln, M. McKee, J. Mackay, R. Magnusson, R. Moodie, M. Mwatsama, S. Nishtar, B. Norrving, D. Patterson, P. Piot, J. Ralston, M. Rani, K. S. Reddy, F. Sassi, N. Sheron, D. Stuckler, I. Suh, J. Torode, C. Varghese and J. Watt, *Lancet*, 2011, **377**, 1438.
4. World Health Organization, *2008–2013 Action Plan for the Global Strategy for the Prevention and Control of Noncommunicable Diseases*, http://whqlibdoc.who.int/publications/2009/9789241597418_eng.pdf (last accessed 18 June, 2012).
5. World Health Organization, *The Global Burden of Disease: 2004 Update*, WHO, Geneva, 2008.
6. J. Ferlay, H. R. Shin, F. Bray, D. Forman, C. Mathers and D. M. Parkin, *Int. J. Cancer*, 2010, **127**, 2893.
7. A. Jemal, R. Siegel, J. Xu and E. Ward, *CA Cancer J. Clin.*, 2010, **60**, 277.
8. F. Bray, A. Jemal, N. Grey, J. Ferlay and D. Forman, *Lancet Oncol.*, 2012, **13**, 790.

9. R. Sankaranarayanan, R. Swaminathan, H. Brenner, K. Chen, K. S. Chia, J. G. Chen, S. C. Law, Y. O. Ahn, Y. B. Xiang, B. B. Yeole, H. R. Shin, V. Shanta, Z. H. Woo, N. Martin, Y. Sumitsawan, H. Sriplung, A. O. Barboza, S. Eser, B. M. Nene, K. Suwanrungruang, P. Jayalekshmi, R. Dikshit, H. Wabinga, D. B. Esteban, A. Laudico, Y. Bhurgri, E. Bah and N. Al-Hamdan, *Lancet Oncol.*, 2010, **11**, 165.

10. World Health Organization, *Fact Sheet No. 297: Cancer*, http://www.who.int/mediacentre/factsheets/fs297/en/index.html (last accessed 18 June 2012).

11. V. T. DeVita Jr, T. S. Lawrence and S. A. Rosenberg, *DeVita, Hellman and Rosenberg's Cancer: Principles and Practice of Oncology*, 8th edn, Lippincott Williams & Wilkins, Philadelphia, PA, 2008.

12. J. T. DiPiro, R. L. Talbert, G. C. Yee, G. R. Matzke, B. G. Wells and L. M. Posey, *Pharmacotherapy: a Pathophysiological Approach*, 7th edn, McGraw-Hill, New York, 2008.

13. P. M. Harari and S. M. Huang, *Semin. Radiat. Oncol.*, 2002, **12**, 21.

14. S. M. Bentzen, P. M. Harari and J. Bernier, *Nat. Clin. Pract. Oncol.*, 2007, **4**, 172.

15. J. A. Bonner, P. M. Harari, J. Giralt, N. Azarnia, D. M. Shin, R. B. Cohen, C. U. Jones, R. Sur, D. Raben, J. Jassem, R. Ove, M. S. Kies, J. Baselga, H. Youssoufian, N. Amellal, E. K. Rowinsky and K. K. Ang, *N. Engl. J. Med.*, 2006, **354**, 567.

16. C. L. Saw and A. N. Kong, *Expert Opin. Ther. Targets*, 2011, **15**, 281.

17. C. L. Saw, Q. Wu and A. N. Kong, *Chin. Med.*, 2010, **5**, 37.

18. I. Berenblum and V. Armuth, *Biochim. Biophys. Acta*, 1981, **651**, 51.

19. C. Heidelberger, A. E. Freeman, R. J. Pienta, A. Sivak, J. S. Bertram, B. C. Casto, V. C. Dunkel, M. W. Francis, T. Kakunaga, J. B. Little and L. M. Schechtman, *Mutat. Res.*, 1983, **114**, 283.

20. F. Weinberg and N. S. Chandel, *Cell Mol. Life Sci.*, 2009, **66**, 3663.

21. A. Mantovani, P. Allavena, A. Sica and F. Balkwill, *Nature*, 2008, **454**, 436.

22. F. Colotta, P. Allavena, A. Sica, C. Garlanda and A. Mantovani, *Carcinogenesis*, 2009, **30**, 1073.

23. A. Mantovani, *Nature*, 2009, **457**, 36.

24. D. Hanahan and R. A. Weinberg, *Cell*, 2000, **100**, 57.

25. D. Hanahan and R. A. Weinberg, *Cell*, 2011, **144**, 646.

26. G. Maciocia, *The Foundations of Chinese Medicine: a Comprehensive Text for Acupuncturists and Herbalists*, 2nd edn, Churchill Livingstone Elsevier, London, 2005.

27. L. Liu and Z. Liu, *Essentials of Chinese Medicine*, Vol. 3, Springer, New York, 2009.

28. B. Peng, *Traditional Chinese Internal Medicine*, 2nd edn, People's Medical Publishing House, Beijing, 2007.

29. P. H. Chiu, H. Y. Hsieh and S. C. Wang, *PLoS One*, 2012, **7**, e31648.

30. C. Zhang, S. Hu, M. Cao, G. Xiao and Y. Li, *Anticancer Drugs*, 2008, **19**, 401.
31. N. Mitani, K. Murakami, T. Yamaura, T. Ikeda and I. Saiki, *Cancer Lett.*, 2001, **165**, 35.
32. P. K. Fu, C. L. Wu, T. H. Tsai and C. L. Hsieh, *Evid. Based Comp. Alt. Med.*, 2012, **2012**, 837513.
33. H. S. Kim, A. R. Kim, J. M. Lee, S. N. Kim, J. H. Choi, D. K. Kim, J. H. Kim, B. Kim, E. Her, Y. M. Yang, Y. M. Kim and W. S. Choi, *J. Pharm. Pharmacol.*, 2012, **64**, 420.
34. Y. T. Tseng, Y. Y. Hsu, Y. T. Shih and Y. C. Lo, *Shock*, 2012, **37**, 312.
35. Q. Du, G. Z. Feng, L. Shen, J. Cui and J. K. Cai, *Can. J. Physiol. Pharmacol.*, 2010, **88**, 1010.
36. H. Li, M. Dai and W. Jia, *Planta Med.*, 2009, **75**, 7.
37. M. Wu and Z. Gu, *Evid. Based Comp. Alt. Med.*, 2009, **6**, 57.
38. B. C. Chan, K. L. Hon, P. C. Leung, S. W. Sam, K. P. Fung, M. Y. Lee and H. Y. Lau, *J. Ethnopharmacol.*, 2008, **120**, 85.
39. Y. C. Oh, J. G. Choi, Y. S. Lee, O. O. Brice, S. C. Lee, H. S. Kwak, Y. H. Byun, O. H. Kang, J. R. Rho, D. W. Shin and D. Y. Kwon, *J. Med. Food*, 2010, **13**, 1125.
40. O. J. Oh, H. Y. Min and S. K. Lee, *Arch. Pharm. Res.*, 2007, **30**, 1236.
41. H. Makabe, N. Maru, A. Kuwabara, T. Kamo and M. Hirota, *Nat. Prod. Res.*, 2006, **20**, 680.
42. D. I. Kim, T. K. Lee, T. H. Jang and C. H. Kim, *Life Sci.*, 2005, **77**, 890.
43. M. K. Jang, H. J. Lee, J. S. Kim and J. H. Ryu, *Arch. Pharm. Res.*, 2004, **27**, 1220.
44. S. K. Lee, C. H. Hong, S. K. Huh, S. S. Kim, O. J. Oh, H. Y. Min, K. K. Park, W. Y. Chung and J. K. Hwang, *J. Environ. Pathol. Toxicol. Oncol.*, 2002, **21**, 141.
45. M. K. Jang, D. H. Sohn and J. H. Ryu, *Planta Med.*, 2001, **67**, 550.
46. Y. K. Park, Y. S. Chung, Y. S. Kim, O. Y. Kwon and T. H. Joh, *Int. Immunopharmacol.*, 2007, **7**, 955.
47. Y. F. Xian, Q. Q. Mao, S. P. Ip, Z. X. Lin and C. T. Che, *J. Ethnopharmacol.*, 2011, **137**, 1425.
48. J. Lu, J. S. Wang and L. Y. Kong, *J. Ethnopharmacol.*, 2011, **134**, 911.
49. N. Li, L. L. Fan, G. P. Sun, X. A. Wan, Z. G. Wang, Q. Wu and H. Wang, *World J. Gastroenterol.*, 2010, **16**, 4483.
50. H. J. Lee, S. A. Kim, H. J. Lee, S. J. Jeong, I. Han, J. H. Jung, E. O. Lee, S. Zhu, C. Y. Chen and S. H. Kim, *PLoS One*, 2010, **5**, e12358.
51. L. Wei, J. Liu, X. C. Le, Y. Han, Y. Tong, A. S. Lau and J. Rong, *Int. J. Oncol.*, 2011, **39**, 735.
52. J. L. Gao, T. C. He, Y. B. Li and Y. T. Wang, *Oncol. Rep.*, 2009, **22**, 1077.
53. S. W. Chen, *Prescriptions from the Taiping Benevolent Pharmaceutical Bureau*, 1151.
54. K. T. Wang, *Standards of Diagnosis and Treatment*, 1602.
55. W. S. Liu, *Liu Wansu's Exposition on Pathogenesis in Nei Jing*, 1186.

56. Q. R. Wang, *Corrections of Medical Errors*, 1830.
57. J. Xue, *Classification and Treatment of Traumatic Injuries*, 1529.
58. P. Q. Lin, *Different Kinds of Diseases*, 1839.
59. Y. H. Yan, *Life-Saving Prescription*, 1253.
60. T. S. Ye, *Secrets of Diagnosis and Treatment on Gynecology by Ye Tianshi*, 1913.
61. Z. J. Zhang, *Synopsis of Golden Chamber, ca* 200.
62. M. X. Lu, *Shibuzhai's Miraculous Prescriptions*, 1866.

CHAPTER 10

Treatment of Stroke with Dan Shen, Salvia miltiorrhiza

JAMES DAVID ADAMS JR

School of Pharmacy, University of Southern California, Los Angeles, CA 90089, USA
E-mail: jadams@usc.edu

10.1 Pathophysiology of Stroke and Its Causes

The majority of strokes are caused by an embolism in one of the cerebral arteries. However, atherosclerosis of the carotid and cerebral arteries is a frequent finding in stroke patients.[1–4] Men are more likely to suffer from atherosclerotic plaques leading to strokes.[5] Women are more likely to suffer from emboli leading to strokes.[5] However, in China, strokes are three times more common than heart attacks and are frequently hemorrhagic, which means they have a much higher rate of mortality than non-hemorrhagic strokes. Devices to remove emboli and stents to facilitate the recanalization of atherosclerotic plaques are treatments sometimes used in stroke patients.[6–10] These devices have safety issues associated with them since they can cause hemorrhage and embolism in their own right. However, patients can benefit from the use of these devices, even when the use is delayed, for as long as 90 days after a stroke. What are the causes of embolism and atherosclerosis in most strokes? Embolism may involve a change in blood coagulation such that clots form more easily. These clots eventually may dislodge and form thrombi in arteries that may occlude blood flow. Hypercoagulability is not a normal condition and involves changes in blood clotting factors or changes in clot breaking factors. Atherosclerosis involves plaque formation in arteries such

RSC Drug Discovery Series No. 31
Traditional Chinese Medicine: Scientific Basis for Its Use
Edited by James David Adams, Jr. and Eric J. Lien
© The Royal Society of Chemistry 2013
Published by the Royal Society of Chemistry, www.rsc.org

that plaque material may occlude arterial flow. Emboli may occur when clots form on plaques or when plaque material breaks off into the bloodstream.

Embolism increases during abdominal obesity associated with visceral fat accumulation, called the metabolic syndrome.[11,12] Obesity occurs in as many as 75% of adults in some countries. Visceral fat secretes adipokines, such as leptin, resistin, tumor necrosis factor alpha (TNF-α), interleukin-6 (IL-6) and visfatin, which are inflammatory proteins.[13] Embolism involves higher levels of C-reactive protein, fibrinogen and factor VIII that are secreted by visceral fat.[12] Another adipokine of interest is plasminogen activator inhibitor type 1.[14] C-reactive protein increases platelet adhesion to endothelial cells. Plasminogen activator inhibitor type 1 increases thrombus and embolus formation since it inhibits fibrinolysis. It does this by inhibiting tissue plasminogen activator and urokinase that are the major enzymes involved in dissolving clots. High fibrinogen levels can lead to high fibrin levels and more coagulation. The conversion of fibrinogen to fibrin is catalyzed by thrombin that increases in patients at risk for a stroke.[15] Factor VIII is a clotting protein that can lead to more clotting. Platelet aggregates in the blood are found at higher levels in patients at risk of stroke.[16]

Atherosclerosis greatly increases during the metabolic syndrome.[14,17,18] Heparin binding epidermal growth factor like growth factor and platelet-derived growth factor are secreted by visceral adipocytes.[19–21] Both of these adipokines stimulate angiogenesis and arterial hypertrophy that are the first steps in atherosclerosis. Toxic lipids, such as ceramide and the endocannabinoids, are produced more abundantly during visceral obesity. Ceramide overproduction is toxic to the kidneys and leads to hypertension. The endocannabinoids cause inflammatory adipokine secretion from visceral fat. These adipokines, such as resistin, TNF-α, IL-6, leptin and angiotensinogen, lead to dysfunction of endothelial cells and wall defects. Platelets and neutrophils adhere to these defects due to the secretion of adhesion molecules. Neutrophils and monocytes are stimulated by visfatin, leptin and oxidized LDL-C to begin an inflammatory reaction in the walls of arteries and produce foam cells. The inflammation and foam cell proliferation produce plaques that grow to occlude arteries. Alternatively, plaques can become unstable and cast emboli due to an increase in matrix metalloproteinase activity in macrophages that is stimulated by C-reactive protein.

There are millions of patients who suffer strokes every year throughout the world. The vast majority of strokes are caused by visceral obesity, as discussed above. Visceral fat accumulation leading to stroke is not always associated with clinically defined obesity, especially in aged patients who are inactive. There are other causes of stroke. Atrial fibrillation is sometimes associated with embolus formation leading to stroke. This may be due to areas of partial blood stagnation in the atria. Atrial fibrillation is most commonly associated with congestive heart failure. Visceral fat accumulation is a cause of congestive heart failure since several adipokines cause cardiac hypertrophy and damage.[22–25] Atrial fibrillation-associated embolism is usually, but not always successfully, prevented with anticoagulants.[26,27] Patients must stay on anticoagulants for the rest of their lives and suffer from the toxicity of these drugs, including hemorrhage.

Smoking can be a cause of stroke since nicotine is toxic to endothelial cells and causes atherosclerosis and embolism.[28] Even second-hand smoke can cause stroke.[29] Marijuana smoking can also cause stroke, since it causes intracranial stenosis.[30] Alcohol consumption can cause stroke since it leads to visceral fat accumulation.[31] Other causes of stroke include hormone replacement therapy[32] and eating high-fat diets, especially diets containing meat.[33,34] Sickle cell disease is an uncommon cause of stroke. Polycythemia and use of erythropoietin can cause strokes in cancer patients. Young athletes who abuse blood packing and erythropoietin-like drugs can suffer from strokes.

The above discussion demonstrates that most strokes can be prevented or at least delayed by adopting lifestyle changes. However, health insurance companies do not usually pay for preventive measures. Governments encourage people to live healthy lifestyles, but pay little towards establishing prevention programs. Modern lifestyles are centered around high-fat fast foods, insufficient exercise, smoking (in some countries) and excessive drinking. Obesity is considered normal and occurs in the majority of adult Americans, as can be seen on the Centers for Disease Control website. In the old days, prior to the 1950s, obesity was uncommon and considered unhealthy. Stroke was much less common in those days and occurred mostly in aged people. In China, it was considered disrespectful to the ancestors to be obese, since each person got their body from their ancestors.

10.2   Prevention of Stroke

"Not smoking, maintaining a healthy weight and controlling blood sugar, blood pressure and cholesterol may add 10 years of life" according to the Centers for Disease Control. In the book released by the Centers for Disease Control, *Prevention Works: CDC Strategies for Heart-Healthy and Stroke-Free America* (available at http://www.cdc.gov/dhdsp/docs/Prevention_works.pdf), strategies for stroke prevention are detailed, namely: control high blood pressure, control high blood cholesterol, know signs and symptoms, call for emergency medical help, improve emergency response, improve quality of care and eliminate disparities. These disparities refer to the fact that in 2002, 41% more black Americans than white Americans died from stroke. The signs of a stroke are FAST: *F*ace – is the smile symmetrical?; *A*rms – can both arms be raised without drooping?; *S*peech – is the speech slurred or strange?; *T*ime – as soon as symptoms are seen, get emergency medical help. The reason why the Centers for Disease Control say that time is important is because thrombolytic therapy, such as tissue plasminogen activator, must be administered within about 3 h to be effective. These same symptoms are used by the National Stroke Association. These measures are a step in the right direction, but fall short of what could be effective prevention.

A more adequate stroke prevention program should involve maintaining low body fat, such as 4% in men and 12% in women. Each patient should find the body composition that allows them to maintain good health. Physical endurance should involve about 1 h of aerobic exercise daily, such as walking, running, bicycling, swimming or other activities. Smoking anything should be avoided. Alcohol consumption should be no more than one drink every day. In addition, a lacto-ovo vegetarian diet should be encouraged in adults. This diet avoids high-fat meats, since fats are addictive and overconsumption of fat leads to visceral fat accumulation.

Tissue plasminogen activator is used in only a few percent of stroke patients since it is difficult for neurologists to visit patients within the 3 h time window. In addition, patients may not visit a hospital within the first 3 h of suffering a stroke. Tissue plasminogen activator can cause hemorrhage in patients and such hemorrhages can be devastating. Its use in patients above the age of 90 years is controversial, since these patients may fare worse with the drug than they would without it.

10.3   *Salvia miltiorrhiza* – **Dan Shen and Its Uses**

Salvia miltiorrhiza (Bunge) Lamiaceae is a deciduous perennial sage found in Asia that grows to 60 cm high. It has pinnately divided leaves that are basal and on the stem. Each leaflet is elliptical and may have serrate edges. The flowers occur in whorls and are blue and claw shaped with an upper and a lower lip. The roots are dark red (dan in Chinese). The plant is found in grassy areas and near streams at low elevations.

The plant contains many compounds, including monoterpenoids, diterpenoids, salvianolic acids and a recently discovered lignan.[35–37] The monoterpenoids include danshensu, caffeic acid, protocatechualdehyde, rosmarinic acid and others (Figure 10.1). Danshensu in Chinese means the active compound in dan shen. It is 3-(3',4'-dihydroxyphenyl)-(2R)-lactic acid. These compounds are water soluble. There are many diterpenoids, such as tanshinone IIA, cryptotanshinone, miltirone, tanshindiol B and others (Figure 10.2). The diterpenoids are not water soluble but are lipid soluble. There are also many salvianolic acids, such as salvianolic acid A, lithospermic acid B and others (Figure 10.3). These compounds are water soluble. The lignin is called XH14, which is 2-(3'-methoxy-4'-hydroxyphenyl)-5-(3-hydroxypropyl)-7-methoxybenzo [*b*]furan-3-carbaldehyde (Figure 10.4). This compound is water soluble. All of these compounds have pharmacological activities.

Most of the active compounds in *S. miltiorrhiza* have poor oral bioavailability. Tanshinones upregulate efflux transporters in the gut, thereby decreasing their own bioavailability.[38] In fact, the elimination of danshensu is hastened by other compounds in *S. miltiorrhiza*.[39] Protocatechuic aldehyde given intravenously increases the area under the curve and decreases the clearance for danshensu, salvianolic acid A and salvianolic acid B.[40] However, protochatechuic aldehyde may have poor oral bioavailability since it is likely a

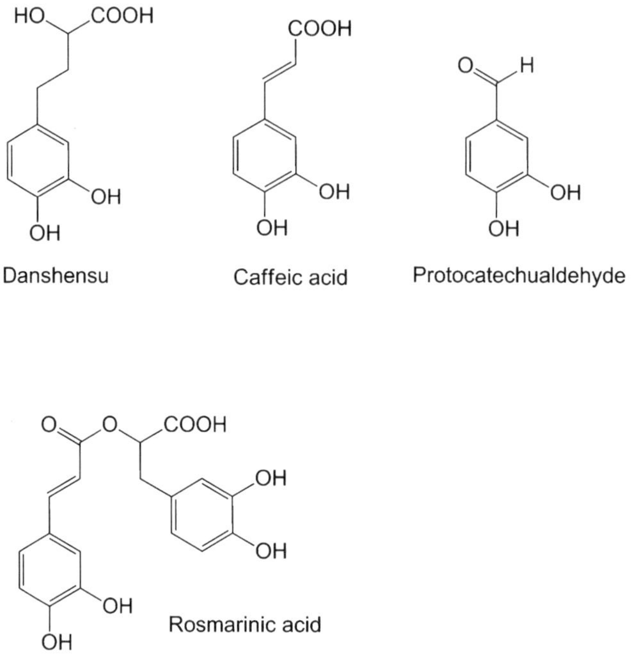

Figure 10.1 Monoterpenoids from *Salvia miltiorrhiza*.

substrate for aldehyde dehydrogenase. Most of the monoterpenoid and tanshinone compounds have half-lives of about 3–4 h.[41] The salvianolic acids have oral half-lives of 3.3 h for salvianolic acid A and 4 h for salvianolic acid B.[42] The oral bioavailability of salvianolic acid B is 2%.[42] Danshensu crosses the blood–brain barrier, but not salvianolic acid B or protocatechuic aldehyde.[43] Tanshinones bind to lipoproteins in the blood and have oral bioavailabilities in rats of about 4–5%.[44] Cryptotanshinone is converted to tanshinone IIA by cytochrome P450. Tanshinone IIA is converted to tanshinone IIB, hydroxytanshinone IIA and other compounds by cytochrome P450.[43] The tanshinones interact with several forms of cytochrome P450 and can inhibit or induce several forms.[44,45] Several potential drug interactions between *S. miltiorrhiza*-derived tanshinones or salvianolic acids and other drugs have been investigated, including midazolam,[46,47] caffeine,[48] metoprolol,[49] tolbutamide[50] and warfarin.[51] None of these potential interactions have been reported in human clinical trials.

The pharmacology of *S. miltiorrhiza* involves interactions with signaling mechanisms, especially extracellular signal regulated kinase (ERK), janus activated kinase (JAK), mitogen activated protein kinase kinase (MEK), activator protein-1 and nuclear factor kappa B (NF-κB) pathways.[42,44] By interacting with these signal transduction pathways, the compounds inhibit the actions of TNF-α, angiotensin-II, IL-6, inducible nitric oxide synthase and peroxisome proliferator activated receptor gamma. The compounds also inhibit

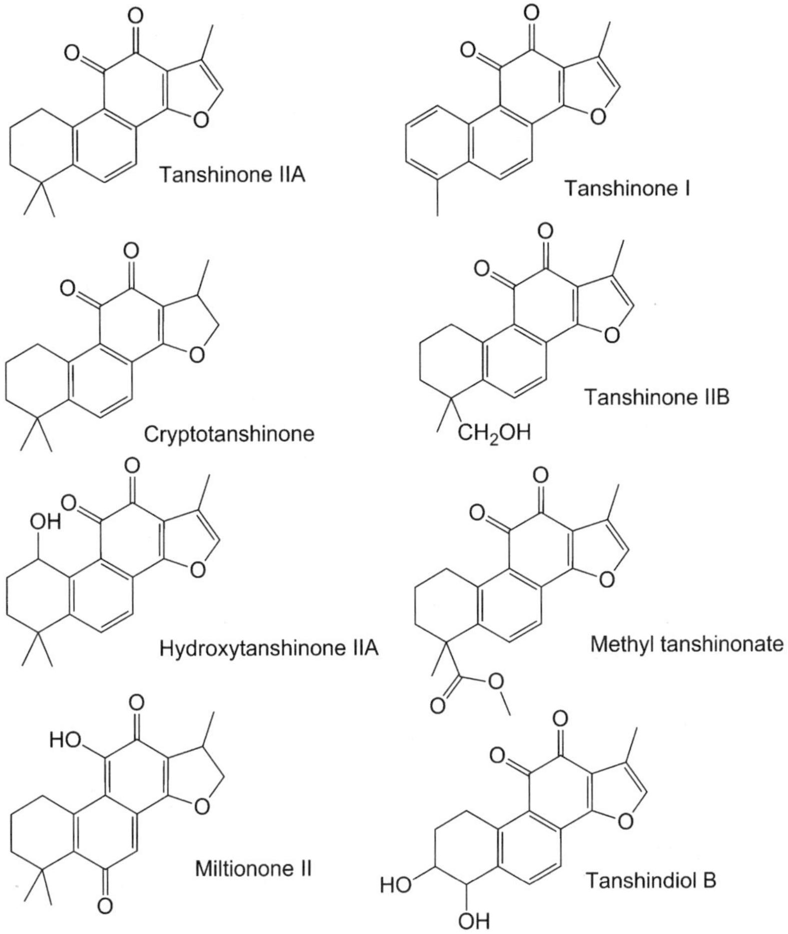

Figure 10.2   Tanshinones from *Salvia miltiorrhiza.*

the expression of plasminogen activator inhibitor type 1, matrix metalloproteinases 2 and 9, cyclooxygenase-2, adhesion molecules (monocyte chemotactic protein-1, VCAM-1 and ICAM-1), vascular endothelial growth factor and other proteins. The *S. miltiorrhiza* compounds involved in these interactions are the tanshinones and salvianolic acids and have similar and overlapping mechanisms of action. These compounds may potentiate or synergize the actions of each other. It is of note that several atherogenic actions of the adipokines discussed above are inhibited by *S. miltiorrhiza* (Figure 10.5).

The lignan from *S. miltiorrhiza,* XH14, inhibits peroxisome proliferator activated receptor gamma induction of the expression of several adipokines, including visfatin and resistin.[52] Similarly, tanshinone IIA and *S. miltiorrhiza* extracts inhibit the expression of several adipokines, including several interleukins, TNF-α, monocyte chemotactic protein-1 and C-reactive protein.[53–55] 15,16-Dihydrotanshinone I is an inhibitor of steroid and mineralocorticoid

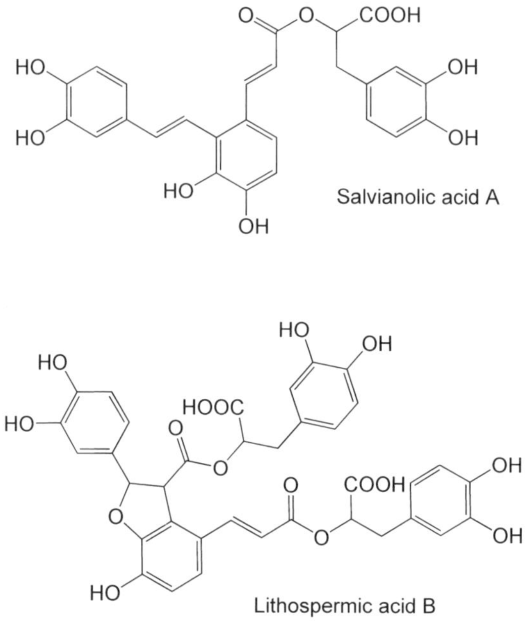

Figure 10.3 Salvianolic acids from *Salvia miltiorrhiza*.

receptors and alters the expression of several genes involved in the metabolic syndrome.[56]

The actions of tanshinone IIA on matrix metalloproteinases appear to be important in the prevention of cardiac fibrosis.[57] Tanshinones also inhibit Bcl-2-induced cardiac myocyte apoptosis.[58] Endothelial cells of the blood–brain barrier are also protected by tanshinone IIA, resulting in less cerebral edema after cerebral artery occlusion.[59] This protection may be from inhibition of adhesion molecule production, activation of calcium influx,[60] activation of potassium channels,[60,61] inhibition of NADPH oxidase or other mechanisms. NADPH oxidase is a transmembrane enzyme found mainly on monocytes and

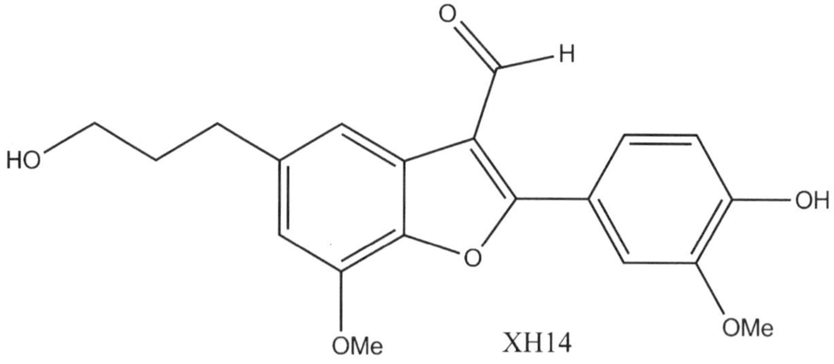

Figure 10.4 Lignan from *Salvia miltiorrhiza*.

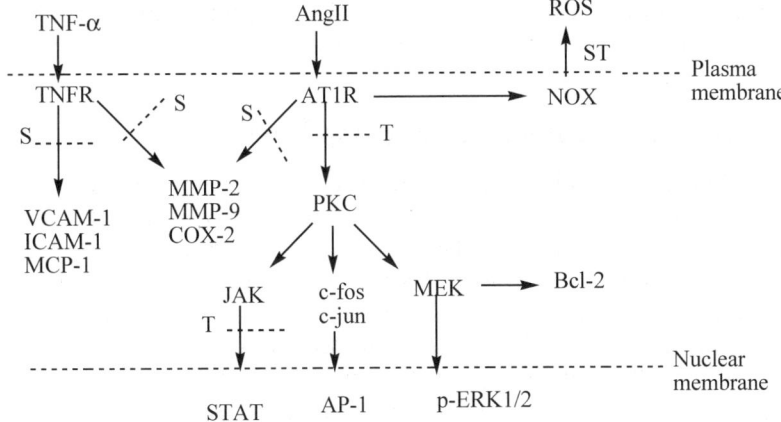

Figure 10.5 Actions of tanshinones (T) and salvianolic acids (S) showing sites of inhibition by dashed lines. TNF-α, tumor necrosis factor alpha; TNFR, TNF-α receptor; VCAM-1, vascular cell adhesion protein-1; ICAM-1, intercellular adhesion molecule-1; MCP-1, monocyte chemotactic protein-1; MMP-2 and -9, matrix metalloproteinase-2 and -9; COX-2, cyclooxygenase-2; JAK, janus kinase; STAT, signal transducer and activator of transcription; AngII, angiotensin-II; AT1R, angiotensin-II receptor 1; PKC, protein kinase C; c-fos, a protein produced by the FOS gene; c-jun, a protein produced by the JUN gene; AP-1, activator protein-1; MEK, mitogen activated protein kinase kinase; p-ERK1/2, phosphorylated extracellular signal regulated kinase 1 and 2; NOX, NADPH oxidase; ROS, reactive oxygen species; Bcl-2, B-cell lymphoma-2 protein.

neutrophils that forms extracellular reactive oxygen radicals that may damage endothelial cells. Activation of calcium and potassium channels appears to be how *S. miltiorrhiza* extracts dilate cerebral arteries.

Salvianolic acid A inhibits platelet aggregation by inhibition of phosphoinositide 3-kinase.[62] This results in less fibrinogen adherence and less clotting. Tanshinone IIA inhibits platelet aggregation by increasing thromboxane B2 levels.[63] Thromboxane A2 is involved in clotting and is converted into thromboxane B2, which is not active. *S. miltiorrhiza* extract downregulates proteinase activated receptor-1 on platelets that is involved in thrombin binding and inhibits aggregation.[64] The mechanisms of action of the salvianolic acids and tanshinones may be additive or synergistic in terms of inhibition of platelet aggregation and blood clotting.

The neuroprotective and stem cell stimulating properties of *S. miltiorrhiza* have recently become of interest. Various studies have found that *S. miltiorrhiza*, salvianolic acid or danshensu can protect the brain from amyloid accumulation.[65–71] The tanshinones have been shown in several publications to inhibit acetylcholinesterase and may be of interest in the treatment of Alzheimer's disease.[66,72] Stem cell survival may be involved in neuroprotective mechanisms in the brain. Salvianolic acid B has been shown to protect stem cells through an MEK/ERK pathway.[73] Salvianolic acids stimulate the growth of endothelial stem cells and neural stem cells.[74,75]

10.4 *Salvia columbariae* **and Its Uses**

Salvia columbariae Benth., chia or golden chia, is an annual sage found in California, Utah and Arizona. It grows to about 1 m high and has basal, pinnate leaves up to 10 cm long. The flowers grow in whorls, are blue, tubular and up to 1 cm long. Seed clusters form from the flowers and contain many black seeds about 1 mm in diameter. The roots are brown. A related species, *Salvia hispanica*, also called chia, grows in Mexico and Guatemala. Mexican chia has been cultivated since pre-Columbian times. The US Department of Agriculture lists the nutrients in 28 g of Mexican chia seeds as 4 g of protein, 9 g of fat and 11 g of fiber. The fat is 57% α-linolenic acid, an omega-3 fatty acid.

S. columbariae is traditionally used as a treatment for stroke and heart attack.[76] The entire plant, roots, stems, leaves and seeds can be used. However, in a patient who has just suffered a stroke, the seeds are easiest to use. This is done by mixing about 10–20 g of seeds into about 100–200 mL of water with stirring for 15 min until a thin gel forms. The gel is given orally to the patient, three times daily for 2 months. The dose can then be decreased to once or twice daily for 4 months. The patient is also instructed to perform mild daily exercise such as walking and careful use of paralyzed limbs for 6 months or more. Unfortunately, *S. columbariae* has not been tested in stroke clinical trials.

The plant contains cryptotanshinone, tanshinone IIA and miltionone II.[76] In addition, β-sitosterol lithospermate is present (Figure 10.6) and is the primary salvianolic acid.[77] Tanshinones are abundant in the plant with cryptotanshinone being the most abundant. The pharmacology of tanshinones is discussed above. The pharmacology of α-linolenic acid is complex and controversial. The clinical evidence for cardiovascular protective effects of α-linolenic acid is mixed, with some trials showing mild effects and others showing no effects.[78] The evidence for protective effects or beneficial effects in stroke are even more inconsistent.[78]

α-Linolenic acid is known to decrease diastolic blood pressure and increase HDL-cholesterol in humans.[79] The fatty acid also protects humans against kidney disease.[80] Low intake of α-linolenic acid appears to be a risk factor for stroke.[81] Soluble IL-6 receptor levels were lower after intake of α-linolenic acid in a twin study.[82] This implies that the action of IL-6 is inhibited or altered by the fatty acid. However, α-linolenic acid administration in humans lowers plasma adiponectin levels.[83] Adiponectin is the protective adipokine that is produced by non-obese visceral fat cells.

Mice and rats fed α-linolenic acid are protected against the development of atherosclerosis and develop cardioprotection.[84,85] This protection is associated with lower levels of TNF-α and IL-6. Of course, α-linolenic acid is a normal human nutrient. The human body has found uses for the nutrient itself and its

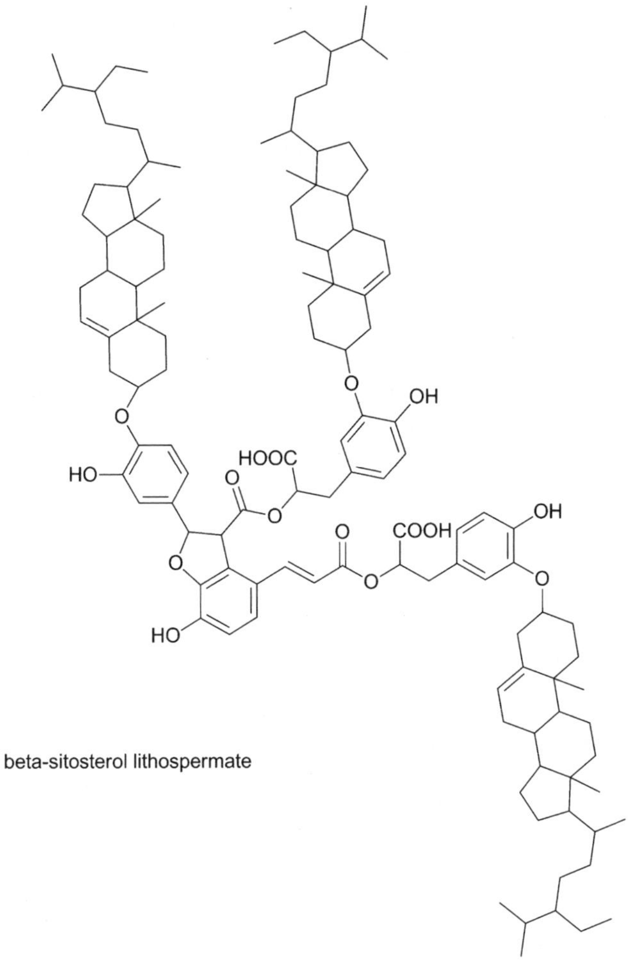

Figure 10.6 Structure of β-sitosterol lithospermate.

metabolites. The metabolites of α-linolenic acid inhibit platelet aggregation,[86] and are involved in altering the expression of fatty acid metabolic enzymes.[87] In addition, metabolites of α-linolenic acid interact with transient receptor potential cation channel vanilloid 1 in the spinal cord and can cause pain.[88] The effects of α-linolenic acid on lipoxins, protectins, resolvins and other beneficial eicosanoids are not known.

S. hispanica, another chia, has been tested in laboratory experiments and clinical trials. The seeds have been found to alter eicosanoid levels and metastasis in cancer cells *in vitro*.[89] The seeds form a gel that can be used in cooking due to the mucoproteins on the seeds that are rich in glutamic acid, arginine and aspartic acid.[90,91] The seeds improve adiposity and normalize hypertriglyceridemia and insulin resistance in obese rats.[92] However, the seeds do not help obese people lose weight or alter inflammatory adipokine levels or blood eicosanoid levels, even though postprandial blood glucose decreases.[91,93] Diabetic patients benefit from *S. hispanica* therapy with decreases in blood pressure, C-reactive protein levels, fibrinogen levels and hemoglobin A1C levels.[94] α-Linolenic acid and eicosapentaenoic acid blood levels increased in the study. It appears that *S. hispanica* should be given in doses large enough to increase blood α-linolenic acid levels in order to achieve a therapeutic effect.

10.5   Clinical Evidence with *Salvia miltiorrhiza* – Dan Shen in Stroke

It is very important to realize that treatment of stroke with any plant medicine, such as *S. miltiorrhiza*, involves concomitant exercise and may require 6 months or more of therapy. In China, exercise typically involves tai chi, although yoga and other mild forms of exercise may be beneficial. *S. miltiorrhiza* therapy typically involves two or more weeks of daily intravenous therapy that is followed by recovery for the next three or more months.

Therapy rarely involves *S. miltiorrhiza* alone, but is usually supplemented with other, beneficial plants. The plants that are added to the therapy are chosen by the physician and are based on what the patient needs and what the physician knows will be beneficial. In other words, the treatment is individualized to fit the needs of the patient. The most common forms of treatment are fufang danshen and gegen danshen. Fufang refers to the addition of *Panax notoginseng* root (also called *Panax pseudoginseng*) or *Dalbergia odorifera*. Gegen refers to *Pueraria lobata* root.

A complete recipe for danshen is as follows:[35] "30 g of the roots of *Astragalus membranaceus* and *Hedysarum coronarium*, 10 g of cinnamon twig (*Cinnamomum verum*), 10 g of peach kernel (*Prunus persica*), 10 g of safflower (*Carthamus tinctorius*), 12 g of *Angelica sinensis* root, 12 g of the rhizome of *Ligusticum chuanxiong*, 12 g of the root of *Paeonia rubra*, 12 g of earthworm (*Lumbricus* sp.), 30 g of the root of *S. miltiorrhiza* and 15 g of the root of *Achyranthis bidentata*. The ingredients are added to water and boiled for an hour or two until the required volume is attained. The preparation is administered orally to the patient. Other plants can be added to this as needed by the patient for severe blood stasis, deficiency of qi and some other conditions." In other words, *Pueraria lobata, Panax pseudoginseng, Dalbergia odorifera* or other plants can be added to suit the patient's needs. *Astragalus membranaceus* is normally used in *S. miltiorrhiza* preparations since it has been shown in clinical trials to improve therapy.

The clinical trials with *S. miltiorrhiza* have been reviewed and even analyzed in meta analysis.[35,95] Many of the 33 published clinical trials found beneficial effects in angina pectoris and other heart patients.[35] Several trials used 50 mg of purified tanshinone IIA injections in heart or stroke patients. These injections were not found to be superior to intravenous injections of *S. miltiorrhiza* extract containing 50 mg of tanshinone IIA. Later work found

Table 10.1 Stroke clinical trials with *S. miltiorrhiza* preparations (from reference 35).

Condition, n	Randomized, double blind	Placebo controlled	Preparation	Dose of tanshinone IIA (i.v.) (mg)	Percentage of patients receiving effective therapy
Stroke, 120	Y, N	N	Danshen	40	88
Subarachnoid bleeding, 42	N, N	N	Danshen	25–75	95
Stroke, 95	N, N	N	Fufang danshen	50 or 100	79 or 95
Stroke, 30	N, N	N	Fufangdanshen	10 intra-arterial	97
Subarachnoid bleeding, 24	N, N	N	Danshen	100	96
Stroke, 48	N, N	N	Danshen	75	100
Stroke, 78	N, N	N	Danshen	40	69
Stroke, 20	N, N	N	Danshen	40	75
Stroke, ?	N, N	N	Danshen	50	55
Stroke, 35	N, N	N	Danshen	?	74
Stroke, 96	N, N	N	Fufangdanshen	40	58
Stroke, 50	N, N	N	Danshen	50	72
Stroke, ?	N, N	N	Danshen	40	90
Stroke, 60	N, N	N	Fufang danshen	50	90
Stroke, 68	N, N	N	Fufang danshen	50	84
Stroke, 60	N, N	N	Fufang danshen	50	33
Stroke, 102	N, N	N	Danshen	50	18
Stroke, 62	N, N	N	Fufangdanshen	40	68
Stroke, 62	Y, N	N	Fufangdanshen	50	69
Stroke, 135	N, N	N	Fufangdanshen	50	65
Stroke, 68	N, N	N	Tanshinone IIA	1000	96
Stroke, 120	N, N	N	Danshen	50	78

that fufang danshen or gegen danshen were as good or better than *S. milti-orrhiza* alone. The 22 stroke clinical trials are summarized in Table 10.1. Few of the clinical trials were placebo controlled because placebos are considered unethical since it is known that *S. miltiorrhiza* is effective. Effective stroke therapy means that patients survived and recovered from paralysis sufficiently to be able to function without assistance.

10.6 Conclusion

Unfortunately, stroke prevention in Europe and the USA is still mired in the use of drugs, endarterectomy and stents.[96] The list of drugs that have been tested in randomized, placebo-controlled, double-blind clinical trials is huge: aspirin, ticlopidine, clopidogrel, sarpogrelate, triflusal, terutroban, cilostazol, warfarin, rivaroxaban, apixaban, ximelagatran, dabigatran, idraparinux and polypills. Polypills are combinations of aspirin, antihypertensives and anti-hyperlipidemic agents. All of these agents suffer from major disadvantages, especially the fact that they are all toxic and must be used for the remainder of the patient's life. In addition, the drugs may work in one gender but not the other. They also suffer from genetic problems as there are great genetic variations in the efficacy and toxicity of these drugs to certain patients. Sadly, medical science in the USA and Europe will continue to search for pills to give patients instead of helping patients with lifestyle issues.

The understanding of stroke is changing outside China. Stroke used to be considered a rapid event that permanently altered a patient's brain with 6 h. However, the use of stents and other intra-arterial devices can be delayed for up to 90 days and still be beneficial to patients. Patients used to be told that paralyzed limbs would never recover. However, physical therapy of up to several months duration has been shown in clinical trials to reverse or partly reverse paralysis after stroke.[97] The Chinese and American Indians have known for many years that exercise can reverse stroke-induced paralysis. How does exercise benefit stroke patients? Exercise stimulates the production of stem cells throughout the body, even the brain.[98] Despite the dogma that brain neurons cannot regenerate, stroke patients can recover from paralysis. The human body has evolved with exercise and continues to need exercise today. Stroke patients need to re-establish their bodies balance by exercising.

Western medicine continues to hinder itself seriously with the pathological insistence on the use of single purified drugs. Plant medicines, such as *S. miltiorrhiza* or *S. columbariae*, are superior to any single purified drug. Plant medicines are called fraudulent or dangerous in the USA by the National Center for Complementary and Alternative Medicine, even though there is a long history of the use of plant medicines in the USA. Of course, it is easier to dose and evaluate drug interactions for single purified drugs. Cancer, AIDS, tuberculosis and other diseases are routinely treated with multiple drug therapy. Even hypertension and hypercholesterolemia are routinely treated with two or three drugs. Clearly, even in the USA, there is an appreciation that healing can benefit from multiple drugs. This should be extended to plant medicines. Funding agencies outside China should seriously consider funding studies to investigate plant medicines in stroke clinical trials.

References

1. L. Wong, *Int. J. Stroke*, 2006, **1**, 158.
2. J. Arenillas, *Stroke*, 2011, **42**, S20.
3. M. Mazighi, J. Labreuche, R. Gongora-Rivera, C. Duyckaerts, J. Hauw and P. Amarenco, *Stroke*, 2008, **39**, 1142.
4. P. Amarenco, P. Lavallee, J. Labreuche, G. Ducrocq, J. Juliard, L. Feldman, L. Cabrejo, E. Meseguer, C. Guidoux, V. Adrai, S. Ratani, J. Kusmierek, B. Lapergue, I. Klein, F. Gongora-Rivera, A. Jaramillo, M. Mazighi, P. Touboul and S. Steg, *Stroke*, 2011, **42**, 22.
5. D. Giralt, S. Domingues-Montanari, M. Mendioroz, L. Ortega, O. Maisterra, M. Perea-Gainza, P. Delgado, A. Rosell and J. Montaner, *Acta Neurol. Scand.*, 2012, **125**, 83.
6. A. Russjan, E. Goebell, S. Havemeister, G. Thomalla, B. Cheng, C. Beck, A. Krutzelmann, J. Fiehler, C. Gerloff and M. Rosenkranz. *Cerebrovasc. Dis.*, 2012, **33**, 30.
7. A. Rouchaud, M. Mazighi, J. Labreuche, E. Meseguer, J. Serfaty, J. Laissy, P. Lavallee, L. Cabrejo, C. Guidoux, B. Lapergue, I. Klein, J. Olivot, H. Abboud, O. Simon, E. Schouman-Claeys and P. Amerenco, *Stroke*, 2011, **42**, 1289.
8. I. Linfante, E. Samaniego, P. Geisbusch and G Dabus, *Stroke*, 2011, **42**, 2636.
9. Y. Loh, D. Kim, Z. Shi, S. Tateshima, P. Vespa, N. Gonzalez, S. Starkman, J. Saver, R. Jahan, D. Liebeskind, G. Duckwiler and F. Vinuela, *Am. J. Neuroradiol.*, 2010, **31**, 1181.
10. R. Nogueira, D. Liebeskind, G. Sung, G. Duckwiler and W. Smith, *Stroke*, 2009, **40**, 3777.
11. A. Vaya, M. Martinez-Triguero, F. Espana, J. Todoli, E. Bonet and D. Corella, *Metab. Syndrome Rel. Dis.*, 2011, **9**, 197.
12. C. Ay, T. Tengler, R. Vormittag, R. Simanek, W. Dorda, T. Vukovich and I. Pabinger. *Haematologica*, 2007, **92**, 374.
13. C. Savopoulos, K. Michalakis, M. Apostolopoulou, A. Miras and A. Hatzitolios, *Maturitas*, 2011, **70**, 322.
14. S. Choi, S. Kwak, Y. Lee, M. Moon, S. Lim, Y. Park, H. Hang and M. Kim, *Clin. Endocrinol.*, 2011, **75**, 628.
15. L. Carcaillon, M. Alhenc-Gelas, Y. Bejot, C. Spaft, P. Ducimetiere, K. Ritchie, J. Dartigues and P. Scarabin, *Arterioscler. Thromb. Vasc. Biol.*, 2011, **31**, 1445.
16. M. Shimizu, M. Yamamoto, S. Takizawa, S. Kohara, S. Takagi and Y. Shinohara, *J. Stroke Cerebrovasc. Dis.*, 2011, **20**, 275.
17. T. Okamura, Y. Kokubo, M. Watanabe, A. Higashiyama, Y. Ono, K. Nishimura, A. Okayama and Y. Miyamoto, *Atherosclerosis*, 2011, **217**, 201.

18. T. Chiu, C. Chen, C. Chen, C. Soon and J. Chen, *Clin. Chim. Acta*, 2012, **413**, 226.

19. Y. Matsuzawa, *FEBS Lett.*, 2006, **580**, 2917.

20. S. Ritchie and J. Connell, *Nutr. Metab. Cardiovasc. Dis.*, 2006, **17**, 319.

21. P. Szmitko, H. Teoh, D. Stewart and S. Verma, *Am. J. Physiol. Heart Circ. Physiol.*, 2007, **292**, H1655.

22. E. Wehr, S. Pilz, B. Boehm, W. Marz and B. Obermayer-Pietsch, *Obesity*, 2011, **19**, 1873.

23. J. Al Suwaidi, M. Zubaid, A. El-Menyar, R. Singh, W. Rashed, M. Ridha, A. Shehab, J. Al-Lawati, H. Amin and A. Al-Mottareb, *J. Clin. Hypertens.*, 2010, **12**, 890.

24. J. Wang, K. Sarnola, S. Ruotsalainen, L. Moilanen, P. Lepisto, M. Laakso and J. Kuusisto, *Atherosclerosis*, 2010, **210**, 237.

25. G. de Simone, R. Devereux, M. Chinali, M. Roman, E. Lee, H. Resnick and B. Howard, *Nutr. Metab. Cardiovasc. Dis.*, 2009, **19**, 98.

26. G. Lip, *Clin. Cardiol.*, 2012, **35**, Suppl. 1, 21.

27. W. Lewis, G. Fonarow, M. Grau-Sepulveda, E. Smith, D. Bhatt, A. Hernandez, D. Olson, E. Peterson and L. Schwamm, *Am. Heart J.*, 2011, **162**, 692.

28. R. Shinton and G. Beevers, *Br. Med. J.*, 1989, **298**, 789.

29. I. Oono, D. Mackay and J. Pell, *J. Public Health*, 2011, **33**, 496.

30. V. Wolff, V. Lauer, O. Rouyer, F. Sellal, N. Meyer, J. Raul, C. Sabourdy, F. Boujan, C. Jahn, R. Beaujeux and C. Marescaux, *Stroke*, 2011, **42**, 1778.

31. P. Gorelick, *Stroke*, 1987, **18**, 268.

32. L. Lisabeth and C. Bushnell, *Lancet Neurol.*, 2012, **11**, 82.

33. Y. Li, Y. He, J. Lai, D. Wang, J. Zhang, P. Fu, X. Yang and L. Qi, *J. Nutr.*, 2011, **141**, 1834.

34. S. Larsson, J. Virtamo and A. Wolk, Am. *J. Clin. Nutr.*, 2011, **94**, 417.

35. J. Adams, R. Wang, J. Yang and E. Lien, *Chin. Med.*, 2006, **1**, 1.

36. H. Sung, J. Jun, S. Kang, H. Kim, D. Shin, I. Kang and Y. Kang, *Chem. -Biol. Interact.*, 2010, **188**, 457.

37. J. Kim and J. Jun, *J. Cell. Biochem.*, 2011, **112**, 3816.

38. Y. Yin, M. Yang, Y. Wang, X. Liu, W. Wu, S. Guan, D. Guo, Y. Cui and B. Jiang, *Am. J. Chin. Med.*, 2010, **38**, 995.

39. J. Wang, Z. Ma, W. Wang, Z. Hong and J. Song, *Zhongguo Zhong Yao Za Zhi*, 2009, **34**, 2943.

40. B. Chang, L. Zhang, W. Cao, Y. Cao, W. Yang, Y. Wang, Y. Chen and X. Liu, *Zongguo Yao Li Xue Bao*, 2010, **31**, 638.

41. Y. Liu, X. Li, Y. Li, L. Wang and M. Xue, *J. Pharm. Biomed. Anal.*, 2010, **53**, 698.

42. J. Ho and C. Hong, *J. Biomed. Sci.*, 2011, **18**, 1.

43. Y. Zhang, L. Wu, Q. Zhang, J. Li, F. Yin and Y. Yuan, *J. Ethnopharmacol.*, 2011, **136**, 129.

44. S. Gao, Z. Liu, H. Li, P. Little, P. Liu and S. Xu, *Atherosclerosis*, 2012, **220**, 3.

45. X. Wang, C. Cheung, W. Lee, P. Or and J. Yeung, *Phytomedicine*, 2010, **17**, 868.
46. X. Wang, W. Lee, X. Xhou, P. Or and J. Yeung, *Phytomedicine*, 2010, **17**, 876.
47. X. Wang and J. Yeung, *Phytother. Res.*, 2011, **25**, 1653.
48. X. Wang and J. Yeung, *J. Pharm. Pharmacol.*, 2010, **62**, 1077.
49. R. Wan, M. Zhou and C. Liu, *Phytother. Res.*, 2010, **24**, 846.
50. X. Wang, W. Lee, P. Or and J. Yeung, *Phytomedicine*, 2010, **17**, 203.
51. W. Wu and J. Yeung, *Phytomedicine*, 2010, **17**, 219.
52. H. Sung, J. Jun, S. Kang, D. Shin, I. Kang and Y. Kang, *Chem.-Biol. Interact.*, 2010, **188**, 457.
53. S. Xu, P. Little, T. Lan, Y. Huang, K. Le, X. Wu, X. Shen, H. Huang, Y. Cai, F. Tang, H. Wang and P. Liu, *Arch. Biochem. Biophys.*, 2011, **515**, 72.
54. W. Wang, B. Yang, L. Wang, R. Liang, C. Chen, N. Hu, L. Cheng, Y. Yang, X. Yin, S. Gao and J. Ye, *Zhongguo Zhong Yao Za Zhi*, 2011, **36**, 784.
55. X. Qin, T. Li, L. Yan, Q. Liu and Y. Tan, *Immunopharmacol. Immunotoxicol.*, 2010, **32**, 51.
56. Q. Liu, Y. Zhang, Z. Lin, H. Shen, L. Chen, L. Hu, H. Jiang and X. Shen, *J. Steroid Biochem. Mol. Biol.*, 2010, **120**, 155.
57. J. Fang, S. Xu, P. Wang, F. Tang, S. Zhou, J. Gao, J. Chen, H. Huang and P. Liu, *Phytomedicine*, 2010, **18**, 58.
58. H. Hong, J. Liu, T. Cheng and P. Chan, *Zhongguo Yao Li Xue Bao*, 2010, **31**, 1569.
59. C. Tang, H. Xue, C. Bai, R. Fu and A. Wu, *Phytomedicine*, 2010, **17**, 1145.
60. F. Lam, S. Deng, E. Ng, J. Yeung, Y. Kwan, C. Lau, J. Koon, L. Zhou, Z. Zuo, P. Leung and K. Fung, *J. Ethnopharmacol.*, 2010, **132**, 186.
61. X. Tan, Y. Yang, J. Cheng, P. Li, I. Inoue and X. Zeng, *Eur. J. Pharmacol.*, 2011, **656**, 27.
62. Z. Huang, C. Zeng, L. Zhu, L. Jiang, N. Li and H. Hu, *J Thromb. Haem.*, 2010, **8**, 1383.
63. J. Liu, T. Lee, M. Miedzyblocki, G. Chan, D. Bigam and P. Cheung, *J. Ethnopharmacol.*, 2011, **137**, 44.
64. Y. Li, L. Shen, R. Chen, J. Li, F. Lu, Y. Qin and J. Liu, *Chin. J. Int. Med.*, 2011, **17**, 625.
65. M. Han, Y. Liu, B. Zhang, J. Qiao, W. Lu, Y. Zhu, Y. Wang and C. Zhao, *Pharm. Biol.*, 2011, **49**, 1008.
66. Y. Zhou, W. Li, L. Xu and L. Chen, *Environ. Toxicol. Pharmacol.*, 2011, **31**, 443.
67. S. Jeon, S. Bose, J. Hur, K. Jun, Y. Kim, K. Cho and B. Koo, *J. Ethnopharmacol.*, 2011, **137**, 783.
68. S. Durairajan, L. Liu, J. Lu, I. Koo, K. Murayama, S. Chung, J. Huang and M. Li, *J. Alzheimers Dis.*, 2011, **25**, 245.
69. T. Liu, J. Jin, Q. Sun, J. Xu and H. Hu, *Neuropharmacology*, 2010, **59**, 595.
70. Z. Mei, B. Situ, X. Tan, S. Zheng, F. Zhang, P. Yan and P. Liu, *Brain Res.*, 2010, **1348**, 165.

71. K. Wong, M. Ho, H. Lin, K. Lau, J. Rudd, R. Chung, K. Fung, P. Shaw and D. Wan, *Planta Med.*, 2010, **76**, 228.

72. K. Wong, J. Ngo, S. Liu, J. Lin, C. Hu, P. Shaw and C. Wan, *Chem.-Biol. Interact.*, 2010, **187**, 335.

73. B. Lu, Z. Ye, Y. Deng, H. Wu and J. Feng, *Cell Biol. Int.*, 2010, **34**, 1063.

74. Y. Li, C. Duan, J. Liu and Y. Xu, *J. Ethnopharmacol.*, 2010, **131**, 562.

75. G. Guo, B. Li, Y. Wang, A. Shan, W. Shen, L. Yuan and S. Zhong, *Sci. Chin.*, 2010, **53**, 653.

76. J. Adams, C. Garcia and M. Wall, *Evid. Based Comp. Alt. Med.*, 2005, **2**, 107.

77. J. Adams, S. Tran, V. Wong, P. Fontaine and J. Petrosyan, *Open Nat. Prod. J.*, 2012, **5**, 7.

78. C. Wang, W. Harris, M. Chung, A. Lichtenstein, E. Balk, B. Kupelnick, H. Jordan and J. Lau, *Am. J. Clin. Nutr.*, 2006, **84**, 5.

79. I. Sioen, M. Hacquebard, G. Hick, V. Maindiaux, Y. Larondelle, Y. Carpentier and S. De Henauw, *Lipids*, 2009, **44**, 603.

80. B. Gopinath, C. Harris, V. Flood, G. Burlutsky and P. Mitchell, *Br. J. Nutr.*, 2011, **105**, 1361.

81. J. de Goede, W. Verschuren, J. Boer, D. Kromhout and J. Geleijnse, *PLoS One*, 2011, **6**, e17967.

82. J. Dai, T. Ziegler, R. Bostick, A. Manatunga, D. Jones, J. Goldberg, A. Miller, G. Vogt, P. Wilson, L. Jones, L. Shallenberger and V. Vaccarino, *Am. J. Clin. Nutr.*, 2010, **92**, 177.

83. T. Nelson, J. Stevens and M. Hickey, *Metab. Clin. Exp.*, 2007, **56**, 1209.

84. C. Bassett, R. McCullough, A. Edel, A. Patenaude, R. LaVallee and G. Pierce, *Am. J. Physiol. Heart Circ. Physiol.*, 2011, **301**, H2220.

85. N. Xie, W. Zhang, J. Li, H. Liang, H. Zhou, W. Duan, X. Xu, S. Yu, H. Zhang and D. Yi, *Arch. Med. Res.*, 2011, **42**, 171.

86. A. Barden, E. Mas, P. Henry, T. Durand, J. Galano, L. Roberts, K. Croft and T. Mori, *Free Rad. Res.*, 2011, **45**, 469.

87. H. Zhu, C. Fan, F. Xu, C. Tian, F. Zhang and K. Qi, *J. Nutr. Biochem.*, 2010, **21**, 954.

88. A. Patwardhan, P. Scotland, A. Akopian and K. Hargreaves, *Proc. Natl. Acad. Sci. U. S. A.*, 2009, **106**, 18435.

89. C. Espada, M. Berra, M. Martinez, A. Eynard and M. Pasqualini, *Prost. Leuk. Essent. Fatty Acids*, 2007, **77**, 21.

90. B. Olivos-Lugo, M. Valdivia-Lopez and A. Tecante, *Food Sci. Technol. Int.*, 2010, **16**, 89.

91. V. Vuksan, A. Jenkins, A. Dias, A. Lee, E. Jovanovski, A. Rogovik and A. Hanna, *Eur. J. Clin. Nutr.*, 2010, **64**, 436.

92. A. Chicco, M. D'Alessandro, G. Hein, M. Oliva and Y. Lombardo, *Br. J. Nutr.*, 2009, **101**, 41.

93. C. Nieman, E. Cayea, M. Austin, D. Henson, S. McAnulty and F. Jin, *Nutr. Res.*, 2009, **29**, 414.

94. V. Vuksan, D. Whitman, J. Sievenpiper, A. Jenkins, A. Rogovik, R. Bazinet, E. Vidgen and A. Hanna, *Diabetes Care*, 2007, **30**, 2804.

95. B. Wu, M. Liu and S. Zhang, *Cochrane Database Syst. Rev.*, 2007, **2**, CD004295.
96. M. Bousser, *Front. Med.*, 2012, **6**, 22.
97. L. Kalra, *Stroke*, 2010, **41**, e88.
98. F. Gage, *Science*, 2000, **287**, 1433.

CHAPTER 11

Drug Discovery from Traditional Chinese Medicine for Neurogenesis: Implications for Stroke and Neurodegenerative Diseases

JIANGANG SHEN* AND XINGMIAO CHEN

School of Chinese Medicine, Research Centre of Heart, Brain, Hormone and Healthy Aging, The University of Hong Kong, Hong Kong SAR, China
*E-mail: shenjg@hkucc.hku.hk

11.1 Background

Stroke and neurodegenerative diseases are aging-related diseases. Stroke is the second most common cause of death and a leading cause of adult disability worldwide.[1] In China, the age-adjusted stroke prevalence ranges from 259.86 to 719 per 100 000 people per year.[2] Neurodegenerative diseases are diseases associated with aging-related progressive neurological function deficits, including Alzheimer's disease (AD), Parkinson's disease (PD), Huntington's disease and multiple sclerosis. AD is the most common neurodegenerative disease and is characterized by progressive cognitive decline leading to a need for complete care within a few years after clinical diagnosis. It is a major cause of dementia. The global prevalence of dementia in people aged over 60 years was estimated to be over 24 million in 2001 and is expected to double every 20

RSC Drug Discovery Series No. 31
Traditional Chinese Medicine: Scientific Basis for Its Use
Edited by James David Adams, Jr. and Eric J. Lien
© The Royal Society of Chemistry 2013
Published by the Royal Society of Chemistry, www.rsc.org

years, affecting more than 80 million people worldwide by 2040.[3–5] The prevalence of senile dementia in China was estimated to be about 6–7 million with an incidence of 5–7% of people over 65 years of age, of which AD accounts for 50–60% cases.[6] PD is the second most common neurodegenerative disease and is characterized by motor deficits, such as resting tremor, bradykinesia, rigidity and postural reflex impairment. Motor deficits in PD are due to the degeneration of the dopaminergic neurons of the pars compacta of the substantia nigra, leading to loss of dopamine in the striatum. Overall incidence rates for PD were estimated at 410–529 per 100 000 at ages over 55 years.[7] The median incidence at 65 years or above in the USA was estimated at 160 per 100 000 person-years.[8] A cross-sectional prevalence study revealed 277 PD cases from a total of 29 454 people aged 55 years or older in Beijing, estimating 1.7 million people with PD in China.[9] With increases in aging populations, it is expected that stroke and neurodegenerative diseases will become major disease burdens in public health worldwide.

Stroke and neurodegenerative diseases share a common feature, the loss of functional neurons and glial cells in the brain and spinal cord. For instance, different types of neurons and glial cells die in a short time period, several days, during acute ischemic stroke. The death of dopamine (DA) neurons gradually occurs in PD and many types of hippocampal and cortical neurons are lost in AD.[10] In recent decades, tremendous efforts have been made to develop therapeutic approaches for improving clinical outcomes for post-stroke disability and neurodegenerative diseases. Nevertheless, clinical outcomes have not improved. For example, recombinant tissue-plasminogen activator (rt-PA) is the only FDA-approved drug, but it has a 3 h golden time window with the potential risk of hemorrhagic transformation. Although many neuroprotective drugs are effective in animal models, they failed to pass clinical trials. Current FDA-approved drugs for AD, including the acetylcholinesterase inhibitors (ACEIs) donepezil, galantamine and rivastigmine and the N-methyl-D-aspartate (NMDA) antagonist memantine, do not halt the progress of the disease. Levodopa is considered the most effective and best tolerated PD drug to date.[11] Levodopa provides good control of the symptoms of PD, but motor complications often develop after long-term use. Other drugs including dopamine agonists, catechol-O-methyltransferase inhibitors or monoamine oxidase type B inhibitors have also been proposed as adjuvant therapies with levodopa, but there have potential adverse events and safety issues for long-term use.[12] Overall, current pharmacological therapies are not able either to arrest or reverse the progression of these diseases. Accordingly, there is an urgent need to develop novel and effective medications for stroke and neurodegenerative diseases.

The discovery of adult neurogenesis shed some light on the development of new therapeutic approaches for stroke and neurodegenerative diseases. neural stem/progenitor cells (NSCs) can potentially develop into new functional neurons in the adult central nervous system (CNS).[13,14] Adult neurogenesis mainly occurs in the subgranular zone (SGZ) in the dentate gyrus (DG) of the hippocampus and the subventricular zone (SVZ) adjacent to the lateral ventricle.[15,16] Enhanced neurogenesis has been reported in hypoxic NSCs *in vitro*[17,18] and in ischemic brains of neonatal mice,[19] adult rats[20] and aged humans *in vivo.*[21] Enhanced neurogenesis by either stem cell transplantation or stimulation of endogenous neurogenesis could partly recover damaged brain function, bringing hope for brain repair treatment.

NSC transplantation is a promising cell therapeutic strategy.[22] Impaired formation of new hippocampal neurons from innate NSCs contributes to memory impairment and cognitive decline in AD.[23] Preclinical evidence indicates that stem cell therapy holds great potential for the replacement of cholinergic cells in AD.[24] Transplantation of stem cell-derived DA neuron precursors also brings substantial benefits for PD.[25,26] Owing to ethical and technical limitations, however, there is still a long way to go for clinical applications of NSC transplantation. The pool of endogenous NSCs of adult brain provides attractive cell sources for regeneration therapy. For drug discovery, most efforts have focused on stimulating endogenous NSCs to form new functional neurons and preventing the death of the neurons in the CNS. For instance, memantine, a drug for AD, was revealed to increase neurogenesis in DG and SVZ with enhanced learning and memory.[27] Therefore, the development of new approaches targeting endogenous NSCs could be an important direction for drug discovery.

Recently, several stem cell-based high-throughput screening platforms have been built for searching for bioactive compounds to improve neurogenesis.[28–31] Many chemically synthesized compounds or natural compounds were screened for promoting NSC proliferation and modulating differentiation. Some of the compounds are promising drug candidates. However, there is still a long way for a drug candidate to travel from bench to bedside. In contrast to Western medicine, traditional Chinese medicine (TCM) has been practiced for more than 2000 years in China. Many medicinal herbs were used for treating stroke, AD and PD patients for centuries with accumulated cases of successfully improved recovery of neurological functions. These medicinal herbs are very efficient resources for seeking drug candidates to promote neurogenesis. In this chapter, we briefly review current progress for drug discovery from Chinese herbal medicines. This review consists of three components: (1) classical TCM theory and scientific studies on Chinese herbal medicine for stroke, AD and PD treatments, (2) current progress with molecular targets and signaling pathways for neurogenesis and (3) progress in exploring active compounds, herbal fractions and formulas for neurogenesis. We hope to provide an overall picture about herbal medicine research for regeneration therapy and shed light on future developments for drug discovery.

11.2 Classical TCM Theory and the Scientific Basis of Chinese Herbal Medicine for Stroke, AD and PD Treatments

TCM is a comprehensive medical system characterized by its own theoretical basis and practical experiences. The records of clinical symptoms related to stroke, AD and PD can be traced back to 2000 years ago, originally from the *Yellow Emperor Canon of Inner Medicine,* the earliest medical textbook in China. According to TCM theory, the dynamic balance of *Yin* and *Yang* is essential for human health, and disharmony of *Yin* and *Yang* is the pathological basis of diseases. Based on the holistic concept of TCM, human health is closely related to the following essential aspects: (1) environmental factors; (2) psychological factors; (3) diet and life style; and (4) constitution, congenital essence and *Qi* (vital energy). Stroke and neurodegenerative diseases are considered pathological states caused by disharmony of *Yin* and *Yang*. To restore balance, TCM practitioners generally integrate different therapeutic approaches and protocols including Chinese herbal medicine, acupuncture, *Qi-gong, Taiji,* mental regulation therapy and others. Here, we briefly introduce classical TCM theory and current progress in Chinese herbal medicine for stroke and PD and AD.

11.2.1 Stroke

In TCM theory, stroke is called *Wind Stroke* and is defined as "a condition mainly characterized by sudden collapse and loss of consciousness, deviation of the tongue and mouth, hemiplegia, slurred speech or only deviation of the tongue and mouth and hemiplegia without collapse." *Wind Stroke* is considered to be the consequence of exposure to mixed pathological states including long-term abnormal diet and life style, stress and abnormal psychological stress. Patients with constitutions deficient in *Gan-Ying* and *Shen-Ying* are susceptible to pathological factors and have greater chances of being attacked by *Wind Stroke.* A *Wind Stroke* stroke is attributed to reversed flow of *Qi* and *Blood,* which produces *Wind, Fire, Phlegm* and *Blood Stasis,* subsequently inducing cerebral thrombosis or cerebral hemorrhage. *Wind Stroke* can be catalogued into *Meridian Stroke* (Superficial and mild), *Zang-fu Stroke* (Deep and severe) and *Sequela.*[32] Therapeutic approaches and formulas are specifically used for different clinical patterns of syndrome differentiation. More than 100 Chinese herbs have been reported for the prevention and treatment of stroke.[33] For example, *Tianma Gouteng Decoction* and *Zhengang Xifen Decoction* can be used for preventing *Wind Stroke* or *Meridian Stroke* at the stages of primary hypertension and transient ischemic attack (TIA). At the acute stage of stroke, *Angong Niuhuang Wan* and *Lingjiao Guoteng Decoction* are suitable for excessive syndrome with *Gang-Yang* and *Gang-Wind* whereas *Shenfu Decoction* can be used for deficiency of *Yang Qi* in *Zang-fu Stroke.* For *Sequela, Jieyu Dan* is used for relieving the symptoms of deviation of the tongue and mouth, hemiplegia and slurred speech with the syndrome of *Phlegm* and *Blood Stasis* whereas *Buyang Huanwu Decoction* is generally applied for neurological deficits and dysfunction with *Qi* deficiency-induced *Blood Stasis*. In recent years, much effort has been expended in exploring the scientific basis of TCM formulas for stroke treatment. For example, *Tianma Gouteng Decoction* (TGD) has been shown to reduce blood pressure, ameliorate cognitive impairment and protect ischemic brains experimentally and clinically.[34–36] Owing to the poor design of clinical trials, however, a recent systematic review challenged its clinical efficacy for primary hypertension.[37] *Buyang Huanwu Decoction* is a classical formula for post-stroke disability. *Buyang Huanwu Decoction* consists of seven items, *Astragalus membranaceus*, *Angelica sinensis*, *Paeonia lactiflora*, *Ligusticum chuanxiong*, *Carthamus tinctorius*, *Prunus persica* and *Lumbricus*, in a ratio of 120:4.5:3:3:3:3:3. The formula was reported to promote neurogenesis, reduce infarction volume and improve neurological function in post-stroke animal models and human subjects.[38–40] Further studies are necessary to explore the scientific basis of these formulas and to conduct well-designed randomized controlled trials (RCTs) to verify their efficacies and safety.

11.2.2   AD

In TCM theory, AD is a type of dementia and is defined as "a kind of mind disease which results from dysfunction of vital activities due to brain atrophy, mainly manifested as dullness and stupidity."[32] Dementia is generally recognized as a pathological state of *Shen-Qi* Deficiency, *Shen Essence* Deficiency, *Pi-Qi* Deficiency or *Xin-Qi* Deficiency and also Stagnation of *Blood* and/or *Phlegm*.[41] AD can be classified into several types of clinical patterns, including *Shen Essence* Deficiency Syndrome, *Gan-Shen Yin* Deficiency Syndrome, *Pi-Shen* Yang Deficiency Syndrome, *Qi* Stagnation and *Blood Stasis* Syndrome, *Phlegm Turbid* Blocking *Orifice* Syndrome and *Xin-Gan Fire* Syndrome. Among them, *Blood* Stagnation and *Shen Qi*/*Shen Essence* Deficiency are common clinical patterns in dementia. *Yigan San* (also named Yukukansan or TJ-54) is a classical TCM formula used for dementia. *Yigan San* consists of seven herbs, *Angelica acutiloba, Atractylodes lancea, Bupleurum falcatum, Poria cocos, Cnidium officinale, Uncaria rhynchophylla* and *Glycyrrhiza uralensis*, in a ratio of 3:4:2:4:3:3:1.5. Clinical RCTs revealed that *Yigan San* improved behavioral and psychological symptoms of dementia,[42,43] which include aggression, agitation, screaming, wandering, hallucinations and delusions. *Yigan San* reduced cholinesterase inhibitor-resistant visual hallucinations in dementia patients.[43] *Yigan San* improved psychiatric symptoms and sleep structure in dementia patients.[44] The mechanisms of action are related to regulating multiple signal pathways, such as the glutamatergic neurotransmitter system, the serotonin receptor and excitotoxicity.[45]

Danggui Shaoyao San (DSS), also called Toki-shakuyaku-san or TJ-23, is another well-known TCM formula for dementia. DSS was recorded in the *Synopsis of Prescriptions of the Golden Chamber*, which was compiled by Zhong-Jing Zhang in the Han dynasty. DSS is composed of six raw herbal materials, *Angelica sinensis* (Oliv.) Diels, *Paeonia lactiflora* Pall., *Ligusticum chuanxiong* Hort., *Poria cocos* (Schw.) Wolf, *Atractylodes macrocephala* Koidz and *Alisma orientalis* (Sam.) Juzep. DSS treatment alleviated β-amyloid (Aβ) cognitive dysfunction,[46] decreased the content and deposition of Aβ, improved neuronal activity and synaptic plasticity and enhanced spatial learning and memory in a spatial cognition of senescence-accelerated mouse prone 8 model, an aged rodent cognitive impairment model.[47] In addition, *Suhexiang Wan* (SHXW) has been used for the treatment of seizures, infantile convulsions, stroke and so forth. KSOP1009, a modified recipe of SHXW, improved memory impairment and reduced Aβ levels in a Tg-APPswe/PS1dE9 AD mouse model. KSOP1009 consists of an ethanol extract of eight herbs, resin of *Liquidambar orientalis* Miller, seed of *Myristica fragrans* Houtt., rhizome of *Cnidium officinale* Makino, lumber of *Santalum album* L., fructus of *Piper longum* L., flower buds of *Eugenia caryophyllata* Merrill et Perry, pollen of *Typha orientalis* Presl. and root of *Salvia miltiorrhiza* Bunge.[48] It is necessary to conduct preclinical experiments and well-designed RCTs to evaluate the efficacies of these formulas on AD patients.

Many herbal extracts and pure compounds have been screened to test their memory-improving effects for drug discovery in recent years. Huperzine A, an alkaloid isolated from *Huperzia serrata* (Qiancengta), was proved to have anti-AChE activity[49] and reduce the formation of Aβ.[50] Clinical trials revealed that huperzine A significantly improved cognitive function in AD patients[51] and vascular dementia patients.[52] A recent review summarized the herbal formulas, herbs and active compounds with potential to improve memory in senile dementia. *Poria cocos*, *Radix et rhizome ginseng*, *Radix polygalae*, *Radix et rhizome glycyrrhizae*, *Radix Angelica sinensis*, *Rhizoma acori tatarinowii*, *Semen ziziphi spinosae*, *Radix rehmanniae*, *Radix ophiopogonis* and *Rhizoma zingiberis* were listed as the top 10 herbs for improving memory. Multiple mechanisms, including estrogen-like, cholinergic, antioxidant, anti-inflammatory, antiapoptotic, neurogenetic and anti-Aβ activities, were discussed. The active compounds, including sinapic acid, tenuifolin, isoliquiritigenin, liquiritigenin, glabridin, ferulic acid, Z-ligustilide, *N*-methyl-β-carboline-3-carboxamide, coniferyl ferulate, 11-angeloylsenkyunolide F and catalpol, revealed potential value for senile dementia.[53] Therefore, Chinese herbal medicine provides vast resources for drug discovery for AD treatment.

11.2.3   PD

PD was first described by James Parkinson in 1817. However, in TCM, precise records for the clinical symptoms of PD and its preliminary treatment can be dated back to 2000 years ago. PD shares similar symptoms to tremor, with the definition of "a disease marked by shaking or tremoring of head or extremities due to brain marrow losing nourishment and tendons, vessels and extremities being out of control."[32] The *Yellow Emperor Canon of Inner Medicine* described the symptoms of tremor and stiffness and the pathogenesis of tremor: "*Wind* refers to the symptoms of shaking and dizziness, resulting from the dysfunction of the *Gan* and the symptoms of stiffness and spasm result from *Wind* or *Dampness*." The symptoms such as limitation of movement, postural disturbances, stiffness and tremor were described in the chapter "*Pulse and Essence Theory*." Zhang Zihe (1156–1228 AD) was the first to describe a typical case of PD in the book *Rumen Shiqin* (*Confucian's Duties to Their Parents*). The reported PD-similar case was 600 years earlier than James Parkinson's report.[54,55] According to TCM theory, PD can be classified by several different clinical patterns, containing "*Gan-Shen* Yin Deficiency Syndrome," "*Qi-Blood* Deficiency Syndrome," "*Qi*-stagnation and *Blood-Stasis* Syndrome," "*Phlegm* Stagnation Syndrome," and "*Shen-Yang* Deficiency Syndrome."[56] There are more than 200 prescriptions with 193 Chinese medicinal materials used for PD treatment in the literature. A multiple-center, randomized, double-blind and placebo-controlled trial revealed that *Bushen Huoxue Decoction* (BHD), a TCM formula for reinforcing *Shen* and activating blood circulation, improved motor function, shortened rise time in the 10 m back and forth test and relieved muscle tension in PD patients.[57] BHD also improved test scores for UPDRS II, PDQ-39 and PDSS and quality of life of PD patients.[58] BHD upregulated expression of the orphan receptor (Nurr1) and tyrosine hydroxylase (TH) in rat brains with PD.[59] Many Chinese herbs or herbal extracts attenuate the degeneration of dopamine neurons and corresponding symptoms in neurotoxin-, 1-methyl-4-phenyl-1,2,3,6-tetrahydropyridine (MPTP)- and 6-hydroxydopamine (6-OHDA)-induced PD models *in vitro* and *in vivo*. A review summarized the herbal extracts and compounds for PD, including green tea polyphenols, catechins, panax ginseng, ginsenoside, EGb 761, polygonum, triptolide from *Tripterygium wilfordii* Hook, polysaccharides from the flowers of *Neriu mindicum*, oil from *Ganoderma lucidum* spores, huperzine, stepholidine, *etc.* These herbs or herbal extracts may promote neuronal survival and neurite growth and facilitate functional recovery by invoking distinct mechanisms such as dopamine transporter inhibitor, monoamine oxidase inhibitor, free radical scavenger, chelator of harmful metal ions, modulator of cell survival genes and signaling, antiapoptosis activity and improving brain blood circulation.[60] *Isorhynchophylline* (IsoRhy), a major tetracyclic oxindole alkaloid isolated from *Uncaria rhynchophylla* (Miq.) Jacks, stimulated autophagy in neuronal cells and reduced pathogenic protein aggregates in neurons.[61] The major chemical structures with anti-Parkinsonian activities include catechols, stilbenoids, flavonoids, phenylpropanoids, lignans, phenylethanoid glycosides and terpenes.[62] More than 20 clinical trials revealed that herbal formulas had therapeutic effects on motor function and daily living activities.[55] However, a systematic review revealed that there is insufficient evidence to prove the efficacy and safety of various treatments with Chinese herbal medicines for PD because of the poor design of clinical trials and potential publication bias.[63]

In summary, many experimental data and clinical trials indicate that Chinese herbal medicine has potential value for stroke and neurodegenerative diseases. However, evidence is still insufficient to support the use of Chinese herbal medicine to improve outcome or slow the progression of those diseases. With increasingly aging populations, there is an urgent need to seek more effective therapeutic strategies for stroke and neurodegenerative diseases. The combination of Chinese herbal medicine and stem cell biology will be a new direction for drug discovery.

11.3 Current Molecular Targets and Signaling Pathways for Neurogenesis

NSCs can potentially develop into new functional neurons in the adult CNS. Enhanced proliferation has been proved in hypoxic NSCs *in vitro*[17,18] and in ischemic brains of neonatal mice,[19] adult rats[20] and aged humans *in vivo*.[21] In post-stroke brains, at 1–2 weeks after transient global ischemia, newly formed neuronal cells from NSCs could migrate into the granule cell area and promote functional recovery.[64] However, endogenous NSCs have only limited capacity for growth and differentiation into new functional neurons and spontaneous brain repair seems to be insufficient to recover from neurological deficits in most cases. The major obstacles include the following: (i) most of the newly proliferated NSCs are unable to form new neurons and integrate into the neurological network; and (ii) there is poor survival of the new neuroblasts 4 weeks after stroke.[65] To overcome those obstacles, pharmacological interventions that stimulate neurogenesis could be a promising strategy for recovery from neurological deficits. For example, combined treatment with basic fibroblast growth factor (bFGF) and epidermal growth factor (EGF) stimulated massive regeneration and triggered repair of hippocampal pyramidal neurons after stroke. Therefore, the development of new molecular targets is critical for drug discovery to improve endogenous neurogenesis.

Adult neurogenesis is a multistep process that requires the proliferation of NSCs, differentiation into the specific neuronal cell type, migration, survival and integration into existing neural networks.[66–68] All of these steps are important for neurogenesis and responsible for the success of regeneration therapy. Signaling pathways involved in the regulation of adult neurogenesis are very complex. The signal molecules come not only from the NSCs themselves but also from the microenvironment or neurogenic niche. The neurogenic niche plays a crucial role in determining where and how neurogenesis can occur. A recent review summarized the cellular signaling involved in modulating adult neurogenesis: Sonic hedgehog (Shh), miR-124, Sox2, Tlx and Wnt–β-catenin signaling pathways are important for regulating proliferation; basic helix–loop–helix (bHLH) transcription factors such as

Asc1, Neurog2 and Tbr2 and epigenetic factors such as Gadd45b, MBD1, MeCP2 and Mll1 are necessary for regulating neuronal differentiation and maturation; IGF-1 and Shh are essential for neuroblast migration; Extrinsic factors including BDNF, FGF-2, GABA, glutamate and NT-3 are necessary for regulating neuronal survival, dendritic arborization, synaptic plasticity and synapse formation; Intrinsic factors including DISC1, Klf-9, NeuroD1, Cdk and cAMP response element (CRE)-binding protein (CREB) participate in dendritic arborization, synaptic integration, neuronal survival and maturation.[69] With progress in stem cell biology, we believe that more and more signaling pathways and their networks will be found and these signaling molecules will be new molecular targets for adult neurogenesis. In the next section, we briefly introduce the roles of Wnt–β-catenin, Notch, phosphatidylinositol 3-kinase (PI3K)–Akt, extracellular signal-regulated kinase (ERK) and caveolin-1 signaling pathways in regulating adult neurogenesis. In addition, we discuss the roles of topical factors and free radicals in modulating adult neurogenesis.

11.3.1   Wnt–β-Catenin Pathway

β-Catenin is a central component of the Wnt–β-catenin signaling pathway. The interaction between Wnt protein and its receptors triggers recruitment of Axin to the plasma membrane and inhibits β-catenin phosphorylation and degradation. The accumulation of β-catenin in the cytoplasm leads β-catenin translocation into the nucleus to form a complex with the T-cell factor/ lymphoid enhancer factor (TCF/LEF) family, subsequently activating target genes. In the activation of the Wnt–β-catenin pathway, the first step is that Wnt interacts with the receptor protein Frizzled (Fzd) and the co-receptor low-density lipoprotein receptor-related protein 5/6 (LRP5/6). The Wnt–Fzd receptor complex forms a ternary cell surface complex with the co-receptor LRP5/6. After translocation into the cytoplasm, Wnt signaling mediates phosphorylation of Dishevelled, which, in turn, inactivates GSK-3β. Wnt signaling regulates GSK-3β activity by physically displacing complex GSK-3β from a member of regulatory binding partners, consequently preventing the phosphorylation and degradation of β-catenin.[70]

The Wnt–β-catenin pathway plays an important role in neural development. It participates in axis formation, midbrain development and oncogenesis.[71–73] Experimental evidence indicates that the Wnt–β-catenin pathway is associated with AD.[74,75] Removing β-catenin resulted in remarkable defects in developing brains whereas overexpressing β-catenin led to enlarged brains and an expanded neural precursor population.[76,77] Blockade of Wnt signaling by either a secreted mutant Wnt protein or a β-catenin inhibitor, Axin, reduced adult hippocampal stem/progenitor cell (AHPs) proliferation and neuronal differentiation *in vitro* and completely abolished neurogenesis.[78,79] Consistently, activation of Wnt signaling promoted AHPs proliferation and neuronal differentiation *in vitro* and increased neurogenesis *in vivo*.[78,79] GSK3β inhibitors were reported to enhance neuronal differentiation of NSCs and promote neurogenesis *in vivo*.[31] Therefore, the Wnt–β-catenin pathway can be a crucial target for screening drugs for neurogenesis.

11.3.2 Notch Signaling Pathways

Notch signaling pathways play critical roles in the maintenance, proliferation and differentiation of NSCs.[80] As cell surface transmembrane receptors, notch proteins mediate multiple cellular functions through cell–cell interactions. Four paralogs of the notch gene (Notch1–4) and five ligands (Jagged1 and 2; Delta1, 2 and 3) have been identified. Notch proteins contain extracellular EGF, which interacts with Delta, Serrate and Lag domains. In the process, Notch is cleaved and releases an intracellular domain (NICD). Upon release, the NICD translocates to the nucleus and represses or activates transcription factors.

Jagged1/Notch1 signaling can regulate stem cell self-renewal, proliferation and differentiation. It maintains NSC self-renewal, induces proliferation and modulates glial and neural fates by activating the downstream pathway NICD, hairy enhancer of Split (Hes1, Hes5) and so on.[81] Reduced Jagged1/Notch1 signaling was reported to decrease NSCs proliferation both *in vitro* and *in vivo* in experimental models.[82] Conversely, administration of Notch ligands activated the Jagged1/Notch1 pathway and increased the proliferation of NSCs in the SVZ.[83] Focal cerebral ischemia can activate Notch1 signaling and promote proliferation of NSCs in the SVZ of adult brains.[84] In addition, Notch1 signaling can mediate crosstalk of NSCs and endothelial cells in microenvironment niches. Endothelial cells can release soluble factors to stimulate the self-renewal of NSCs and inhibit their differentiation *via* activating Notch and Hes1.[85] Notch signaling modulates glial and neural fates in a stepwise manner, first by inhibiting neuronal fate and promoting glial fate and second by inducing astrocyte differentiation. The Notch-mediated inhibition of neuronal differentiation is achieved by inhibiting neighboring cells from becoming the same cell type. Taken together, these findings indicate that Notch signaling is important for adult neurogenesis.

A high-throughput screening study has been performed to seek Hes1 dimer inhibitors for neuronal differentiation.[86] Several natural compounds have been found to reduce Hes1 dimer formation, including lindbladione, lycogarubin B, lycogaric acid A, lyoniside and nudiposide.[86] It is necessary to investigate how these compounds regulate and reduce Hes1 dimer formation and evaluate further their effects on promoting neurogenesis in different experimental animal models of stroke and neurodegenerative diseases.

11.3.3 Phosphatidylinositol 3-Kinase (PI3K)–Akt and Extracellular Signal-Regulated Kinase (ERK) Pathways

PI3K–Akt and ERK signaling pathways regulate diverse cellular processes including proliferation, survival and malignant transformation. Akt and ERK

signaling promote proliferation, migration and maturation of neural progenitor cells via regulating their downstream targets, including HIF-1α,
GSK-3β and CREB. HIF-1α, a principal mediator of hypoxic adaptation,
modulates Wnt–β-catenin signaling by enhancing β-catenin activation and
expression of the downstream effectors LEF-1 and TCF-1 in hypoxic
embryonic stem (ES) cells and NSCs. HIF-1α deletion impairs hippocampal
Wnt-dependent processes, including NSC proliferation, differentiation and
neuronal maturation.[87] GSK-3β can function as signaling nodes to regulate
and orchestrate the diverse cellular responses involved in the process of
neurogenesis via affecting its downstream signaling components including
HIF-1α, HIF-2α and β-catenin. CREB is required for EGF-induced cell
proliferation and serum response element activation in NSCs isolated from
the forebrain subventricular zone of adult mice.[88] Experimental studies
revealed that inactivation of Akt–CREB by either pharmacological inhibition
or the expression of a dominant negative Akt inhibited AHPs proliferation.[89]
Vanadium, a stimulator of PI3K–Akt and ERK, increased NSC proliferation
in DG and promoted the migration of newborn neurons in ischemic brain.[90]
In addition, PI3K–Akt and ERK also regulate many mitogens, including
basic fibroblast growth factor (FGF-2), Shh and insulin-like growth factor 1
(IGF-1). Thus, PI3K–Akt and ERK signal pathways can be important
molecule targets for adult neurogenesis.

11.3.4   Trophic Factors

NSCs are in close proximity to blood vessels and are surrounded by glial cells
to form a microenvironmental niche. The proliferation and differentiation of
NSCs and neural growth depend on microenvironment niche signals, including
many trophic factors and cytokines. For example, EGF and FGF-2 increase
neurogenesis *in vitro* and *in vivo*.[91,92] Intraventricular administration of EGF
improved forelimb asymmetry with increased neuronal differentiation and cell
migration from SVZ to cortex in different stroke models.[93,94] Overexpression
of FGF-2 can improve the standardized disability scale with increased
neurogenesis in a permanent ischemic stroke model.[95] Another important
factor is vascular endothelial growth factor (VEGF), which improves recovery
from neurological deficits through enhancing neurogenesis, neuromigration
and angiogenesis in ischemic brain injury.[96–99] Brain-derived neurotrophic
factor (BDNF) promotes adult NSC survival *in vitro* and its overexpression
increases newly generated neurons in the olfactory bulb (OB).[100] IGF-1
promotes neuron and oligodendrocyte differentiation of adult NSCs *in
vitro*[101,102] and stimulates neurogenesis *in vivo*.[103] Transforming growth factor
alpha (TGF-α) infused into rat striatum led to migration of NSCs from SVZ to
the infusion sites.[104] Platelet-derived growth factor (PDGF) promotes
neuronal differentiation,[105] while ciliary neurotrophic factor (CNTF) increases
astroglial differentiation.[60] Drug discovery targeting these factors and their

downstream signaling pathways or directly regulating their release will be beneficial for promoting neurogenesis.

11.3.5   Free Radicals

Free radicals, including reactive oxygen species (ROS) and reactive nitrogen species (RNS), play important roles in the pathological process of stroke and neurodegenerative diseases. ROS are believed to be cytotoxic factors.[106–108] However, recent studies showed that low-concentration H_2O_2 promoted NSC self-renewal and neuronal differentiation. Genetic or pharmacological strategies to decrease endogenous ROS formation reduced cell proliferation in SVZ.[109,110] As many compounds from TCM have polyphenol structures to harbor antioxidant effects, which also possess pro-oxidant effects by generating ROS under some conditions, it will be of interest to explore potential concentration-related effects of these compounds in regulating adult neurogenesis.

RNS, including NO and ONOO⁻, also participate in modulating neurogenesis under both physiological and pathological conditions. NO appears to play dual roles in adult neurogenesis. A non-selective NO synthase (NOS) inhibitor, N^G-nitro-L-arginine methyl ester (L-NAME), and an nNOS inhibitor, 7-nitroindazole (7-NI), increased NSC proliferation and neuronal differentiation; while an NO donor blocked NSC proliferation and promoted astrocyte differentiation.[111,112] nNOS knockout mice showed more NSC proliferation and higher survival rates of newly formed mature neurons than wild-type mice.[113] Conversely, NO derived from iNOS or eNOS promoted neurogenesis in ischemic brain.[114,115] A recent study reported that nNOS derived from neurons and neural stem cells had opposite roles in regulating neurogenesis. The NSCs treated with the nNOS inhibitor N^5-(1-iminobut-3-enyl)-L-ornithine (L-VNIO) or nNOS gene deletion exhibited significantly decreased proliferation and neuronal differentiation, indicating that NSC-derived nNOS is essential for neurogenesis. The NSCs co-cultured with neurons displayed a significantly decreased proliferation. Deleting nNOS gene in neurons or scavenging extracellular NO abolished the effects of co-culture, suggesting that neuron-derived nNOS, a source of exogenous NO for NSCs, exerts a negative control on neurogenesis. The study proposed that NSC-derived nNOS stimulates neurogenesis *via* activating telomerase, whereas neuron-derived nNOS could repress neurogenesis by supplying exogenous NO that hinders CREB activation, in turn reducing nNOS expression in NSCs.[116] The neurogenesis-promoting or cytotoxic effects appear to be mainly dependent on NO concentration. The NO donor NOC-18 at 10 μM increased cell proliferation, whereas at 100 μM it inhibited cell proliferation.[117] In a neurovascular niche, NO-induced VEGF signaling mediates an inducible positive feedback signaling loop between NSCs and brain microvascular endothelial cells, providing for homeostasis of the resident NSCs and BMECs.[118]

In ischemic brain, when NO and O_2^- are simultaneously produced, they react together extremely rapidly, at a rate close to the diffusion limit, to form peroxynitrite ($k_2 = 4.7 \times 10^9$ M^{-1} s^{-1}). Peroxynitrite is far more active than its precursors.[119] The reaction rate of NO and O_2^- is much higher than in their reactions with other biomeolecules. Peroxynitrite is well known to be a cytotoxic factor in many pathological conditions including stroke and neurodegenerative diseases. However, recent studies indicated that peroxynitrite stimulated endothelial cell proliferation and tube formation and new blood vessel growth.[120] Peroxynitrite was also responsible for the hypoxia-induced increase in pulmonary artery smooth muscle cell proliferation.[121] The peroxynitrite donor 3-morpholinylsydnoneimine chloride (SIN-1) promoted NSC proliferation.[122] The mechanisms include $ONOO^-$ mimicking VEGF activity,[123] increasing VEGF mRNA[124] and activating ERK1/2 pathway.[125] However, how peroxynitrite switches its cell proliferation functions into cytotoxicity remains to be addressed. Given that many herbal products for promoting neurogenesis contain many antioxidant ingredients, it is important to understand how ROS and RNS affect adult neurogenesis and further investigate whether the antioxidant properties of the herbal products contribute to their modulating effects on adult neurogenesis.

11.3.6 Caveolin-1

Caveolins are 22 kDa proteins found in caveolae, plasma membrane invaginations (50–100 nm). Caveolins consist of three subtypes, caveolin-1 (Cav-1), caveolin-2 (Cav-2) and caveolin-3 (Cav-3). The subtypes of Cav-1 and Cav-2 are widely expressed in neuronal cell types and brain regions,[126–128] whereas cav-3 is muscle specific.[129] As integral membrane proteins, caveolins are the structural proteins of caveolae and are essential to caveolae structure formation. Caveolins physically interact with a large number of proteins *via* the caveolin scaffold domain (CSD).[130] The binding proteins contain the CAV-binding site, "φXφXXXXφ" or "φXXφXXXXφ," where φ is phenylalanine, tyrosine or tryptophan and X is any amino acid residue. Proteins with these character domains include eNOS, iNOS, nNOS, matrix metalloproteinases (MMPs), EGF receptor and aquaporin. Caveolins can also negatively regulate a variety of cellular signal pathways, such as G proteins, Src tyrosine kinases, ras, estrogen receptors, protein kinase C, integrins, MAP kinase and EGF-R.[131,132] Caveolins can bind and regulate these proteins and participate in multiple cellular activities, including mitogenic signaling,[133] apoptosis,[134,135] cholesterol transport,[136] cancer progression and metastasis,[137,138] and cardiovascular functions.[139]

Cav-1 plays diverse roles in proliferation and differentiation of various cell systems. Cav-1 inhibits cellular proliferation and differentiation in many cell types, including embryonic fibroblasts,[140] mesangial cells,[132] endothelial cells,[141] neuroblastoma cells[142] and cancer cells.[143] Cav-1 promotes liver regeneration by regulating lipid metabolism[144] and has a dual role in the proliferation of vascular smooth muscle cells.[145] Cav-1 participates in post-injury reactive neural plasticity. Cav-1 decreases neurite outgrowth and branching and neurite density in injured differentiated PC12 cells[142] and blocks the formation of neurites and phosphorylation of Erk upon bFGF treatment in N2a cells.[146] Cav-1 appears to be a critical protein in cell fate decision of neural progenitor cells for neurogenesis. Cav-1 deficiency mice showed increased NSC proliferation in the SVZ of adult mouse brain.[147] Our recent studies suggest that Cav-1 can inhibit neuronal differentiation by reducing expression of VEGF and phosphorylation of signal transducer and activator of transcription 3 (p-STAT3), Akt and MAPK in both post-ischemic brains *in vivo* and hypoxic NSCs *in vitro*.[148] In addition, Cav-1 can inhibit oligodendroglial differentiation through modulating β-catenin expression and promote astroglial differentiation of NPCs through regulating Notch1/NICD and Hes1 expression in NSCs.[149,150] Therefore, Cav-1 could be a new molecular target for drug discovery to regulate the proliferation and cell fate decision of NSCs.

11.4 Drug Discovery from TCM for Improving Neurogenesis

Experimental and clinical studies revealed that many compounds, herbal extracts and TCM formulas have potential therapeutic value for attenuating the progressive loss of functional neurons and restoring neurological function in stroke and neurodegenerative diseases. They could promote neuronal survival and neurite growth and facilitate neurological functional recovery by acting on different cellular signaling pathways and regulating diverse molecular targets. In this section, we particularly focus on research strategies and current progress on studies of the active compounds, herbal extracts and TCM formulas with potential for promoting neurogenesis and their relevant molecular targets and mechanisms.

11.4.1 Screening Active Compounds and Extracts from Herbs for Improving Neurogenesis with *In Vitro* Cultured Stem/ Progenitor Cell Models

The *in vitro* cultured stem/progenitor cell models provide a high-throughput platform for initially screening drug candidates. Embryonic stem cells, adult neural stem/progenitor cells from different species and immortalized NPCs have been applied for drug screening. There are two general strategies and criteria for the selection of drug candidates. One is based on previous experimental and clinical reports or the direct experience of TCM practitioners in the treatment of stroke, AD and PD in human subjects; the other is to screen the potential drug candidates by targeting particular signaling molecules with experimental or/and bioinformatics approaches. Drug screening based on the first strategy appears to be more efficient and widely accepted for studies of herbal medicine since multiple signaling molecules and targets are involved in the regulation of neurogenesis. In the literature, the identified active compounds and fractions are mostly from herbs used for stroke and neurodegenerative diseases with previous experimental and clinical trial reports. MLC901 is a simplified formula from a State Food and Drug Administration (SFDA)-registered botanical drug, MLC 601, originally developed from a TCM formula named "*Danqi Piantang Jiaonang*,"[151] which contains *Radix astragali*, *Radix salvia miltiorrhizae*, *Radix paeoniae rubra*, *Rhizoma chuanxiong*, *Radix angelicae sinensis*, *Carthamus tinctorius*, *Prunus persica*, *Radix polygalae* and *Rhizoma acori tatarinowii*. Previously double-blind randomized, controlled trials showed that *Danqi Piantang Jiaonang* could improve the recovery of neurological function in stroke patients. MLC901 was reported to increase the proliferationof human ESC-derived progenitors.[152] The active compounds contributing to those activities remain to be addressed.

Tables 11.1–11.3 summarize current progress on the identified compounds, fractions and herbal formulas for improving neurogenesis by various stem/progenitor cell models. For example, salvianolic acid B (Sal B), an active compound from *Salvia miltiorrhiza*, was initially screened from 45 medicinal herbs used for stroke, AD and PD. In an *in vitro* cellular experiment, Sal B was found to promote NPC self-renewal and proliferation of rat embryonic neural stem cells through modulating the PI3K–Akt signaling pathway.[153] Further studies on an ischemic stroke rat model confirmed its effects, showing that even delayed post-ischemic treatment with Sal B improved cognitive impairment in ischemic rat brain.[154] EGb761 is a standard extract of *Ginkgo biloba*, a medicinal herb for the treatment of stroke, AD and PD. Bilobalide is a unique ingredient of *Ginkgo biloba* whereas quercetin is a representative antioxidant component. A recent study revealed that bilobalide, quercetin and EGb 761 all promoted proliferation of rat hippocampal progenitor cells and embryonic brain cells. They share the same signaling pathway by mediating phosphorylation of CREB.[155,156] Animal experiments in AD models consistently revealed that EGb 761 increased cell proliferation and promoted new mature neuron formation in the DG and SVZ in TgAPP/PS1 mice via restoring CREB phosphorylation.[156] Baicalin is a flavonoid compound isolated from *Scutellaria baicalensis G*. Baicalin was previously reported to promote the differentiation of human umbilical cord blood mesenchymal stem cells and rat bone marrow stromal cells into neurons.[157,158] Recently, we found that baicalin down-regulated p-stat3 and Hes1 and upregulated NeuroD1 and Mash1 (Ascl) in cultured embryonic NPCs. Furthermore, baicalin promoted neural differentiation but inhibited glial formation, indicating that baicalin could regulate cell fate decision in embryonic NPCs.[159] Tenuigenin and *Polygala tenuifolia* root extract are isolated and extracted from *Polygala tenuifolia*, a medicinal herb used for improving insomnia and memory. They were reported to improve NSC proliferation and neuronal differentiation.[160,161] Similar cases also include ginsenosides Rg5[162] and *Panax notoginseng* saponins,[163] wolfberry polysaccharides[164] and epime-dium flavonoids.[165] Nevertheless, most of these studies are very primary.

Table 11.1 Single compounds with the effects of improving neurogenesis in cellular models.

Compound	Plant origin	Effect on neurogenesis	Cells	Potential pathway	Ref.
Salvianolic acid B	*Salvia miltiorrhiza*	Proliferation (BrdU+) ↑ MTS assay ↑	Rat embryonic neural stem/progenitor cells	PI3K–Akt ↑	153, 154
Baicalin	*Scutellaria baicalensis G*	Neurosphere formation ↑ Neuronal differentiation (MAP2+) ↑	Rat embryonic neural stem/progenitor cells	p-Stat3 ↓ Hes1 ↓, NeuroD1 ↑ Mash1 ↑	159
Curcumin	*Curcuma longa*	Glial formation (GFAP) ↓ Proliferation (BrdU+) ↑ MTT assay ↑	C17.2 neural progenitor cell line	pERK ↑ p-p38 ↑	166
Tetramethylpyrazine	*Ligusticum chuanxiong*	Proliferation (BrdU+) ↑ MTT assay ↑ Cyclin-D1 ↑ Neurosphere formation ↑ Neuronal differentiation (TUJ-1+) ↑ Glial formation (GFAP+) ↓	Hypoxic rat embryonic neural stem/progenitor cells	pERK ↑	167
Gentisides C–K Ginsenosides Rg5	*Gentiana rigescens* Franch. *Panax notoginseng*	Neurite outgrowth ↑ Neuronal differentiation (TUJ/MAP2+) ↑	PC12 Mouse embryonic neural stem cells	$[Ca^{2+}]$ ↑	176 162
Bilobalide and quercetin	*Ginkgo biloba*	Proliferation (BrdU+) ↑ Dendritic process (MAP2+) ↑ Restored amyloid-β oligomers	Rat hippocampal progenitor cells.	BDNF ↑ pCREB ↑	155
Tenuigenin	*Polygala tenuifolia*	Proliferation (BrdU+) ↑ Neurosphere formation ↑ Neuronal differentiation (TUJ-1+) ↑ glial formation (GFAP) ↑	Rat embryonic neural stem/progenitor cells		160

Table 11.2 Fractions of single herbs with the effect of promoting neurogenesis in cellular models.

Name	Plant origin	Effect on neurogenesis	Cells	Potential pathway	Ref.
Panax notoginseng saponins	*Panax notoginseng*	Proliferation (BrdU+) ↑ Neuronal differentiation (TUJ1+) ↑	Rat hippocampal neural stem cells		163
Standard *Ginkgo biloba* extract EGb 761	*Ginkgo biloba*	Proliferation (BrdU+) ↑	Embryonic brain cells	pCREB ↑	156
Polygala tenuifolia root extract	*Polygala tenuifolia*	Neuronal differentiation (Tuj-1) ↑	Neuronal precursor cells (HiB5)		161
Polysaccharides from wolfberry	*Lyceum barbarum*	Proliferation (BrdU+) ↑	C17.2 neural progenitor cell line		164
Epimedium flavonoids	*Epimedium grandiflorum*	Proliferation (BrdU+) ↑ Neurosphere volume ↑ Neuronal differentiation (NF-200+) ↑	Rat hippocampal neural stem cells		165

Table 11.3 TCM formulas with the effect of promoting neurogenesis in cellular models.

Formula	Plant origin	Effect on neurogenesis	Cells	Potential pathway	Ref.
MLC901	Radix astragali, Radix salvia miltiorrhizae, Radix paeoniae rubra, Rhizoma chuanxiong, Radix angelicae sinensis, Carthamus tinctorius, Prunus persica, Radix polygalae, Rhizoma acori tatarinowii	Proliferation (cell density) ↑	Human ESC-derived progenitors		152
Buyang Huanwu Decoction-containing serum	Radix astragali membranaceus, Radix angelicae ainensis, radix Paeonia rubra, Rhizoma chuanxiong, Semen persicae, Flos carthami, Lumbricus	Neurite outgrowth ↑ Neuronal differentiation (NF+)	Rat embryonic neural stem/progenitor cells		168

Several experiments addressed the potential molecular targets and signal pathways, such as PI3K–Akt,[153] Notch,[159] ERK,[166,167] and BDNF.[155] How those compounds regulate signaling and molecule targets and whether they directly bind these proteins or indirectly affect their activities and expression remain to be addressed. Bioinformatics and databases appear to be important tools for addressing those questions. After conducting initial screening of the active compounds in cellular models, it is a promising strategy to use bioinformatics approaches and database systems for understanding their molecular targets. More importantly, *in vivo* animal experiments are essential to confirm their bioactivities and evaluate the potential value of drug candidates.

11.4.2 Applications of Stroke, AD and PD Models for Identifying Active Compounds, Fractions and Formulas for Promoting Neurogenesis

In recent decades, efforts have been made to seek active compounds and fractions from herbs and herbal formulas for improving neurogenesis with different animal models of stroke, AD and PD. However, progress appears to be unsatisfactory. Although some compounds are promising for promoting neurogenesis in *in vitro* cellular experiments, they are not always effective in animal experiments owing to bioavailability, pharmacokinetic and pharmacodynamic problems. Few well-designed studies on the effects and mechanisms of isolated compounds, fractions and herbs or herbal formulas for promoting neurogenesis have been reported in the literature.

Tables 11.4–11.6 summarize the compounds, fractions and herbal formulas for promoting neurogenesis in animal models *in vivo*. For example, *Buyang Huanwu Decoction* (BYHWD) has been a classical formula for post-stroke disability for 300 years. Previous cellular experiments revealed that NSCs treated with BYHWD-containing serum had increased neuronal differentiation and neurite outgrowth.[168] To demonstrate the efficacy of BYHWD on improving neurogenesis and recovering neurological function, we investigated the effects of the whole formula and its herbal components on neurological behavior performance and infarction volume in focal cerebral ischemic rats. The neurological deficit scores and infarction volumes were measured at days 3, 7 and 14 after 30 min of middle cerebral artery occlusion. The results showed that BYHWD and its herbal components significantly improved neurological behavior performance and reduced infarction volume in the ischemic brains. By immunohistochemical staining with the thymidine analog 5-bromo-2′-deoxyuridine (BrdU), we found that the formula stimulated the proliferation of the progenitors in the hippocampus and SVZ in the ischemic brains. Its mechanisms were related to regulating VEGF and its receptor Flk-1 in ischemic brains.[40,169] We are working to identify the active compounds and the synergic effects of its active compounds for regulating adult neurogenesis and exploring their potential molecular targets. *Fuzhi San* is a herbal formula used for improving learning and memory in AD. A recent study revealed that increased proliferation of neural progenitor cells and prolonged survival of the newborn cells in the hippocampal DG of SAMP-8 aging mice improved learning and memory.[170] *Fuzi* polysaccharide-1, a water-soluble polysaccharide isolated from *Fuzi*, promoted NSC proliferation and differentiation into new neurons and had antidepressant effects through regulating BDNF signaling in aged mice.[171] Different herbal extracts such as the aqueous extract of *Liuwei Dihuang Tang* and *Nelumbo nucifera* rhizome and the methanol extract of *Zizyphus* also promoted proliferation and increased spatial learning ability.[172] Icariin, an active component extracted from *Herba epimedii*, enhanced neurogenesis and improved spatial learning and memory activities.[173] Overall, current progress indicates that Chinese herbal medicine is an important resource for drug discovery to promote adult neurogenesis and improve the recovery of neurological function in stroke and neurodegenerative diseases. Although the precise mechanisms and molecular targets of herbal medicines are still unclear, current studies are providing novel insight into drug discovery and representing a step forwards in fighting aging-related diseases.

Table 11.4 Single compounds with the effects of improving neurogenesis in stroke and neurodegenerative diseases models.

Compound	Plant origin	Model	Effect on neurogenesis	Effects on functional recovery	Potential pathway	Ref.
Curcumin	*Curcuma longa*	Mice	Proliferation (BrdU+) ↑ I n DG and SVZ			166
Icariin	*Herba epimedii*	Aging rats	Proliferation (BrdU+/GFAP+, BrdU+/PSA-NCAM+ and BrdU+/Olig2+) ↑ in DG	Improved Morris water maze		173
Quercetin	Various herbs	Rats/MCAO ischemic stroke model	Proliferation (BrdU+) ↑ in SVZ			177
Salvianolic acid B	*Salvia miltiorrhiza*	Rats/ischemic stroke model by electro-coagulation	Proliferation (BrdU+) ↑ in hippocampus	Improved Morris water maze		153

Table 11.5 Fractions of single herbs with the effects of improving neurogenesis in stroke and neurodegenerative diseases models.

Name	Plant origin	Model	Effect on neurogenesis	Effects on functional recovery	Potential pathway	Ref.
Fuzi polysaccharide-1	Aconitum carmichaeli Debx.	Mice	Proliferation (BrdU+) ↑ in DG; New mature neuron (NeuN+/BrdU+) ↑	Reduces immobility in the forced swim test and latency in the suppressed-feeding test	BDNF ↑	171
Zizyphus methanol extract	Zizyphus jujuba	Mice	Proliferation (Ki67+) ↑ in DG; Neuroblast (DCX+) ↑ in DG			178
Lemon balm extract	Melissa officinalis L.	Mice	Proliferation (Ki67+) ↑, (Brdu+) ↑ in DG; Neuroblast (DCX+) ↑ in DG		GABA level ↑	179
Platycodon grandiflorum aqueous extract	Platycodon grandiflorum	Mice	Proliferation (Ki67+) ↑, (BrdU+) ↑ in DG; Neuroblast (DCX+) ↑ in DG			180
Ginkgo biloba extract	Ginkgo biloba	Mice	Proliferation (Ki67+) ↑ in DG; Neuroblast (DCX+) ↑ in DG			181
Standard Ginkgo biloba extract EGb 761	Ginkgo biloba	TgAPP/PS1 mice	Proliferation (BrdU+) ↑ in DG; New mature neuron (NeuN+/BrdU+) ↑		Aβ ↓; pCREB ↑	156
Polygala tenuifolia root extract	Polygala tenuifolia	Rats	Proliferation (BrdU+) ↑, (Nestin+) ↑ in hippocampal CA1; Neuronal differentiation (Tuj-1+/BrdU+) ↑			161
Nelumbo nucifera rhizome extract	Nelumbo nucifera	Rats	Proliferation (Ki67+) ↑, (BrdU+) ↑ in DG; Neuroblast (DCX+) ↑ in DG	Increased latency of reaction in passive avoidance test		182
Cornel iridoid glycoside	Cornus officinalis	Rats/MCAO	Proliferation (BrdU+) ↑, (Nestin+) ↑ in SVZ; New mature neuron (NeuN+/BrdU+) ↑	Improved neurological severity score	VEGF ↑; Flk-1 ↑	183

Table 11.6 TCM formulas with the effect of improving neurogenesis in stroke and neurodegenerative disease models.

Formula	Plant origin	Model	Effect on neurogenesis	Effects on functional recovery	Potential pathway	Ref.
Kami-ondam-tang (Jiawei wen-dam-tang)	Pinelliae rhizoma, Bambusae caulis, Aurantii immaturus fructus, Poria, Citri reticulatae pericarpium, Ginseng radix, Glycyrrhizae radix, Polygalae radix, Jujubae fructus, Scrophulariae radix, Rehmanniae radix, Zizyphi spinosae semen, Zingiberis rhizoma	Mice	Neuroblast (DCX+) ↑ in hippocampus	Increased step-through latency in the passive avoidance task	pAkt ↑, BDNF ↑, pCREB ↑	184
Aqueous extract of Liuwei dihuang-tang	Rehmanniae radix, Dioscorae radix, Corni fructus, Alimatis rhizoma, Hoelen, Moutan cortex radicis, Maximowicziae fructus, Cervi cornu	Rats	Proliferation (BrdU+) ↑ in DG	Increased spatial learning ability by radial-arm maze test		172
Fuzhi San	Panax ginseng, Scutellaria baicalensis, Acorus talarinowi, Glycyrrhiza uralensis	SAMP-8 mice	Proliferation (BrdU+) ↑, (PCNA+) ↑ in DG and SVZ	Improved Morris water maze		170
MLC 601	Radix astragali, Radix salviae miltiorrhizae, Radix paeoniae rubra, Rhizoma chuanxiong, Radix angelicae sinensis, Carthamus tinctorius, Prunus persica, Radix polygalae, Rhizoma acori tatarinowii and 5 animal components (Hirudo, Eupolyphaga seu steleophaga, Calculus bovisartifactus, Buthus martensii, Cornu saigae tataricae)	Mice/MCAO ischemia stroke model	Proliferation (BrdU+) ↑ Neuronal differentiation (BrdU+/DCX+, BrdU+/NeuN+) ↑	Improved rotorod and locomotor activity	BDNF ↑	152

Table 11.6 (*Continued*)

Formula	Plant origin	Model	Effect on neurogenesis	Effects on functional recovery	Potential pathway	Ref.
MLC 901	*Radix astragali, Radix salvia miltiorrhizae, Radix paeoniae rubra, Rhizoma chuanxiong, Radix angelicae sinensis, Carthamus tinctorius, Prunus persica, Radix polygalae, Rhizoma acori tatarinowii*	Rat four-vessel occlusion model	Proliferation (BrdU+) ↑ in DG Neuronal differentiation (BrdU+/DCX+, BrdU+/NeuN+) ↑	Improved Morris water maze Improved grip strength test		185
Buyang Huanwu Decoction	Huangqi (*Radix astragalis seu hedysari*), Danggui (*Guiwei*) (*Radix angelicae sinensis*), Chishao (*Radix Paeoniae rubra*), Chuanxiong (*Rhizoma ligustici chuanxiong*), Honghua (*Flos carthami*), Taoren (*Semen persicae*), Dilong (*Pheretima*)	Rats/ ischemic stroke models	Proliferation (BrdU+) ↑ in DG and SVZ	Improved neurological function deficit scores	VEGF ↑ Flk-1 ↑	40,169

11.5 Conclusion and Perspectives

In this chapter, an attempt has been made to provide a comprehensive picture of classical therapeutic strategies of TCM, current studies on molecular targets for drug discovery and research progress on the active compounds, fractions and formulas for stroke and neurodegenerative diseases. Chinese herbal medicine represents a large chemical library for seeking single molecules that are being finally developed into new drugs for the treatment of stroke, AD and PD. Many compounds or extracts from Chinese herbal medicine have been studied for their effects on neurogenesis and neurological function with both *in vitro* and *in vivo* experimental systems. Current progress indicates that some active compounds, fractions and even herbal formulas hold potential for this purpose. Although the strategy may be promising and efficient, it is still premature for some herbal products to be developed into new drugs. Much effort needs to be devoted to evaluating the efficacy and safety of the herbal products, to explore the active components and clarify their mechanisms of action.

With the globally increasing popularity of and expenditure on herbal therapies recently, drug discovery from these natural extracts has attracted more attention. TCM herbs offer a great and unique source of both single compounds and complex combinations of compounds for drug screening. US and Chinese scientists have jointly established a library of 202 authenticated medicinal plant and fungal species and about 10 000 standard fractions from these materials.[174] Several natural products from plants have become frontline drugs for diseases, such as paclitaxel, vinblastine and camptothecin for cancer therapy.[175] We believe that similar cases will occur for drug discovery for the treatment of stroke and neurodegenerative diseases.

Finally, we should keep clearly in mind that adult neurogenesis involves multiple signal networks, and many of these are as yet unknown. The complex network for controlling neurogenesis makes drug development extremely difficult. How can simply targeting several signaling pathways and molecules achieve the goals of globally adjusting the abnormal network systems? TCM practitioners generally use herbal formulas as a holistic treatment by targeting multiple signaling pathways. Without considering the synergic effects of different compounds and fractions, we would shift the direction from the right track and make it difficult to achieve our goals. On the other hand, Chinese medicine aims to restore the harmony of the whole body rather than only targeting the brain. Further studies should be conducted to explain how Chinese medicine works to restore the balance of Yin and Yang on a scientific basis. With the development of proteomics, metabonomics and bioinformatics, we have powerful tools to open up this complex world. We believe the development of Chinese herbal medicine will be beneficial for the therapy of stroke and neurodegenerative diseases.

Abbreviations

Aβ	β-amyloid
ACEI	acetylcholinesterase inhibitor
AD	Alzheimer's disease
AHP	adult hippocampal stem/progenitor cell
BDNF	brain-derived neurotrophic factor
bFGF	basic fibroblast growth factor
BHLH	basic helix–loop–helix
BrdU	thymidine analog 5-bromo-2′-deoxyuridine
Cav	caveolin
CNS	central nervous system
CREB	cAMP response element (CRE)-binding protein
CNTF	ciliary neurotrophic factor
EGF	epidermal growth factor
ERK	extracellular signal-regulated kinase
ES	embryonic stem cells
DG	dentate gyrus
FGF-2	fibroblast growth factor 2
Fzd	Frizzled
HES	hairy enhancer of Split
PI3K	phosphatidylinositol 3-kinase
IGF-1	insulin-like growth factor 1
IsoRhy	isorhynchophylline
L-NAME	N^G-nitro-L-arginine methyl ester
L-VNIO	N^5-(1-iminobut-3-enyl)-L-ornithine
MMP	matrix metalloproteinase
MPTP	1-methyl-4-phenyl-1,2,3,6-tetrahydropyridine
6-OHDA	6-hydroxydopamine
7-NI	7-nitroindazole
NMDA	N-methyl-D-aspartate
NO	nitric oxide
NOS	nitric oxide synthase
NSC	neural stem/progenitor cell
OB	olfactory bulb
PD	Parkinson's disease
PDGF	platelet-derived growth factor
RCT	randomized controlled trial
ROS	reactive oxygen species
RNS	reactive nitrogen species
rt-PA	recombinant tissue-plasminogen activator
SGZ	subgranular zone
Shh	Sonic hedgehog
SIN-1	3-morpholinylsydnoneimine chloride
STAT3	signal transducer and activator of transcription 3
SVZ	subventricular zone

TCM   traditional Chinese medicine
TCF/LEF   T-cell factor/lymphoid enhancer factor
TGF-α   transforming growth factor alpha
TH   tyrosine hydroxylase
VEGF   vascular endothelial growth factor

Acknowledgements

This work was supported by the Hong Kong RGC General Research Fund (GRF No. 777610M, 777611M, 776512M) and the Seed Funding Programme for Basic Research of the University of Hong Kong (201111159021).

References

1. R. Bonita, S. Mendis, T. Truelsen, J. Bogousslavsky, J. Toole and F. Yatsu, *Lancet Neurol.*, 2004, **3**, 391.
2. M. Liu, B. Wu, W. Z. Wang, L. M. Lee, S. H. Zhang and L. Z. Kong, *Lancet Neurol.*, 2007, **6**, 456.
3. C. P. Ferri, M. Prince, C. Brayne, H. Brodaty, L. Fratiglioni, M. Ganguli, K. Hall, K. Hasegawa, H. Hendrie, Y. Huang, A. Jorm, C. Mathers, P. R. Menezes, E. Rimmer and M. Scazufca, *Lancet*, 2005, **366**, 2112.
4. C. Reitz, C. Brayne and R. Mayeux, *Nat. Rev. Neurol.*, 2011, **7**, 137.
5. H. Hampel, D. Prvulovic, S. Teipel, F. Jessen, C. Luckhaus, L. Frolich, M. W. Riepe, R. Dodel, T. Leyhe, L. Bertram, W. Hoffmann and F. Faltraco, *Prog. Neurobiol.*, 2011, **95**, 718.
6. M. J. Dong, B. Peng, X. T. Lin, J. Zhao, Y. R. Zhou and R. H. Wang, *Age Ageing*, 2007, **36**, 619.
7. K. Wirdefeldt, H. O. Adami, P. Cole, D. Trichopoulos and J. Mandel, *Eur. J. Epidemiol.*, 2011, **26**, Suppl 1, S1.
8. D. Hirtz, D. J. Thurman, K. Gwinn-Hardy, M. Mohamed, A. R. Chaudhuri and R. Zalutsky, *Neurology*, 2007, **68**, 326.
9. Z. X. Zhang, G. C. Roman, Z. Hong, C. B. Wu, Q. M. Qu, J. B. Huang, B. Zhou, Z. P. Geng, J. X. Wu, H. B. Wen, H. Zhao and G. E. Zahner, *Lancet*, 2005, **365**, 595.
10. V. L. Roger, A. S. Go, D. M. Lloyd-Jones, R. J. Adams, J. D. Berry, T. M. Brown, M. R. Carnethon, S. Dai, G. de Simone, E. S. Ford, C. S. Fox, H. J. Fullerton, C. Gillespie, K. J. Greenlund, S. M. Hailpern, J. A. Heit, P. M. Ho, V. J. Howard, B. M. Kissela, S. J. Kittner, D. T. Lackland, J. H. Lichtman, L. D. Lisabeth, D. M. Makuc, G. M. Marcus, A. Marelli, D. B. Matchar, M. M. McDermott, J. B. Meigs, C. S. Moy, D. Mozaffarian, M. E. Mussolino, G. Nichol, N. P. Paynter, W. D. Rosamond, P. D. Sorlie, R. S. Stafford, T. N. Turan, M. B. Turner, N. D. Wong and J. Wylie-Rosett, *Circulation*, 2011, **123**, e18.

11. A. Vlaar, A. Hovestadt, T. van Laar and B. R. Bloem, *Pract. Neurol.*, 2011, **11**, 145.

12. J. Kulisevsky and J. Pagonabarraga, *Drug Saf.*, 2010, **33**, 147.

13. C. G. Gross, *Nat. Rev. Neurosci.*, 2000, **1**, 67.

14. S. Temple, *Nature*, 2001, **414**, 112.

15. A. Alvarez-Buylla and J. M. Garcia-Verdugo, *J. Neurosci.*, 2002, **22**, 629.

16. H. A. Cameron and R. D. McKay, *J. Comp. Neurol.*, 2001, **435**, 406.

17. N. Picard-Riera, B. Nait-Oumesmar and A. Baron-Van Evercooren, *J. Neurosci. Res.*, 2004, **76**, 223.

18. Y. Tanaka, H. Kanno, M. Dezawa, T. Mimura, A. Kubo and I. Yamamoto, *Neurosci. Lett.*, 2005, **383**, 28.

19. J. M. Plane, R. Liu, T. W. Wang, F. S. Silverstein and J. M. Parent, *Neurobiol. Dis.*, 2004, **16**, 585.

20. P. Thored, J. Wood, A. Arvidsson, J. Cammenga, Z. Kokaia and O. Lindvall, *Stroke*, 2007, **38**, 3032.

21. J. Macas, C. Nern, K. H. Plate and S. Momma, *J. Neurosci.*, 2006, **26**, 13114.

22. V. Bonnamain, I. Neveu and P. Naveilhan, *Front. Cell. Neurosci.*, 2012, **6**, 17.

23. C. Zhao, W. Deng and F. H. Gage, *Cell*, 2008, **132**, 645.

24. C. V. Borlongan, *Exp. Neurol.*, 2012, **237**, 142.

25. L. M. Bjorklund, R. Sanchez-Pernaute, S. Chung, T. Andersson, I. Y. Chen, K. S. McNaught, A. L. Brownell, B. G. Jenkins, C. Wahlestedt, K. S. Kim and O. Isacson, *Proc. Natl. Acad. Sci. U. S. A.*, 2002, **99**, 2344.

26. M. Dezawa, H. Kanno, M. Hoshino, H. Cho, N. Matsumoto, Y. Itokazu, N. Tajima, H. Yamada, H. Sawada, H. Ishikawa, T. Mimura, M. Kitada, Y. Suzuki and C. Ide, *J. Clin. Invest.*, 2004, **113**, 1701.

27. K. Jin, L. Xie, X. O. Mao and D. A. Greenberg, *Brain Res.*, 2006, **1085**, 183.

28. Y. Liu, R. Lacson, J. Cassaday, D. A. Ross, A. Kreamer, E. Hudak, R. Peltier, D. McLaren, I. Munoz-Sanjuan, F. Santini, B. Strulovici and M. Ferrer, *J. Biomol. Screen.*, 2009, **14**, 319.

29. H. J. Kim and C. Y. Jin, *Korean J. Physiol. Pharmacol.*, 2012, **16**, 1.

30. M. A. Arai, *Chem. Pharm. Bull.*, 2011, **59**, 417.

31. K. J. Kim, J. Wang, X. Xu, S. Wu, W. Zhang, Z. Qin, F. Wu, A. Liu, Y. Zhao, H. Fang, M. Zhu, J. Zhao and Z. Zhong, *J. Biomol. Screen.*, 2012, **17**, 129.

32. X. Wang, *Internal Medicine of Traditional Chinese Medicine*, Higher Education Press, Beijing, 2007.

33. X. Gong and N. J. Sucher, *Phytomedicine*, 2002, **9**, 478.

34. C. Y. Zhang, G. Y. Du, W. Wang, Z. G. Ye, D. Q. Wang, X. F. Sun and D. Z. Zhao, *Zhongguo Zhong Yao Za Zhi*, 2004, **29**, 1061.

35. S. C. Ho, Y. F. Ho, T. H. Lai, T. H. Liu, S. Y. Su and R. Y. Wu, *Am. J. Chin. Med.*, 2008, **36**, 593.

36. Y. H. Zhao, Y. D. Liu, Y. Guan and N. W. Liu, *J. Tradit. Chin. Med.*, 2010, **30**, 171.
37. H. W. Zhang, J. Tong, G. Zhou, H. Jia and J. Y. Jiang, *Cochrane Database Syst. Rev.*, 2012, (6), CD008166.
38. G. X. Cai and B. Y. Liu, *Zhongguo Wei Zhong Bing Ji Jiu Yi Xue*, 2010, **22**, 591.
39. X. M. Li, X. C. Bai, L. N. Qin, H. Huang, Z. J. Xiao and T. M. Gao, *Neurosci. Lett.*, 2003, **346**, 29.
40. G. Cai, B. Liu, W. Liu, X. Tan, J. Rong, X. Chen, L. Tong and J. Shen, *J. Ethnopharmacol.*, 2007, **113**, 292.
41. Fu, R. J., *J. Tradit. Chin. Med.*, 1991, **2**, 56.
42. K. Iwasaki, T. Satoh-Nakagawa, M. Maruyama, Y. Monma, M. Nemoto, N. Tomita, H. Tanji, H. Fujiwara, T. Seki, M. Fujii, H. Arai and H. Sasaki, *J. Clin. Psychiatry*, 2005, **66**, 248.
43. K. Iwasaki, M. Maruyama, N. Tomita, K. Furukawa, M. Nemoto, H. Fujiwara, T. Seki, M. Fujii, M. Kodama and H. Arai, *J. Clin. Psychiatry*, 2005, **66**, 1612.
44. H. Shinno, Y. Inami, T. Inagaki, Y. Nakamura and J. Horiguchi, *Prog. Neuropsychopharmacol. Biol. Psychiatry*, 2008, **32**, 881.
45. Y. S. Ho, K. F. So and R. C. Chang, *Chin. Med.*, 2011, **6**, 15.
46. Z. Y. Hu, G. Liu, H. Yuan, S. Yang, W. X. Zhou, Y. X. Zhang and S. Y. Qiao, *J. Ethnopharmacol.*, 2010, **128**, 365.
47. Z. Y. Hu, G. Liu, X. R. Cheng, Y. Huang, S. Yang, S. Y. Qiao, L. Sun, W. X. Zhou and Y. X. Zhang, *Exp. Gerontol.*, 2012, **47**, 14.
48. S. Jeon, S. Bose, J. Hur, K. Jun, Y. K. Kim, K. S. Cho and B. S. Koo, *J. Ethnopharmacol.*, 2011, **137**, 783.
49. Y. Q. Liang and X. C. Tang, *Acta Pharmacol. Sin.*, 2006, **27**, 1127.
50. H. Y. Zhang, H. Yan and X. C. Tang, *Neurosci. Lett.*, 2004, **360**, 21.
51. B. S. Wang, H. Wang, Z. H. Wei, Y. Y. Song, L. Zhang and H. Z. Chen, *J. Neural Transm.*, 2009, **116**, 457.
52. R. Wang, H. Yan and X. C. Tang, *Acta Pharmacol. Sin.*, 2006, **27**, 1.
53. Z. Lin, J. Gu, J. Xiu, T. Mi, J. Dong and J. K. Tiwari, *Evid. Based Comp. Alt. Med.*, 2012, **2012**, 692621.
54. Z. X. Zhang, Z. H. Dong and G. C. Roman, *Arch. Neurol.*, 2006, **63**, 782.
55. G. Q. Zheng, *J. Alt. Comp. Med.*, 2009, **15**, 1223.
56. Q. Li, D. Zhao and E. Bezard, *Behav. Pharmacol.*, 2006, **17**, 403.
57. M. H. Yang, M. Li, Y. Q. Dou, Y. Liu, X. D. Luo, J. Z. Chen and H. J. Shi, *Zhong Xi Yi Jie He Xue Bao*, 2010, **8**, 231.
58. M. Li, M. H. Yang and Y. Liu, *Zhong Xi Yi Jie He Xue Bao*, 2012, **10**, 310.
59. M. H. Yang, H. M. Wang and Y. Liu, *Chin. J. Integr. Med.*, 2011, **17**, 43.
60. L. W. Chen, Y. Q. Wang, L. C. Wei, M. Shi and Y. S. Chan, *CNS Neurol. Disord. Drug Targets*, 2007, **6**, 273.
61. J. H. Lu, J. Q. Tan, S. S. Durairajan, L. F. Liu, Z. H. Zhang, L. Ma, H. M. Shen, H. Y. Chan and M. Li, *Autophagy*, 2012, **8**, 98.

62. J. X. Song, S. C. Sze, T. B. Ng, C. K. Lee, G. P. Leung, P. C. Shaw, Y. Tong and Y. B. Zhang, *J. Ethnopharmacol.*, 2012, **139**, 698.

63. V. Chung, L. Liu, Z. Bian, Z. Zhao, W. Leuk Fong, W. F. Kum, J. Gao and M. Li, *Mov. Disord.*, 2006, **21**, 1709.

64. J. Liu, K. Solway, R. O. Messing and F. R. Sharp, *J. Neurosci.*, 1998, **18**, 7768.

65. O. Lindvall and Z. Kokaia, in *Adult Neurogenesis*, ed. K. G. Gage and F. H. Song, Cold Spring Harbor Laboratory Press, Cold Spring Harbor, NY, 2008, 549.

66. A. Alvarez-Buylla, J. M. Garcia-Verdugo and A. D. Tramontin, *Nat. Rev. Neurosci.*, 2001, **2**, 287.

67. P. Rakic, *Prog. Brain Res.*, 2002, **138**, 3.

68. P. Taupin and F. H. Gage, *J. Neurosci. Res.*, 2002, **69**, 745.

69. Y. Mu, S. W. Lee and F. H. Gage, *Curr. Opin. Neurobiol.*, 2010, **20**, 416.

70. L. Zhang, X. Yang, S. Yang and J. Zhang, *Eur. J. Neurosci.*, 2011, **33**, 1.

71. Q. Tao, C. Yokota, H. Puck, M. Kofron, B. Birsoy, D. Yan, M. Asashima, C. C. Wylie, X. Lin and J. Heasman, *Cell*, 2005, **120**, 857.

72. M. Kunz, M. Herrmann, D. Wedlich and D. Gradl, *Dev. Biol.*, 2004, **273**, 390.

73. P. Polakis, *Curr. Opin. Genet. Dev.*, 2007, **17**, 45.

74. G. V. De Ferrari and N. C. Inestrosa, *Brain Res. Brain Res. Rev.*, 2000, **33**, 1.

75. G. V. De Ferrari, A. Papassotiropoulos, T. Biechele, F. Wavrant De-Vrieze, M. E. Avila, M. B. Major, A. Myers, K. Saez, J. P. Henriquez, A. Zhao, M. A. Wollmer, R. M. Nitsch, C. Hock, C. M. Morris, J. Hardy and R. T. Moon, *Proc. Natl. Acad. Sci. U. S. A.*, 2007, **104**, 9434.

76. V. Brault, R. Moore, S. Kutsch, M. Ishibashi, D. H. Rowitch, A. P. McMahon, L. Sommer, O. Boussadia and R. Kemler, *Development*, 2001, **128**, 1253.

77. A. Chenn and C. A. Walsh, *Science*, 2002, **297**, 365.

78. D. C. Lie, S. A. Colamarino, H. J. Song, L. Desire, H. Mira, A. Consiglio, E. S. Lein, S. Jessberger, H. Lansford, A. R. Dearie and F. H. Gage, *Nature*, 2005, **437**, 1370.

79. Q. Qu, G. Sun, W. Li, S. Yang, P. Ye, C. Zhao, R. T. Yu, F. H. Gage, R. M. Evans and Y. Shi, *Nat. Cell Biol.*, 2010, **12**, 31.

80. J. D. Lathia, M. P. Mattson and A. Cheng, *J. Neurochem.*, 2008, **107**, 1471.

81. N. Gaiano and G. Fishell, *Annu. Rev. Neurosci.*, 2002, **25**, 471.

82. Y. Nyfeler, R. D. Kirch, N. Mantei, D. P. Leone, F. Radtke, U. Suter and V. Taylor, *EMBO J.*, 2005, **24**, 3504.

83. A. Androutsellis-Theotokis, R. R. Leker, F. Soldner, D. J. Hoeppner, R. Ravin, S. W. Poser, M. A. Rueger, S. K. Bae, R. Kittappa and R. D. McKay, *Nature*, 2006, **442**, 823.

84. X. Wang, X. Mao, L. Xie, D. A. Greenberg and K. Jin, *J. Cereb. Blood Flow Metab.*, 2009, **29**, 1644.

85. Q. Shen, S. K. Goderie, L. Jin, N. Karanth, Y. Sun, N. Abramova, P. Vincent, K. Pumiglia and S. Temple, *Science*, 2004, **304**, 1338.
86. M. A. Arai, A. Masada, T. Ohtsuka, R. Kageyama and M. Ishibashi, *Bioorg. Med. Chem. Lett.*, 2009, **19**, 5778.
87. J. Mazumdar, W. T. O'Brien, R. S. Johnson, J. C. LaManna, J. C. Chavez, P. S. Klein and M. C. Simon, *Nat. Cell Biol.*, 2010, **12**, 1007.
88. H. Iguchi, T. Mitsui, M. Ishida, S. Kanba and J. Arita, *Endocr. J.*, 2011, **58**, 747.
89. J. Peltier, A. O'Neill and D. V. Schaffer, *Dev. Neurobiol.*, 2007, **67**, 1348.
90. N. Shioda, F. Han and K. Fukunaga, *Int. Rev. Neurobiol.*, 2009, **85**, 375.
91. C. G. Craig, V. Tropepe, C. M. Morshead, B. A. Reynolds, S. Weiss and D. van der Kooy, *J. Neurosci.*, 1996, **16**, 2649.
92. J. P. Wagner, I. B. Black and E. DiCicco-Bloom, *J. Neurosci.*, 1999, **19**, 6006.
93. B. Kolb, C. Morshead, C. Gonzalez, M. Kim, C. Gregg, T. Shingo and S. Weiss, *J. Cereb. Blood Flow Metab.*, 2007, **27**, 983.
94. T. Teramoto, J. Qiu, J. C. Plumier and M. A. Moskowitz, *J. Clin. Invest.*, 2003, **111**, 1125.
95. R. R. Leker, F. Soldner, I. Velasco, D. K. Gavin, A. Androutsellis-Theotokis and R. D. McKay, *Stroke*, 2007, **38**, 153.
96. A. Schanzer, F. P. Wachs, D. Wilhelm, T. Acker, C. Cooper-Kuhn, H. Beck, J. Winkler, L. Aigner, K. H. Plate and H. G. Kuhn, *Brain Pathol.*, 2004, **14**, 237.
97. K. Jin, Y. Zhu, Y. Sun, X. O. Mao, L. Xie and D. A. Greenberg, *Proc. Natl. Acad. Sci. U. S. A.*, 2002, **99**, 11946.
98. L. Cao, X. Jiao, D. S. Zuzga, Y. Liu, D. M. Fong, D. Young and M. J. During, *Nat. Genet.*, 2004, **36**, 827.
99. Y. Wang, K. Jin, X. O. Mao, L. Xie, S. Banwait, H. H. Marti and D. A. Greenberg, *J. Neurosci. Res.*, 2007, **85**, 740.
100. B. Kirschenbaum and S. A. Goldman, *Proc. Natl. Acad. Sci. U. S. A.*, 1995, **92**, 210.
101. G. J. Brooker, M. Kalloniatis, V. C. Russo, M. Murphy, G. A. Werther and P. F. Bartlett, *J. Neurosci. Res.*, 2000, **59**, 332.
102. J. Hsieh, J. B. Aimone, B. K. Kaspar, T. Kuwabara, K. Nakashima and F. H. Gage, *J. Cell Biol.*, 2004, **164**, 111.
103. M. A. Aberg, N. D. Aberg, H. Hedbacker, J. Oscarsson and P. S. Eriksson, *J. Neurosci.*, 2000, **20**, 2896.
104. H. Wichterle, J. M. Garcia-Verdugo and A. Alvarez-Buylla, *Neuron*, 1997, **18**, 779.
105. P. Mohapel, H. Frielingsdorf, J. Haggblad, O. Zachrisson and P. Brundin, *Neuroscience*, 2005, **132**, 767.
106. R. H. Fabian, D. S. DeWitt and T. A. Kent, *J. Cereb. Blood Flow Metab.*, 1995, **15**, 242.
107. O. Peters, T. Back, U. Lindauer, C. Busch, D. Megow, J. Dreier and U. Dirnagl, *J. Cereb. Blood Flow Metab.*, 1998, **18**, 196.

108. G. W. Kim, T. Kondo, N. Noshita and P. H. Chan, Stroke, 2002, **33**, 809.
109. J. E. Le Belle, N. M. Orozco, A. A. Paucar, J. P. Saxe, J. Mottahedeh, A. D. Pyle, H. Wu and H. I. Kornblum, *Cell Stem Cell*, 2011, **8**, 59.
110. B. C. Dickinson, J. Peltier, D. Stone, D. V. Schaffer and C. J. Chang, *Nat. Chem. Biol.*, 2011, **7**, 106.
111. M. A. Packer, Y. Stasiv, A. Benraiss, E. Chmielnicki, A. Grinberg, H. Westphal, S. A. Goldman and G. Enikolopov, *Proc. Natl. Acad. Sci. U. S. A.*, 2003, **100**, 9566.
112. B. Moreno-Lopez, C. Romero-Grimaldi, J. A. Noval, M. Murillo-Carretero, E. R. Matarredona and C. Estrada, *J. Neurosci.*, 2004, **24**, 85.
113. S. Fritzen, A. Schmitt, K. Koth, C. Sommer, K. P. Lesch and A. Reif, *Mol. Cell. Neurosci.*, 2007, **35**, 261.
114. D. Y. Zhu, S. H. Liu, H. S. Sun and Y. M. Lu, *J. Neurosci.*, 2003, **23**, 223.
115. R. Zhang, L. Zhang, Z. Zhang, Y. Wang, M. Lu, M. Lapointe and M. Chopp, *Ann. Neurol.*, 2001, **50**, 602.
116. C. X. Luo, X. Jin, C. C. Cao, M. M. Zhu, B. Wang, L. Chang, Q. G. Zhou, H. Y. Wu and D. Y. Zhu, *Stem Cells*, 2010, **28**, 2041.
117. B. P. Carreira, M. I. Morte, A. Inacio, G. Costa, J. Rosmaninho-Salgado, F. Agasse, A. Carmo, P. Couceiro, P. Brundin, A. F. Ambrosio, C. M. Carvalho and I. M. Araujo, *Stem Cells*, 2010, **28**, 1219.
118. Q. Li, M. C. Ford, E. B. Lavik and J. A. Madri, *J. Neurosci. Res.*, 2006, **84**, 1656.
119. M. P. Murphy, M. A. Packer, J. L. Scarlett and S. W. Martin, *Gen. Pharmacol.*, 1998, **31**, 179.
120. R. Rietze, P. Poulin and S. Weiss, *J. Comp. Neurol.*, 2000, **424**, 397.
121. E. O. Agbani, P. Coats and R. M. Wadsworth, *J. Cardiovasc. Pharmacol.*, 2011, **57**, 584.
122. M. Yoneyama, K. Kawada, Y. Gotoh, T. Shiba and K. Ogita, *Neurochem. Int.*, 2010, **56**, 740.
123. D. H. Platt, M. Bartoli, A. B. El-Remessy, M. Al-Shabrawey, T. Lemtalsi, D. Fulton and R. B. Caldwell, *Free Rad. Biol. Med.*, 2005, **39**, 1353.
124. R. K. Upmacis, R. S. Deeb, M. J. Resnick, R. Lindenbaum, C. Gamss, D. Mittar and D. P. Hajjar, *Am. J. Physiol. Cell Physiol.*, 2004, **286**, C1271.
125. C. Zouki, S. L. Zhang, J. S. Chan and J. G. Filep, *FASEB J.*, 2001, **15**, 25.
126. T. Ikezu, H. Ueda, B. D. Trapp, K. Nishiyama, J. F. Sha, D. Volonte, F. Galbiati, A. L. Byrd, G. Bassell, H. Serizawa, W. S. Lane, M. P. Lisanti and T. Okamoto, *Brain Res.*, 1998, **804**, 177.
127. D. D. Mikol, S. S. Scherer, S. J. Duckett, H. L. Hong and E. L. Feldman, *Glia*, 2002, **38**, 191.
128. D. N. Arvanitis, H. Wang, R. D. Bagshaw, J. W. Callahan and J. M. Boggs, *J. Neurosci. Res.*, 2004, **75**, 603.

129. J. G. Shen, W. S. Lee, J. P. Chen and D. P. Yang, *Am. J. Biomed. Sci.*, 2011, **3**, 126.

130. K. G. Rothberg, J. E. Heuser, W. C. Donzell, Y. S. Ying, J. R. Glenney and R. G. Anderson, *Cell*, 1992, **68**, 673.

131. C. Schwencke, R. C. Braun-Dullaeus, C. Wunderlich and R. H. Strasser, *Cardiovasc. Res.*, 2006, **70**, 42.

132. Y. Fujita, S. Maruyama, H. Kogo, S. Matsuo and T. Fujimoto, *Kidney Int.*, 2004, **66**, 1794.

133. J. Thyberg, *Arterioscler. Thromb. Vasc. Biol.*, 2003, **23**, 1481.

134. T. E. Peterson, M. E. Guicciardi, R. Gulati, L. S. Kleppe, C. S. Mueske, M. Mookadam, G. Sowa, G. J. Gores, W. C. Sessa and R. D. Simari, *Arterioscler. Thromb. Vasc. Biol.*, 2003, **23**, 1521.

135. P. Gargalovic and L. Dory, *J. Lipid Res.*, 2003, **44**, 1622.

136. C. J. Fielding and P. E. Fielding, *Adv. Drug Deliv. Rev.*, 2001, **49**, 251.

137. E. Burgermeister, M. Liscovitch, C. Rocken, R. M. Schmid and M. P. Ebert, *Cancer Lett.*, 2008, **268**, 187.

138. J. G. Goetz, P. Lajoie, S. M. Wiseman and I. R. Nabi, *Cancer Metastasis Rev.*, 2008, **27**, 715.

139. P. G. Frank, G. S. Hassan, J. A. Rodriguez-Feo and M. P. Lisanti, *Curr. Pharm. Des.*, 2007, **13**, 1761.

140. B. Razani, J. A. Engelman, X. B. Wang, W. Schubert, X. L. Zhang, C. B. Marks, F. Macaluso, R. G. Russell, M. Li, R. G. Pestell, D. Di Vizio, H. Hou Jr, B. Kneitz, G. Lagaud, G. J. Christ, W. Edelmann and M. P. Lisanti, *J. Biol. Chem.*, 2001, **276**, 38121.

141. K. Fang, W. Fu, A. R. Beardsley, X. Sun, M. P. Lisanti and J. Liu, *Cell Cycle*, 2007, **6**, 199.

142. S. B. Gaudreault, J. F. Blain, J. P. Gratton and J. Poirier, *J. Neurochem.*, 2005, **92**, 831.

143. M. Shatz and M. Liscovitch, *Int. J. Radiat. Biol.*, 2008, **84**, 177.

144. M. A. Fernandez, C. Albor, M. Ingelmo-Torres, S. J. Nixon, C. Ferguson, T. Kurzchalia, F. Tebar, C. Enrich, R. G. Parton and A. Pol, *Science*, 2006, **313**, 1628.

145. D. G. Sedding and R. C. Braun-Dullaeus, *Trends Cardiovasc. Med.*, 2006, **16**, 50.

146. M. J. Kang, J. S. Seo and W. Y. Park, *Neuroreport*, 2006, **17**, 823.

147. J. F. Jasmin, M. Yang, L. Iacovitti and M. P. Lisanti, *Cell Cycle*, 2009, **8**, 3978.

148. Y. Li, J. Luo, W. M. Lau, G. Zheng, S. Fu, T. T. Wang, H. P. Zeng, K. F. So, S. K. Chung, Y. Tong, K. Liu and J. Shen, *PLoS One*, 2011, **6**, e22901.

149. Y. Li, W. M. Lau, K. F. So, Y. Tong and J. Shen, *Neurochem. Int.*, 2011, **59**, 114.

150. Y. Li, W. M. Lau, K. F. So, Y. Tong and J. Shen, *Biochem. Biophys. Res. Commun.*, 2011, **407**, 517.

151. C. Chen, N. Venketasubramanian, R. N. Gan, C. Lambert, D. Picard, B. P. Chan, E. Chan, M. G. Bousser and S. Xuemin, *Stroke*, 2009, **40**, 859.

152. C. Heurteaux, C. Gandin, M. Borsotto, C. Widmann, F. Brau, M. Lhuillier, B. Onteniente and M. Lazdunski, *Neuropharmacology*, 2010, **58**, 987.

153. P. Zhuang, Y. Zhang, G. Cui, Y. Bian, M. Zhang, J. Zhang, Y. Liu, X. Yang, A. O. Isaiah, Y. Lin and Y. Jiang, *PLoS One*, 2012, **7**, e35636.

154. G. Guo, B. Li, Y. Wang, A. Shan, W. Shen, L. Yuan and S. Zhong, *Sci. China Life Sci.*, 2010, **53**, 653.

155. F. Tchantchou, P. N. Lacor, Z. Cao, L. Lao, Y. Hou, C. Cui, W. L. Klein and Y. Luo, *J. Alzheimers Dis.*, 2009, **18**, 787.

156. F. Tchantchou, Y. Xu, Y. Wu, Y. Christen and Y. Luo, *FASEB J.*, 2007, **21**, 2400.

157. Y. Jia, Y. Yang, Y. Zhou, Y. Song, L. Liu, J. Song, X. Wang, L. Zhong and X. Yu, *Zhonghua Yi Xue Za Zhi*, 2002, **82**, 1337.

158. X. H. Yan and R. B. Huang, *Zhonghua Er Ke Za Zhi*, 2006, **44**, 214.

159. Y. Li, P. Zhuang, B. Shen, Y. Zhang and J. Shen, *Brain Res.*, 2012, **1429**, 36.

160. Y. Chen, X. Huang, W. Chen, N. Wang and L. Li, *Neurochem. Res.*, 2012, **37**, 771.

161. H. J. Park, K. Lee, H. Heo, M. Lee, J. W. Kim, W. W. Whang, Y. K. Kwon and H. Kwon, *Phytother. Res.*, 2008, **22**, 1324.

162. J. W. Liu, S. J. Tian, J. de Barry and B. Luu, *J. Nat. Prod.*, 2007, **70**, 1329.

163. Y. C. Si, J. P. Zhang, C. E. Xie, L. J. Zhang and X. N. Jiang, *Am. J. Chin. Med.*, 2011, **39**, 999.

164. B. W. Lau, J. C. Lee, Y. Li, S. M. Fung, Y. H. Sang, J. Shen, R. C. Chang and K. F. So, *PLoS One*, 2012, **7**, e33374.

165. R. Yao, L. Zhang, X. Li and L. Li, *Neurol. Res.*, 2010, **32**, 736.

166. S. J. Kim, T. G. Son, H. R. Park, M. Park, M. S. Kim, H. S. Kim, H. Y. Chung, M. P. Mattson and J. Lee, *J. Biol. Chem.*, 2008, **283**, 14497.

167. Y. Tian, Y. Liu, X. Chen, H. Zhang, Q. Shi, J. Zhang and P. Yang, *Neurosci. Lett.*, 2010, **474**, 26.

168. J. Sun, Y. Bi, L. Guo, X. Qi, J. Zhang, G. Li, G. Tian, F. Ren and Z. Li, *J. Ethnopharmacol.*, 2007, **113**, 199.

169. L. Tong, X. H. Tan and J. G. Shen, *Zhongguo Zhong Xi Yi Jie He Za Zhi*, 2007, **27**, 519.

170. H. Yang, S. R. Wen, G. W. Zhang, T. G. Wang, F. X. Hu, X. L. Li and D. S. Wang, *Exp. Gerontol.*, 2011, **46**, 628.

171. H. C. Yan, H. D. Qu, L. R. Sun, S. J. Li, X. Cao, Y. Y. Fang, W. Jie, J. C. Bean, W. K. Wu, X. H. Zhu and T. M. Gao, *Int. J. Neuropsychopharmacol.*, 2010, **13**, 623.

172. K. S. Lee, B. V. Lim, H. K. Chang, H. Y. Yang, G. H. Bahn, E. K. Paik, J. H. Park and C. J. Kim, *Fitoterapia*, 2005, **76**, 514.

173. B. Wu, Y. Chen, J. Huang, Y. Ning, Q. Bian, Y. Shan, W. Cai, X. Zhang and Z. Shen, *J. Ethnopharmacol.*, 2012, **142**, 746.

174. D. M. Eisenberg, E. S. Harris, B. A. Littlefield, S. Cao, J. A. Craycroft, R. Scholten, P. Bayliss, Y. Fu, W. Wang, Y. Qiao, Z. Zhao, H. Chen, Y. Liu, T. Kaptchuk, W. C. Hahn, X. Wang, T. Roberts, C. E. Shamu and J. Clardy, *Fitoterapia*, 2011, **82**, 17.

175. G. M. Cragg, M. R. Boyd, J. H. Cardellina II, D. J. Newman, K. M. Snader and T. G. McCloud, *Ciba Found. Symp.*, 1994, **185**, 178; discussion 190.

176. L. Gao, L. Xiang, Y. Luo, G. Wang, J. Li and J. Qi, *Bioorg. Med. Chem.*, 2010, **18**, 6995.

177. L. L. Zhang, Q. Cao, Z. Y. Hu, X. H. Yan and B. Y. Wu, *Nan Fang Yi Ke Da Xue Xue Bao*, 2011, **31**, 1200.

178. I. K. Hwang, K. Y. Yoo, D. Y. Yoo, J. H. Choi, C. H. Lee, I. J. Kang, D. Y. Kwon, Y. S. Kim, D. W. Kim and M. H. Won, *J. Med. Food*, 2011, **14**, 195.

179. D. Y. Yoo, J. H. Choi, W. Kim, K. Y. Yoo, C. H. Lee, Y. S. Yoon, M. H. Won and I. K. Hwang, *Neurochem. Res.*, 2011, **36**, 250.

180. K. Y. Yoo, O. K. Park, I. K. Hwang, H. Li, S. Y. Ryu, I. J. Kang, J. S. Yi, Y. S. Bae, J. Park, Y. S. Kim and M. H. Won, *Neurosci. Lett.*, 2008, **444**, 97.

181. D. Y. Yoo, Y. Nam, W. Kim, K. Y. Yoo, J. Park, C. H. Lee, J. H. Choi, Y. S. Yoon, D. W. Kim, M. H. Won and I. K. Hwang, *J. Vet. Med. Sci.*, 2011, **73**, 71.

182. W. M. Yang, K. J. Shim, M. J. Choi, S. Y. Park, B. J. Choi, M. S. Chang and S. K. Park, *Neurosci. Lett.*, 2008, **443**, 104.

183. R. Q. Yao, L. Zhang, W. Wang and L. Li, *Brain Res. Bull.*, 2009, **79**, 69.

184. J. G. Hong, D. H. Kim, S. J. Park, J. M. Kim, M. Cai, X. Liu, C. H. Lee and J. H. Ryu, *J. Ethnopharmacol.*, 2011, **137**, 251.

185. H. Quintard, M. Borsotto, J. Veyssiere, C. Gandin, F. Labbal, C. Widmann, M. Lazdunski and C. Heurteaux, *Neuropharmacology*, 2011, **61**, 622.

CHAPTER 12

Medicinal Uses of Seaweed in Traditional Chinese Medicine

ERI OSHIMA

Consultant, San Diego, CA, USA
E-mail: eoshima@usc.edu

12.1 Background

12.1.1 Introduction

Throughout the coastal regions of the world, diverse species of edible seaweed are harvested, and from ancient times seaweed has been part of the daily diet in many cultures. It is a popular food ingredient consumed by hundreds of millions of people in China, Korea, Japan, the Philippines, Malaysia, Indonesia, Vietnam and Polynesia, and also Ireland, the North Sea regions and maritime Canada. In Asia, seaweed is a popular daily food ingredient. In Europe and Canada, seaweed has been known as an important source of nutrition such as laverbread (*Porphyra* sp.) from Wales, dulse (*Palmaria palmata*) from Iceland and "Irish moss" (*Chondrus crispus*) from maritime Canada.[1]

Traditional Chinese medicine (TCM) has recorded numerous species of seaweed in the Chinese *materia medica*. Medicinal uses of seaweed are known globally. Traditional healers from ancient Greece, South America and Polynesia used seaweed as herbal remedies.[1,2] Today, the Chumash Native American healer from Southern California uses kelp (*Fucus vesiculosus*) for alleviating edema and pain from the lower extremities by wrapping them with large fronds of kelp.[3] Epidemiological observations demonstrated potential

RSC Drug Discovery Series No. 31
Traditional Chinese Medicine: Scientific Basis for Its Use
Edited by James David Adams, Jr. and Eric J. Lien
© The Royal Society of Chemistry 2013
Published by the Royal Society of Chemistry, www.rsc.org

links between dietary seaweed and health benefits, which has inspired the global community to look for biochemical explanations.

This chapter provides an overview of TCM uses of seaweed. Because preventive medicine is an important teaching of TCM and seaweed has been a food ingredient for as long as the history of TCM, both aspects of seaweed are discussed. The focus is on the macroalgae species commonly found in Chinese *materia medica* (Table 12.1).

First, the TCM classics and TCM historical background (Table 12.2) are outlined, with explanations of basic TCM theory and TCM diagnostic methods based on systematic evaluations of disease symptoms. Next, TCM herbal theory is explained, followed by therapeutic uses of seaweed. The discussion of modern research starts with an overview of bioactive substances. It is a condensed compilation of excellent reviews by many authors (Tables 12.3 and 12.4). In the clinical section, several controlled trials and case reports involving dietary seaweed are discussed (Table 12.5). The last section focuses on TCM hospital-initiated clinical trials, highlighting the recent increase in reporting activity and quality assessment. Reports of human TCM studies with seaweed are outlined. Finally, safety information is provided, followed by a concluding section summarizing the totality of evidence (Table 12.6).

12.1.2 Uses in Industry

Seaweed is cultivated and harvested on a large scale. According to the 2003 report from the Food and Agricultural Organization of the United Nations, 7.5–8 million tons of wet seaweed were used worldwide annually.[4] The total annual value was US $5.5–6 billion, of which $5 billion was from direct human consumption. The popularity of dietary seaweed has been growing. China and Korea are the largest producers of brown seaweed and Japan of *nori* (*Porphyria* sp.). Worldwide, seaweed ingredients are important for the food processing industry. The commercially important polysaccharides are agar, alginates and carrageenans used as food binding, gelling and texturing agents and also food stabilizers and emulsifiers. Furthermore, polysaccharides are versatile agents in research laboratories. Agar, for example, is a well-known media material for microbiology. Agar and carrageenans are extracted from red algae and alginates from brown algae. In agriculture, seaweed has been used as fertilizer in northern Europe. The kelp ash made from bladderwrack (*Fucus vesiculosus*) is well known. For animal and fish farming, seaweed was added to the feed to increase the nutrient content and promote growth. The pigments from red algae, phycobiliproteins, are used in the food industry as natural colorants.[4]

In the healthcare product industry, countless numbers of marketed dietary supplements are manufactured from ingredients extracted from seaweed, such as dietary fibers, antioxidants, minerals and micronutrients. Cosmetic lotions and creams may contain hydrocolloidal polysaccharides extracted from seaweed. Pharmaceutical research has been testing the dietary seaweed constituents in various disease models in an effort to identify potential therapeutic agents. A

Table 12.1   Common seaweed in Chinese **materia medica**.

Botanical name Latin binomials	Chinese name	Other names	Notes
Brown algae: Phaeophyceae (Class)			
Undaria pinnatifida (Harvey) Suringar	裙带菜 Qún dài cài	わかめ Wakame Mieok	Popular food
Laminaria japonica Areschoug	海带 Hǎi dài	昆布 Kombu Kelp	

Ecklonia kurome
  Okamura
	昆布
	Kūn bù
	鹅掌菜
	É zhǎng cài
		クロメ
		Kurome
			Hǎi dài, Kūn bù
			  and Hǎi zǎo are
			  common herbal
			  ingredients

Sargassum spp.:
 	
1. *S. pallidum* (Turner)
  C. Agardh
2. *S. kjellmanianum*
  Yendo
3. *S. fusiforme* (Harvey)
  Setchell
4. *S. thunbergii* (Mert.)
  O. Kuntze
5. *S. tortile* C. Agardh
	海藻
	Hǎi zǎo
	1. 馬尾藻
	Mǎ wěi zǎo
	2. 海黍子
	hǎi shǔ zǐ
	3. 鹿尾菜
	Lù wěi cài
	羊栖菜
	Yáng qī cài
	4. 鼠尾藻
	Shǔ wěi zǎo
		ホンダワラ類
		Hondawara-rui
		3. *Hizikia*
		  fusiformis
		ひじき
		Hijiki
		4. 海虎尾
		Umitoranoo
		5. ヤレモク
		Yaremoku
			common herbal
			  ingredients
			Kombu is a source
			  of *umami* flavor

Nemacystus decipiens (Suringar) Kuckuck	水云 Shuǐ yún	水雲 Mozuku	Popular food
Red algae: Rodophyta (Phylum)			
Porphyra spp.: *P. tenera* Kjellman	紫藻 Zǐ zǎo 紫菜 Zǐ cài 甘紫菜 Gān zǐ cài	Nori Gim	Paper-thin sheet used for wrapping sushi
Caloglossa leprieurii (Montagne) J. Agardh	鸪鸪菜 Zhè gū cài 美舌藻 Měi shé zǎo		Anthelmintic Kainic acid
Digenea simplex	海人草 Hǎi rén cǎo	海人草 Kai nin so マクリ Makuri	Anthelmintic Kainic acid
Eucheuma gelatinae	麒麟菜 Qí lín cài	キリン藻 Kirin so	
Gelidium amansii	石花菜 Shí huā cài	天草 Tengusa	Laxative 寒天, kanten (agar agar) food additive

Table 12.1   (*Continued*)

Botanical name Latin binomials	Chinese name	Other names	Notes
Gracilaria spp.	龙鬚菜 Lóng xū cài	龍鬚菜 しらも Shiramo	Source of agar

Green algae: Chlorophyta (Phylum)

Ulva lactuca Linnaeus	石蓴 Shí chún	Sea lettuce アオサ Aosa	Popular food ingredient
Enteromorpha spp.: 1. *E. clathrata* 2. *E. prolifera* 3. *E. compressa* 4. *E. intestinalis*	干苔 Gān tái 1. 条浒苔 Tiáo hǔ tái 2. 浒苔 Hǔ tái 3. 扁浒苔 Biǎn hǔ tái 4. 肠浒苔 Cháng hǔ tái	Dried moss 青海苔 Aonori	Industry-scale use as additives for farm animals and fish feed; human food additives; dietary supplements
Codium fragile (Suringar) Hariot	水松 Shuǐ sōng 刺松藻 Cì sōng zǎo	ミル Miru 海松 Umimatsu	Anthelmintic

Blue–green algae: Cyanophyceae (Class)

1. *Nostoc commune* var. *flagelliforme* 2. *Nostoc commune* Vaucher	1. 发菜 Fà cài	1. 髮菜 Hassai 2. 石くらげ Ishikurage	1. Filamentous seaweed, delicacy food

Plants

Zostera marina Linnaeus	大叶藻 Dà yè zǎo	Sea grass Common eel grass	Flowering plants

Sources (websites last accessed 9 July 2012):

National University of Ireland, Galway, *AlgaeBase*, http://www.algaebase.org.

Complementary and Alternative Healing University, *Chinese Herb Dictionary*, http://alternative-healing.org/.

J. Zhou, G. Xie and X. Yan, *Encyclopedia of Traditional Chinese Medicines*, Springer, Heidelberg, 2001, Vol. 5, p. 601.

National Taiwan Museum, *Seaweeds Net of Taiwan*, http://web.ntm.gov.tw/seaweeds/home.asp.[17]

小药箱, www.xiaoyaoxiang.com (in Chinese).

医学百科, http://www.wiki8.com (in Chinese).

中药剂_中医世家, http://www.zysj.com.cn/zhongyaocai/ (in Chinese).

State Food and Nutrition Consultant Committee, 浅论浒苔的开发与利用, http://www.sfncc.org.cn/Z_Show.asp?ArticleID=2033 (in Chinese).[18]

浙江海洋药用藻类资源调查研究初报, X. Chen, *J. Zhejiang Coll. Tradit. Chin. Med.*, 2000, 24 (2), 65–68 (in Chinese).

和漢三才圖會 卷第九十七 水草 藻類 苔類 寺島良安, *Yabtyan 20081124*, http://homepage2.nifty.com/onibi/wakan97.html (in Japanese).

product based on carrageenans was tested in large human clinical trials for viral illness. The results were negative for efficacy, but the product was shown to be safe.[5] The anticoagulant activity of fucoidan has been intensively studied because of its potential as a therapeutic agent.[6]

12.1.3 General Characteristics

Dietary macroalgae used for TCM are classified into brown, red, green and blue–green algae, taxonomically called Phaeophyceae, Rhodophyta, Chlorophyta and Cyanophyceae, respectively. This classification is based on the photosynthetic pigments. Such pigments are evolutionarily significant because they enable the algae to absorb sunlight in its aquatic habitat where varying degrees of the light spectrum are available. The major classes of photosynthetic pigments are chlorophylls, carotenoids and phycobiliproteins. Seaweed extracts iodine and bromine from seawater. Iodine is an essential nutrient for humans to synthesize thyroid hormones. Seaweed probably uses such halide molecules for defending itself from the harsh environment.[7] The surface of seaweed is covered with viscous materials consisting of polysaccharides that aid in defending itself from mechanical damage and infections. The seaweed described in Chinese *materia medica* has diverse morphology from several meters long to a few centimeters growth in brown and green algae, respectively. AlgaeBase (www.algaebase.org) is an excellent source of information about the taxonomy, morphology and worldwide distribution of seaweed.

12.1.4 Nomenclature

The seaweed from Chinese *materia medica* has multiple names: the Latin binomials, the names in Chinese, colloquial English and other languages. For example, in the USA, often the Japanese name is printed on food product packages, such as *nori*, *wakame* and *kombu*. The corresponding Chinese names are *zǐ cài*, *qún dài cài* and *hǎi dài*, respectively, with the Latin binomials *Porphyra tenera* Kjellman, *Undaria pinnatifida* (Harvey) Suringar and *Laminaria japonica* Areschoug, respectively. Moreover, *kombu* in Japanese and *kūn bù* in Chinese refer to different brown algae even though the written Chinese characters are the same and the pronunciation is very similar. Table 12.1 sets out the Latin binomials, Chinese, English and other names in a side-by-side listing.[8]

12.2 TCM Theory

12.2.1 Classics

In the beginning, knowledge of herbal medicine was passed on to generations through verbal teaching. Around 300 AD, *Shén Nóng Běn Cǎo Jīng, The God Emperor Shennon's Classic Book of Plants*, was written (Table 12.2).[9–14] It is known as the first Chinese *materia medica*. Probably in the later Han dynasty,

Table 12.2 TCM classics.[9–14]

Title	Year written Author	Significance
神农本草经 *Shén Nóng Běn Cǎo Jīng* *The God Emperor Shennon's* *Classic Book of Plants*	300s AD Anonymous	First written Chinese *materia medica*; compilation of verbal knowledge passed down over generations
黄帝内经 *Huáng Dì Nèi Jīng* *Yellow Emperor's Inner Classic*	Probably later Han dynasty (206 BC–220 AD)[a] Anonymous	One of the earliest descriptions of the fundamental philosophy of TCM
伤寒论 *Shāng Hán Lùn* *Treatise on Cold Damage* *Disorders*	~220 AD Zhāng Jī Zhāng Zhòng Jīng	One of the earliest textbooks describing the fundamental principles of TCM treatments by analysis of symptoms and determination of imbalance in the body
本草纲目 Běn cǎo gāng mù Chinese *materia medica*	Ming dynasty (1590 AD) Lǐ Shí Zhēn	Well-quoted Chinese *materia medica* to this day

[a]Metropolitan Museum of Art, *Heilbrunn Timeline of Art History: Han Dynasty*, http://www. metmuseum.org/toah/hd/hand/hd_hand.htm (last accessed 9 July 2012).

the first volume of *Huáng Dì Nèi Jīng*, *The Yellow Emperor's Inner Classic*, was written. It is considered the first written textbook explaining the fundamental theory of TCM. Among the many volumes, *Sù Wèn, The Plain Questions*, is the most famous. Basic TCM theory is that a person has an interactive relationship with the energy from the Universe, the five elements of the Earth, the four seasons, the internal body organs, the body surface, everyday diet and the mind's emotional state. A healthy person is in a state of good balance in this dynamic network. The goal of TCM treatment is to maintain balance and to recover from imbalance.[11–14]

From the beginning, disease prevention was recognized as one of the main tasks of TCM.[11–14] *Huáng Dì Nèi Jīng* explains the term *wèi bìng* as referring to the state when the disease symptoms are not yet manifested, but a person may have a propensity or risk factors for developing illness. In contrast, *jì bìng* is a state where a person has already developed the symptoms. It is important to take care of *wèi bìng* in order to prevent the onset of disease symptoms and also to treat *jì bìng* to avoid further exacerbation of symptoms. For disease prevention, TCM regards daily diet and medicine as sharing the same therapeutic function.[15]

The *Treatise on Cold Damage Disorders*, *Shāng Hán Lùn*, was written around 200 AD.[9–14] It is the first TCM textbook that explains the systemic evaluation of disease symptoms. It is believed that "cold damage" refers to acute epidemic diseases common to that time, such as influenza or bacterial infections with fever and diarrhea transmitted through drinking water. It describes the methods of analyzing the patient's physical signs and symptoms

Table 12.3 Useful reviews: bioactive compounds in dietary seaweed.

Title	Highlights	Ref.
Marine algal constituents	Comprehensive nutritional analysis; tables of nutrient values comparing brown, red and green algae	Yuan 2008[1]
Nutritional value of edible seaweeds	Comparison of nutritional values of seaweed with other natural food items	MacArtain et al. 2007[40]
Bioactive compounds in seaweed: functional food applications and legislation	Comprehensive detailed review including tables of useful references for each category of bioactivity	Holdt and Kraan 2011[26]
Medicinal and pharmaceutical uses of seaweed natural products	Therapeutics-focused review including chemical structures	Smit 2004[35]
Antioxidants from macroalgae: potential applications in human health and nutrition	Useful tables of antioxidant compounds	Cornish and Garbary 2010[48]
Dietary seaweed and human health	Table of observational studies showing possible link between dietary seaweed and disease risk reduction	Brownlee et al. 2011[25]
Anti-inflammatory compounds of macro algae origin	Comparison of various seaweed extraction methods	Jaswir and Monsur 2011[49]
Fucoidan structure and bioactivity	Comprehensive and useful review	Li et al. 2008[6]
Therapies from fucoidan multifunctional marine polymers	Recent update and therapeutic applications discussed	Fitton 2011[29]

systematically by examining several categories of TCM entities. The principal concept is that disease manifestations are influenced by *Yin* and *Yang*, the flow and quality of *Qi*, the meridian channels running through the body and the condition of the outer and inner body. The *Qi* can arise from the solid organs or from circulating blood; it may be flowing smoothly or constrained, nutritive or defensive or having *Yin* or *Yang* quality. Through careful evaluations of the patient, the patterns of symptoms called *zhèn* are determined. A total of eight patterns, *bā zhèng*, are systemically reviewed in order to determine the nature of the imbalance. In *Shāng Hán Lùn*, a symptom pattern-based diagnostic method was first described as *bā gāng biàn zhèng*, *i.e.*, differentiation of the patterns or syndromes based on the eight principles.[11–14]

Following the publication of the above classics, TCM textbooks were published one after another: for example, the new interpretations and additional volumes to

Table 12.4 Summary of nutrients and bioactive compounds in dietary seaweed[1,26]

Nutrients			
Bioactive compounds	Brown algae	Red algae	Green algae
Polysaccharides Soluble dietary fiber	Alginic acid Fucoidan Laminarin	Agar Carrageenans	Polyuronides (highly branched polysaccharides)
Minerals I, K, Ca, Mg, Fe, Zn, Cu, Mn	Mineral content up to 39.8% dry weight High iodine	Mineral content up to 43.6% dry weight	Mineral content up to 29.2% dry weight
Proteins Essential amino acids Taurine	*U. pinnatifida* Protein: 18% dry weight	*P. tenera* Protein: 38 % dry weight Lectins	*E. compressa* Protein: 32% dry weight Lectins
Omega-3 fatty acids EPA Sterols	Stearidonic acid Fucosterol	High EPA DHA Desmosterol	DHA Mixture of sterols
Photosynthetic pigments	Chlorophyll Fucoxanthin	Chlorophyll Phycoerythrin	Chlorophyll
B vitamins Polyphenols Choline	Phlorotannins	Vitamin B_{12}	Vitamin B_{12}

the classics; elaborations of diagnostic methods; subspecialties covering gynecological, pediatrics, epidemic diseases and other special cases; updated *materia medica*; herbal medicine textbooks; and acupuncture and moxibustion textbooks.[9,10] In the 1500s, Lǐ Shí Zhēn thoroughly studied volumes of *materia medica* written before his time. He then compiled the old master's knowledge together with his own and authored *Běn Cǎo Gāng Mù* in 1590. It is considered one of the most comprehensive books of Chinese *materia medica*.[11–14]

TCM *materia medica* is digitized; several sites are listed in Table 12.1. Well-known herbs and herbal formulas can be found in numerous TCM herbal dictionaries, glossaries and inventories accessible through Chinese and bilingual sites.

12.2.2 Concept of Interdependence and Balance

The TCM philosophy of the Universe is that all phenomena belong to either *Yin* or *Yang*.[11–14] The positivity of *Yang* and negativity of *Yin* together create a state of good balance. The Earth consists of five elements: Fire, Wood, Metal, Earth and Water. TCM has a solid organ-like concept, *zàng*. There are five *zàng* including the heart, liver, lungs, spleen and kidneys. The *fǔ* organs are equivalent to the hollow organs in the body *i.e.*, the large and small intestines, gall bladder, urinary bladder and stomach. The five *zàng* and *fǔ* organs, the

Table 12.5 Human controlled clinical studies and case reports with dietary seaweed.

Health focus	Subjects	Seaweed, source	Ref.
Postprandial glycemic response	12 healthy women in Spain	*Porphyra tenera* Northern Spain	Goñi *et al.* 2000[51]
Type 2 diabetes mellitus	20 Korean subjects with type II diabetes	Sea tangle and sea mustard (brown algae) Korea	Kim *et al.* 2008[52]
Metabolic syndrome	17 subjects in Ecuador with metabolic syndrome	*Undaria pinnatifida* (Harv.) Suringar Patagonian coast of Argentina	Teas *et al.* 2009[53]
Estrogen progesterone metabolism	3 pre-menopausal women in northern California with abnormal menstrual cycles	*Fucus vesiculosus* North Atlantic	Skibola 2004[54]
Estrogen phytoestrogen metabolism	15 healthy women of European descent in Massachusetts	*Alaria esculenta* (L.) Greville Northeastern shores of North America	Teas *et al.* 2009[55]
Herpes viruses Active and latent infection	15 subjects with active or latent infection with herpes virus	*Undaria pinnatifida* Tasmania, Australia	Cooper *et al.* 2002[56]
HIV-AIDS	11 subjects with HIV in South Carolina	*Undaria pinnatifida* (Harvey) Suringar Patagonian coast of Argentina Spirulina (*Arthrospira platensis*) California	Teas and Irhimeh 2012[57]

five elements of the Earth, the planets in the Universe, the four seasons and *Yin* and *Yang* are all interconnected and influence each other. The interdependent relationship is Fire–Mars–summer–heart–small intestine; Wood–Jupiter–spring–liver–gall bladder; Metal–Venus–autumn–lung and large intestine; Earth–Saturn–intermediate season–spleen and stomach; and Water–Mercury–winter–kidney and urinary bladder.[11–14] Within the body, the *Qi*, *xuě* (blood) and *jīn yè* (all other body fluids) travel through multiple channels called *jīng luò* or meridians into and around five *zàng* and six *fŭ* anatomical organ-like entities.[11–14]

A person exists in a dynamic network of *Yin* and *Yang* interdependent states. TCM views that an imbalance of *Yin* and Yang can result in disease

Table 12.6 Totality of evidence: TCM medicinal uses of seaweed.

Source	Findings
TCM herbal theory	Herbal properties of seaweed: Cold, salty and/or bitter; drains fluids downwards, softens and eliminates hardness, dissipates nodules and disperses accumulations; enters liver, stomach, kidney meridian channels; *Yang* invigorating (Section 12.2.4)
Chinese *materia medica*	Indications: Chronic inflammation; fever, coughs, bronchitis, sticky sputum, phlegm stagnation, simple goiter, cervical lymphadenopathy, tumor, hepatocirrhosis, blood stasis, food stagnation, fluid accumulation with edema, vitamin deficiency, constipation and scrofula, antiparasitic agents; for obstetrics, labor induction (Section 12.2.4)
Translational: Positive research data *in vitro* and animal studies	Bioactive chemicals and nutrients: Dietary fibers, sulfated polysaccharides, fucoidan, carrageenans, alginates; iodine and other minerals; proteins and essential amino acids; omega-3 fatty acids and sterols; chlorophyll, fucoxanthin and phycobiliproteins; polyphenols and phlorotannins (Section 12.3.1 and Table 12.4)
Epidemiological observations	Association with reduced risk: Colon, rectal and prostate cancer, type II diabetes, obesity, osteoporosis, breast cancer (Section 12.3.2)
Human controlled clinical studies (Non-TCM)	Positive effects from small studies: Estrogen and progesterone levels, viral infection and symptoms, postprandial glycemic response, metabolic syndrome and type II diabetes (Section 12.3.2 and Table 12.5)
TCM human clinical studies	Positive results are difficult to interpret; unique TCM RCT design is needed (Sections 12.3.3.1 and 12.3.3.2)

symptoms. Diseases usually manifest as *zhèng*, *i.e.*, disease symptom patterns similar to a "syndrome." There are eight guiding principles to evaluate disease patterns, *i.e.*, *bā gāng biàn zhèng*.[11–14] A combination of eight factors is based on excess or deficiency, hot or cold, exterior or interior of the body and *Yin* or *Yang*. Identification of *zhèng* is achieved by asking questions, listening, observing and examining the patient to detect various external and internal factors such as excessive eating or drinking habits, fever, erythema or pallor, presence of pains, abdominal mass, irregularity or weakness in the rhythm, strength and pattern of the peripheral pulses and changes on the tongue surface. Also, the quality and strength of voice, breath pattern and odor, energy level, mental acuity and state of mind such as agitation and irritability *versus* calmness are evaluated. Manifestation of disease symptoms is organized in eight patterns, called *bā zhèng*. Based on the examination, the pattern of symptoms is determined and the patient's imbalance is diagnosed. The TCM practitioner appropriately selects the herbal prescription that will help counteract the factors causing imbalance.[11–14]

12.2.3 Herbal Theory

TCM herbal medicine is the practice of selecting the right principal herbs and compatible herbal combinations that can correct imbalance as determined by an evaluation of symptoms. Profound knowledge of the herb's therapeutic properties and the disease syndrome is required. The textbook by Yang offers an excellent teaching of herbal theory including the therapeutic properties of seaweed.[16] Herbs possess four main temperatures, hot, warm, cold or cool, plus neutral temperature. According to each practitioner's experience, the less clear cut temperature ranges may be assigned to a herb; for instance, thin or thick hot and thin or thick cold refer to the temperature quality from the body's superficial surface and deeper parts. Herbs are assigned flavor groups: pungent, sweet, bitter, sour, salty and bland, based on the property of action in the body rather than the actual taste in the mouth. Additionally, aromatic and astringent flavors exist. One herb may have two or three flavors. By applying the knowledge of diverse flavor and temperature patterns of the herbs, an herbalist can create unique sets of herbal combinations to treat various disease symptoms. Herbal prescriptions are individualized. The unique combination of herbs is applied to the individual patient's unique case. Therefore, it is possible that two herbalists' prescriptions for the same patient may not be exactly identical. [11–14,16]

One of the important properties of herbs is the direction of travel in the body. Herbs enter the body and travel through the meridian channels. Each herb possesses a tendency to enter one main meridian or sometimes a secondary meridian. Each herb has a direction of travel in the body such as upwards, downwards, outwards or inwards. The direction of movement is an important property of herbs. It is important to correct stagnation of the flow of *Qi*, blood, other body fluids and phlegm as well as water metabolism and non-functioning meridians.[11–14,16]

In TCM, a combination of herbs is usually prescribed. Selected first is the principal herb with therapeutic action matched to the particular patterns of *zhèng*, the syndrome. Together, supportive herbs that can perform synergistic effects may be prescribed. To alleviate the side effects or sometimes to improve the overall taste, additional herbs may be added.[16] Usually, a mixture of dried herbs is prescribed as a decoction. They are first soaked in water, then boiled and strained. The liquid is the decoction that is divided into portions and ingested over the course of a day. In some cases, the herbs are formulated into a pill form made from an edible binder material.

12.2.4 Medicinal Uses of Seaweed

Although diverse species of seaweed are listed in the Chinese *materia medica*, the TCM descriptions of therapeutic actions are very similar except that a few of them possess anthelmintic property.[16–18] TCM characterizes seaweed as having a salty flavor and cold temperature. Some seaweed species have both salty and bitter flavors. Seaweed in general enters the body through the liver, stomach and the kidney meridian channels. *Hǎi z*ǎo and *kūn bù* used together may also enter the lung meridians. The therapeutic action in the body is to drain fluids downwards, soften and eliminate hardness, dissipate nodules and disperse accumulations.[16] It has *Yang* invigorating quality. It has been prescribed for many conditions relating to chronic inflammation, stagnation, swelling, tumors and water imbalance. Some of the conditions are fever, coughs, bronchitis, sticky sputum, phlegm stagnation, simple goiter, swelling in the neck with cervical lymphadenopathy, tumor, hepatocirrhosis, blood stasis, food stagnation, fluid accumulation with edema, vitamin deficiency (scurvy, beriberi), constipation and swelling of the scrotum (scrofula).[16–19] Furthermore, well-known anti-parasitic seaweeds are *zhè gū cài* and *makuri* (*Cologlossa* sp. and *Digenea* sp., respectively). For obstetrics, *Laminaria* sp. has been used for induction of labor or as an agent to enhance uterine contractions to induce placental detachment after birth.[19]

With the combination of salty flavor and cold temperature, seaweed has the ability to transform phlegm. In TCM, phlegm is a pathological fluid product that can cause stagnation of *Qi*, resulting in *Yang Qi* deficiency. According to TCM phlegm theory explained by Clavey, it can be visible and lodge in the respiratory tract as a sticky fluid or it can be invisible and internal.[20]

Invisible phlegm can accumulate in any part of the body and result in complicated symptoms. For instance, in the chest and ribs, local obstruction may cause pain and uncomfortable sensations in the area surrounding the heart to the back, or it may block upward and downward movement of *Yin* and *Yang*, causing imbalance in the heart and kidneys with insomnia, palpitations and lower back pain, or in the stomach it may cause nausea and vomiting, loss of appetite and distension in the upper abdomen.[20] Chronic stagnation of internal phlegm may lead to the formation of a mass or tumor if other pathological factors exist.[16] An important advantage of seaweed is that it can alleviate phlegm stagnation that may cause many complicated disease symptoms.

Using seaweed as the principle ingredient, a herbalist can add other compatible herbs to make a special prescription tailored to the individual patient's needs. The brown algae *hǎi zǎo* (*Sargassum* sp.), *kūn bù* (*Ecklonia* sp.) and *hǎi dài* (*Laminaria* sp.) are common seaweeds used in a prescription. Often, the full Latin binomials are not mentioned. *Hǎi zǎo* may have stronger actions for dissipating nodules whereas *kūn bù* may have stronger action in softening hardness. *Hǎi zǎo* may be recommended for treating goiter and scrofula and *kūn bù* for hepatosplenomegaly, liver cirrhosis and tumors.[16] The common herbal combination for goiter is *hǎi zǎo yù hú tāng* (*Sargassum* decoction for the Jade hot pot soup) with *hǎi zǎo* and *kūn bù* as the principle ingredients. Another less common prescription is *jú hé wán* (Tangerine seed pill) for unilateral scrotal swelling in which all three above seaweeds are included.[16]

According to classical teaching, *hǎi zǎo* (*Sargassum* sp.) and *Radix glycyrrhizae* (licorice root) are an incompatible combination and should not be used together. In practice, however, the combination is sometimes prescribed. A recent article from the TCM University reviewed the literature from 1979 to 2010 where *hǎi zǎo yù hù tāng* decoction was used with licorice.[21] The study concluded that in practice, *hǎi zǎo yù hù tāng* decoction and licorice root have been used together and no reports of adverse effects were found.

As a general precaution, TCM advises not to use downward-draining herbs in certain situations. Because downward-draining herbs may drain *Qi* and blood circulation downwards, they should not be prescribed as the only medication for diarrhea, heavy menstruation and pregnancy. For serious abdominal conditions that require immediate attention such as intestinal obstruction, appendicitis, cholecystitis and pancreatitis, patients must be evaluated by Western medical examinations first. TCM advises that only after that, a TCM practitioner can prescribe downward-draining herbs.[16]

12.3   Modern Research

12.3.1   Bioactive Substances in Seaweed

Throughout the history of TCM, seaweed has been used both as a medicine and as a health food. Cultures where TCM has been practiced have incorporated seaweed as part of the daily diet. Many epidemiological studies have shown that dietary seaweed is associated with reduced risk of chronic diseases associated with diet.[22-25] Such observations inspired global research interest in dietary seaweed. The following is a compilation of several excellent reviews and biochemical research studies focusing on the constituents in dietary seaweed commonly found in Chinese *materia medica* (Table 12.3). The table of nutrients and bioactive compounds was prepared highlighting the uniqueness of each of the brown, red and green algae group (Table 12.4).

12.3.1.1   Dietary Fiber

Seaweed has a high dietary fiber content. The total dietary fiber is classified as polysaccharides. According to the proximate composition analysis by Yuan, total dietary fiber ranged from 30.8 to 74.6% dry weight in brown algae, from 24.7 to 59.4 % in red algae and from 28.5 to 55.4% in green algae.[1] The health benefits of dietary fiber are well known. Insoluble dietary fiber functions as a bulking agent in the gut similarly to a laxative. Soluble dietary fiber (SDF) has the capacity to hold water and form colloidal gel in the gut. This viscous material can absorb luminal lipids and cholesterol dissolved in the bile acids. This process can eventually lead to a decrease in serum cholesterol and glucose levels.[1,26] Examples of such hydrocolloids are agar, alginates and carrageenans. Agar, isolated from red algae, is a food additive and a molding agent for Japanese confectionaries (*kanten*). Alginates extracted from brown algae are used in various industries from food processing, pharmaceuticals, chemicals, animal feeds to cosmetics. Carrageenans are sulfated polysaccharides extracted from red algae species, with a long history of use in the food industry as food additives and binding agents.

Digestion of SDF takes place in the colon by the colonic microflora.[27] The bacterial fermentation process enables the host to extract energy from SDF by producing short-chain fatty acids. In the case of sulfated polysaccharides, the digestion mechanism is not well known. Recently, a study implicated possible horizontal gene transfer from the marine bacteria to the host microflora isolated from Japanese individuals.[28] The genes coding for porphyranases that act on porphyran, a sulfated polysaccharide of red algae, may have been transferred from the marine bacteria attached to dietary red algae such as *nori* to the Japanese colonic bacteria.

12.3.1.2 Sulfated Polysaccharides

Among the sulfated polysaccharides, fucoidans are isolated from brown algae. They are composed of sulfated fucose-rich polysaccharides. Fucoidans have been tested in various cell lines and animal disease models relating to inflammation, neoplasia, oxidation, coagulation, viral infection and other diseases. The groups of Li, Holdt and Fitton reported numerous positive results implicating fucoidan's important role in health maintenance and prevention from harmful effects.[6,26,29]

The anti-inflammatory function of fucoidan is well known. Fucoidan has been shown to interact with the cell surface adhesion molecules called selectins that aid lymphocytes to migrate during inflammation.[29] By blocking selectins, fucoidan can inhibit leukocyte migration and adhesion to inflammation sites. In another study, fucoidan-containing extracts were obtained from *Sargassum hemiphyllum* and tested in the mouse macrophage cell line activated by lipopolysaccharide (LPS). The sulfated polysaccharide extract reduced the secretion of pro-inflammatory mediators including nitric oxide (NOS), IL-1β, IL-6 and TNF-α. Also, a reduction was observed in the mRNA expressions of IL-1β, inducible NOS and cyclooxygenase-2 and the protein levels of nuclear factor κB.[30] Furthermore, studies have shown that fucoidan can block the complement cascade system of immunity resulting in the inhibition of pro-inflammatory anaphylatoxin synthesis. It is speculated that this may be the mechanism for the antiallergic activity of fucoidan.[29]

Antioxidant activity of sulfated polysaccharides has been demonstrated in *Laminaria japonica*, *Fucus vesiculosus*, *Ecklonia kurome*, *Porphyra* sp. and *Ulva* sp.[26] Fucoidan from *L. japonica* was shown to significantly inhibit the increase in lipid peroxides in the serum, liver and spleen of diabetic mice; a correlation may also exist between the molecular weight, the sulfate content and the antioxidant activity.[6]

In viral infection, fucoidan has been shown to block viral replication by blocking the binding of viral particles to the host cells. Antiviral activity against herpes simplex viruses, HSV-1 and HSV-2 and other enveloped viruses including human immunodeficiency virus (HIV) and human cytomegalovirus (CMV) were demonstrated. Recently, a small clinical study was conducted with 13 Japanese patients suffering from human T-lymphotropic virus type-1 (HTLV-1) infection with HTLV-1 associated myelopathy/tropical spastic paraparesis (HAM/TSP).[31] The patients received 6 g per day of fucoidan obtained from *Cladosiphon okamuranus* for 6–13 months. The results showed a 42.4 % decrease in the HTLV-1 proviral load without affecting the host immune cells. Proviral load is an important indicator of clinical outcome. In the same study, it was also demonstrated *in-vitro* that fucoidan inhibited the cell-to-cell transmission of HTLV-1; however, the viability of infected cell line was maintained.

The antitumor activity of fucoidan is well known, but the detailed mechanism of action is not yet known. It is speculated that immunomodulation may play a role in the antitumor effects. Studies have shown that fucoidan can interact with the enzymes involved in metastasis, reduce solid tumor size in animal models and directly kill tumor cells *in vitro*.[6,26] Fucoidan may inhibit tumor metastasis by inhibiting the enzymes involved in tumor cell migration and implantation processes such as matrix metalloproteases, hyaluronidases and elastases. The antitumor action may also be due to the inhibition of tumor cell adhesion to various substrates. In animal models of solid tumors, tumor size reduction may be secondary to antiangiogenesis activity. Fucoidan exhibited direct tumor killing of human HS-sultan cells *via* the caspase and ERK signaling pathway.[6,26] In immunomodulation, tumor cell destruction may be mediated by enhancement of the type 1 T-helper cell and natural killer cell responses. In gamma-ray irradiated rats, fucoidan enhanced the recovery of immunological function by inhibiting apoptosis in splenic lymphocytes.[32]

The anticoagulation properties of fucoidan have been intensively studied because of the potential to develop a natural oral anticoagulant. Anticoagulation activity is mediated by the inactivation of thrombin and binding to the heparin cofactor.[6,29]

In the stomach, fucoidan demonstrated protective effects on the human gastric mucosa infected with *Helicobacter pylori* by preventing the adhesion of bacteria to the gastric mucosa.[6,33] In the liver, fucoidan exhibited antifibrosis activity in the rodent liver fibrosis model.[29] For lipid metabolism, fucoidans improved rat hypercholesterolemia and hyperlipidemia.[6] Neuroprotective effects from amyloid toxicity and radiation were reported *in vitro* with rat neurons and neuronal stem cells respectively. Neuroprotection was also observed in the mouse Parkinson's disease model.[29]

Carrageenans are sulfated linear galactans extracted from red algae species. Three forms, kappa, lambda and iota, exist with characteristic gel-forming properties.[26] Carrageenans possess anticoagulant activity through antithrombotic effects. Carrageenans demonstrated antiviral activity *in vitro* by blocking the adsorption stage of HSV virus infection. Carrageenan gels demonstrated *in vitro* anti-HIV activity. The product, Carraguard, underwent a Phase III clinical trial in South Africa enrolling 6000 non-pregnant HIV-negative women. However, the study concluded that the product was not effective in preventing HIV transmission from male to female although it was safe for vaginal use.[5] Carrageenans are "generally regarded as safe" (GRAS) by the US Food and Drug Administration (FDA). The pro-inflammatory molecules of carrageenans used in biology research probably have different molecular weights from the therapeutic forms.[26]

12.3.1.3 Minerals

Well-known TCM herbal remedies for goiter use brown algae as the main herbs. Historically, the common cause of goiter was iodine deficiency. Seaweed is known for its high content of iodine. The brown algae in general have the highest iodine content. The analysis by Yuan[1] showed that in *Laminaria* spp. and *Sargassum* spp., the iodine content ranged from 2 to 10,000 µg g^{-1} dry weight and from 30 to 3,000 µg g^{-1} respectively. In comparison, the values for *Ulva* spp. (green algae) were 20–250 µg g^{-1}. One study averaged the values from ten *Laminaria* species obtained worldwide, which showed 1,542 µg g^{-1} dry weight.[34] In this study, the estimated iodine contents of Japanese *kombu* and *wakame* were 2,353 and 42 µg g^{-1}, respectively. The daily iodine intake in Japan was estimated to be 1,000 to 3,000 µg per day. In comparison, the United States Recommended Daily Allowance of iodine for people over 14 years of age and not pregnant is 150 µg per day. Seaweed also contains other important minerals, including potassium, calcium, magnesium, iron, zinc, copper and manganese. The mineral contents fluctuate seasonally.[1]

12.3.1.4 Proteins

Protein content was in general higher in red algae, with *Palmaria tenera* having 37.8% dry weight, *Enteromorpha compressa* (green algae) 32.4% and *Undaria pinnatifida* (brown algae) 18%.[1] Most species contained all of the essential amino acids. Within the non-essential amino acids, the proportions of aspartate and glutamate were relatively high. In *P. palmate*, aspartate content was 18.5% and glutamate content was 9.9% of the total amino acids. In *Sargassum* spp., the proportions were 9.99% and 11.4%, respectively. Aspartate and glutamate are the chemical basis of *umami* in seaweed used for food flavoring. Seaweed also contains taurine, a natural chemical similar to the amino acids. Taurine is important in the production of bile acids. Its antihypertensive, hypocholesterolemic and antioxidant activities were reported.[26]

Lectins are carbohydrate-binding proteins of non-immune origin. Bioactive lectins have been isolated from various red and green algae, including *Ulva* spp., *Gracilaria* spp. and *Eucheuma* spp. Lectins are known to play a role in various protein–carbohydrate interactions. Bioactivities have been demonstrated in many studies relating to host–pathogen interaction, cell–cell recognition, induction of apoptosis, antiadhesion, metastasis and differentiation, recognition and binding of carbohydrates including viruses, bacteria, fungi and parasites.[26]

Kahalide F is a depsipeptide isolated from green algae, *Bryopsis* spp. It showed antitumor activities in the lung, colon and prostate cancer cell lines.[26,35] It has a potentially unique mechanism of action as an antitumor agent by acting on lysosomal membranes. The compound is undergoing clinical development as a therapeutic agent for lung carcinoma.

Peptides isolated from *Porphyra yezonensis* and *Hizikia fusiformis* showed angiotensin-converting enzyme (ACE) inhibitory activity. Extracts of *U. pinnatifida* exhibited hypotensive effects in a rat hypertensive model. In a study comparing ACE inhibitory activities among various species of brown and green algae, *Ecklonia cava* (brown algae) showed the highest ACE inhibitory activity.[26]

12.3.1.5   Lipids

Seaweed has in general low lipid contents ranging from as low as 0.3% to 7.2% dry weight.[1] Seaweed supplies essential polyunsaturated fatty acids (PUFAs). According to Yuan's analysis, the contents of C18:2n-6, C18:3n-3, C20:5n-3 [eicosapentaenoic acid (EPA)] and C22:6n-3 [docosahexaenoic acid (DHA)] were 0.69–10.03, 0.23–11.97, 1.01–33.1 and 0.80–12.9% of total fatty acids, respectively.[1] EPA was the highest in red algae, *Gracilaria changgi* (33.1%) and *Palmaria* spp. (24.05%). Brown algae had an EPA range of 5.50–9.43%. In this analysis, DHA was detected only in *U. lactuca* (0.80%) and *G. changgi* (12.9%) and not in the brown algae. The beneficial health effects of PUFAs are well known. PUFAs have been shown to reduce pro-inflammatory cytokine production in inflammatory immune responses. PUFAs have antiatherogenic and antithrombotic activity, which can exert cardioprotective effects.[36]

Seaweed contains various types of sterols such as cholesterol and fucosterol from brown algae and desmosterol from red algae. Fucosterol and desmosterol showed hypocholesterolemic effects in animal models.[1,26] Fucosterol had anti-inflammatory properties in LPS-induced macrophages.[37]

12.3.1.6   Pigments

Seaweed contains chlorophylls and other unique photosynthetic pigments. Among the carotenoids, fucoxanthin, isolated from brown algae, is known for its various biological activities. Antioxidant activities were demonstrated by many studies which reported that fucoxanthin was a strong scavenger of reactive oxygen species.[38] In the inflammatory reaction, fucoxanthins reduced the production of inflammatory cytokines and mediators. Anti-neoplastic activity of fucoxanthin was shown as an apoptosis-inducing property.[38] In another study, antiobesity effects were observed in mice, probably due to the regulation of mitochondrial uncoupling protein 1 in white adipose tissues.[39]

The unique photosynthetic pigments isolated from red algae belong to the phycobiliproteins. Red algae contain mainly phycoerythrin. Various bioactivities were reported in inflammation, cardiovascular protection, antitumor effects and antioxidation.[26]

12.3.1.7 Micronutrients

Seaweed contains both water- and fat soluble-vitamins. Variable amounts of vitamins A, B, C and E (tocols and tocotrienols) are present.[1] Seaweed is a non-animal and non-fish source of vitamin B_{12}. *Porphyra* spp. contained 0.323 µg of vitamin B_{12} per gram dry weight and *Enteromorpha* spp. contained 0.636 µg.[1] Another study showed that *Ulva lactuca* can provide 0.625 µg of vitamin B_{12} per gram dry weight.[40] Seaweed contains the essential nutrient choline and also polyphenols, including catechins and phlorotannins. Brown seaweed is a rich source of phlorotannins. The phlorotannins obtained from *Ecklonia cava* from Korea showed cytotoxic activities against cultured human tumor cell lines and ACE inhibitory activity.[41,42] Also, phlorotannins showed inhibitory effects on ADP-induced platelet aggregation in *Sargassum thunbergii* from China.[43]

12.3.1.8 Toxins

Kainic and domoic acids are potent neuronal agonists and excitatory amino acids. They are used in neurophysiological research and are also known for antiroundworm toxicity. The kainoids are structurally related to glutamic and aspartic acids. TCM uses seaweed containing kainoids, *i.e.*, *Caloglossa leprineurii* and *Digenea simplex*, as anthelmintic agents.[35] A human study from China published in 1965 reported that an oral dose, probably a decoction, of *Caloglossa leprineurii* was effective in elimination of ascarids.[44]

Poisoning from seaweed is rare. A review by Smit described reported cases of poisonings from red algae species.[35] In 1991, poisoning from polycavernoside A was documented in Guam after consumption of *Polycavernosa tsudai*.[45] According to the literature, *P. tsudai* is closely related to *Gracilaria edulis*. In Japan, a fatal poisoning was reported in 1993 due to prostaglandin E_2 from consumption of *ogonori*, *G. verrucosa*. In 1994, non-fatal poisoning was reported in Hawaii from aplysiatoxins found in *G. coronopifolia*.[46]

12.3.1.9 Variations

Among the various species of seaweed, the biochemical constituents are not uniform. Wide variations exist among the species and the red, green and brown algae groups. Even within the same species, seasonal variations are observed. The minerals, proteins, fatty acids, photosynthetic pigments, phytosterols and vitamins all showed such fluctuations.[1] Furthermore, coastal habitats and geographical locations influence environmental factors, which may lead to biochemical variations even within the same species. Any of the above factors can influence the nutritive value and biological potency of the whole seaweed or the extracts. It is therefore important to record not only the species identity but also, if possible, the date and costal location of the harvest. The study from Alexandria, Egypt, illustrates this point. Seasonal variations in antioxidant activity were measured in *Enteromorpha compressa* extracts prepared by several extraction methods. The study showed antioxidant activity measured by DPPH methods was highest in the spring in crude ethanolic extracts.[47]

12.3.2   Human Clinical Studies with Dietary Seaweed

As outlined in 12.3.1, translational research continues to provide positive results from bioactive substances in dietary seaweed. Recent observational studies from Korea and Japan showed associations between the consumption of dietary seaweed and positive health effects in type II diabetes, osteoporosis, obesity, cardiovascular mortality, allergic rhinosinusitis and breast cancer.[25] Another recent observational study showed that even in young Japanese children, an inverse relationship was observed between high intake of dietary seaweed and blood pressure.[50]

It is important to understand the strength and weaknesses of various human research methods intended to show a substance–disease relationship. The FDA Guidance for Industry: Evidence-Based Review System for the Scientific Evaluation of Health Claims – Final 2009, written in English, Chinese and Hindi, offers guidance in properly designing observational studies and randomized controlled clinical trials.

Stronger data from translational research and observational studies support TCM uses of seaweed; however, it is not well known how the bioactive substances are absorbed and function positively in the human body. Recently, several controlled clinical studies were conducted with dietary seaweed interventions. Even though the studies were too small to draw any general conclusions about efficacy, causality or safety, they may trigger further interest in initiating larger trials. The study subjects consumed either brown or red seaweed and the investigators measured parameters related to glycemic response, type II diabetes mellitus, metabolic syndrome, estrogen metabolism and viral infection (Table 12.5).

A study conducted in Spain evaluated dietary seaweed and glycemic response in 12 healthy 20 to 24-year-old female subjects.[51] The *nori* alga (*Porphyra tenera*) from northern Spain was powdered and encapsulated at 500 mg per capsule. First the baseline glycemic response was obtained by measuring postprandial blood glucose levels after ingestion of white bread containing 50 g of starch. The next day, each of the subjects ingested six capsules with white bread containing 50 g of starch. The results showed a sharp decrease in postprandial glucose peaks.

A study from Korea reported the effects of dietary brown seaweed in patients with type II diabetes mellitus.[52] Twenty diabetic subjects were randomized to either a seaweed supplementation group or a control group. Dried brown seaweed was powdered and formed into pills and taken orally by the subjects three times per day, equivalent to 48 g of seaweed per day. After 4 weeks, fasting blood glucose and the 2 hour postprandial blood glucose decreased significantly in the group who had taken seaweed pills compared with the group without seaweed. The serum total triglycerides also decreased significantly, with a significant increase in high-density lipoprotein cholesterol; however, the total cholesterol and low-density lipoprotein values were not affected. The effects may have been due to the 2.5-fold increase in soluble dietary fiber intake compared with the control group. Blood analysis showed an elevation in antioxidant enzyme activities in blood samples obtained from the seaweed supplementation group.

The effects of dietary seaweed in the metabolic syndrome were evaluated in a double-blind crossover study.[53] The subjects were 27 men and women from Ecuador who were non-seaweed consumers and had at least one parameter of the metabolic syndrome. *Wakame, Undaria pinnatifida*, harvested in Argentina was powdered and encapsulated at 500 mg per capsule. Subjects were assigned to one of two groups. Group 1 received 4 weeks of placebo followed by 4 weeks of 4 g per day of seaweed (four capsules twice per day). Group 2 took 4 weeks of 4 g per day of seaweed followed by 4 weeks of 6 g per day (six capsules twice per day). The study concluded that daily intake of 6 g per day, which is similar to the average daily intake in Japan, showed positive effects with a reduction in systolic blood pressure in one hypertensive man and a reduction in the waist circumference of one woman.

The next two studies reported the effects of dietary seaweed on estrogen and progesterone metabolism. The case report from Northern California included three premenopausal women with abnormal menstrual cycles.[54] Daily intake of either 700 mg or 1.4 g of encapsulated powdered brown seaweed, *Fucus vesiculosus,* showed improvement in menstruation with a prolongation of the duration of cycles. Also, antiestrogenic effects were observed with a reduction in the serum 17β-estradiol levels and an increase in progesterone levels. This pilot study led to further investigations on the binding affinity of *F. vesiculosus* to the estrogen and progesterone receptors.[55]

The role of dietary seaweed in estrogen and phytoestrogen metabolism was evaluated in a double-blind crossover study enrolling 15 healthy postmeno-pausal women of European descent in Massachusetts, USA.[56] The subjects were randomized to two groups. One group took seaweed for 7 weeks, and during the seventh week a daily soy protein isolate, isoflavones, was also con-sumed. Another group took placebo for 7 weeks with a soy protein challenge, *i.e.*, daily isoflavones during the seventh week. After a 3 week washout period, the subjects were crossed over to the alternate group. The seaweed dosage was 5 g per day (10 capsules) prepared from powdered brown alga, *Alaria* sp., harvested along the northeastern shores of North America. The results showed that of the five subjects who were equol excreters, soy increased urinary equol excretion (equol is a metabolite of soy isoflavones; it is produced by intestinal bacteria in some people). The combination of seaweed and soy further

increased equol excretion. The results showed that the seaweed supplementation dose was inversely associated with serum estradiol levels. The investigators speculated on a possible role of seaweed in modulating colonic bacteria.

The last two studies reported clinical cases of viral illness and positive effects
of dietary seaweed. The first study reported beneficial effects of *Undaria
pinnatifida* from Tasmania.[57] The formulation was ingested by patients who
presented with acute or latent infections from the herpes group virus including
herpes simplex virus type 1, type 2, Epstein–Barr virus and herpes zoster virus.
Fifteen subjects with active infection and six patients with latent infection were
enrolled. The dosage for 15 patients with active herpetic viral infections was
2.24 g (four capsules) per day, which was within the range of average daily
seaweed consumption in Japan. Six patients with latent infection were given
1.12 g (two capsules) per day, except one patient who took four capsules per
day. The treatment was associated with increased healing rates in patients with
active infection. *In vitro* experiments showed that seaweed had mitogenic
effects on human T-cells.

A recent study in South Carolina evaluated the effects of dietary seaweed in
patients with HIV infection.[58] Patients were not required to be on anti-
retroviral therapy and were referred to the study because of declining CD4 cell
counts and increasing viral load. The first part of the study was a 3 month
safety study. After 3 months, additional patients were recruited for a longer
duration of therapy. Five and six patients participated in the shorter and
longer studies, respectively. Patients were randomly assigned to one of the
three treatment arms: *Undaria pinnatifida* from the Patagonian coast of
Argentina or spirulina (*Arthrospira platensis*) from California or both. The
daily dose regimens were 5 g of *Undaria* (10 capsules); 6 g of spirulina (10
capsules) or a combination (10 capsules with 2.5 g of *Undaria* + 3 g of
spirulina). The results showed that dietary seaweed was non-toxic to the
subjects with HIV infection who had never consumed seaweed. The *Undaria*-
only and *Undaria* and spirulina combination groups demonstrated some
effects. Quality of life indicators showed improvement at 3 weeks. The CD4
cells and HIV-1 viral loads remained stable. One subject who continued in the
study for 13 months showed significant improvement in CD4 counts and
decreased HIV viral load. The investigators speculated that dietary seaweed
may offer immediate support for people with HIV.

The above controlled studies and case reports indicated potential causality
between dietary seaweed intake and health benefits. Combined with the results
from observational studies, positive human research data indicating health
benefits of dietary seaweed have been accumulating. In contrast, TCM has a
unique standard of practice in evaluating and treating illness in which the
principal herb is prescribed with multiple other herbs with synergistic effects.
The TCM standard practice makes it challenging to quantify the effectiveness
of medicinal uses of an herbal ingredient. In recent years, rapidly increasing
numbers of TCM clinical studies and case reports have been published in
Chinese and international journals.

12.3.3 TCM Clinical Studies – Herbal Uses of Seaweed

Researchers at Sīchuān, China, analyzed 3,159 TCM randomized controlled trials (RCTs) conducted in China, Hong Kong and Taiwan during 1965–2008 and published in either Chinese or foreign journals.[59] The meta-analysis performed quality assessment of the clinical trial design, focusing on the randomization and blinding methods. From 1993 to 2007, Chinese and foreign journals published 2,535 and 374 RCTs, respectively. The number of reports published between 1965 and 1993 was relatively small. For the RCTs published in Chinese journals, the proportion of reports with adequate randomization was increased, but not that of blinding methods. In contrast, TCM RCTs reported in foreign journals showed a large improvement in both parameters. The survey concluded that many of the TCM RCTs reported in Chinese journals during 1993–2007 lacked key information and it was difficult to interpret the studies. The results of this survey showed the limitations of applying conventional RCT design standards to the TCM system.

12.3.3.1 Seaweed in TCM Clinical Studies

The following are five human studies from 2002 to 2010 that reported positive effects of seaweed used as herbal therapy. A 2002 clinical study in Běijīng examined endothelial cell injury in atherosclerosis and protective effects of a herbal liquid formulation taken daily for 2 months.[60] Subjects were patients with a diagnosis of coronary artery disease with stable angina pectoris with hypertension, mild to moderate cerebral arteriosclerosis or primary hyperlipidemia. Thirty subjects received the test formulation, and 31 subjects in the positive control group received 20 mg per day of lovastatin, a cholesterol-lowering drug. The TCM intervention was a liquid formulation containing *Radix ginseng rubra* as the principal herb with *Fructus crataegi* and *Sargassum* sp., a brown alga, as the supporting herbs. The TCM rationale for the formula was to reinforce deficient spleen *Qi*, eliminate phlegm and dampness, remove blood stasis and eventually to prevent the formation of atherosclerosis. The results showed that both the herbal formulation and lovastatin significantly decreased total cholesterol, low-density lipoprotein cholesterol, apolipoprotein ApoB-100 and Apo-AI. Triglyceride levels decreased in the test group only. Also, the level of plasma endothelin decreased significantly in the test group. Adverse reactions were unremarkable.

In 2002, investigators from Zhèjiāng treated patients with chronic gastritis for 1 month with a capsule formulation containing *Shí huā cài*, a red alga *Gelidium* sp.[61] A total of 140 patients with chronic gastritis were randomized to the test group of 100 patients and the control group of 40 patients who received *Yang- Wei-Chong- Ji*, a commercial herbal prescription used for gastritis. The results showed the test and the control group had similar effectivenesses of 95 and 92.5%, respectively.

A 2007 case report from Guǎngdōng and Jiāngxī discussed the effectiveness of *Hǎi zǎo*, *Sargassum* sp. (brown alga) and licorice in treating 120 women with mammary gland hyperplasia.[62] The prescription was a combination of nine herbs including *Hǎi zǎo* and licorice. The study design allowed the investigators to prescribe any of the ten additional herbs as needed according to the individual patient's needs. The results showed 92.5% total effectiveness. No toxic side effects or allergic reactions were observed.

A 2006 case report from Shǎnxī included 34 patients with uterine muscle tumor who were treated with brown algae, *Hǎi zǎo* and *Kūn bù* (*Ecklonia* sp.) and nine other herbs in a decoction.[63] Additional herbs were prescribed as needed according to each individual situation. The treatment showed an effectiveness of 82.35% after three courses of therapy.

A 2010 study in Guǎngdōng investigated 108 patients with thyroid enlargement (goiter) treated with *Hǎi zǎo yù hù tāng* herbal prescription, the main ingredients of which are brown algae, *Hǎi zǎo* and *Kūn bù*.[64] The comparator group consisted of 116 patients who took a non-seaweed-containing herbal mixture, *Xiāo yáo sǎn*. The parameters used for determination of efficacy were the reduction of the enlarged glands, progression to surgery, blood tests of thyroid function, clinical examination of the mass and overall symptoms. The study design allowed patients with hyperthyroidism to receive additional herbs as needed. The study concluded that *Hǎi zǎo yù hù tā ng* showed only a short-term efficacy and cautioned the use of iodine-containing *Hǎi zǎo* and *Kūn bù* for patients with hyperthyroidism and goiter when no surgery was planned.

The above studies illustrated some of the challenges of designing TCM controlled trials. For example, seaweed was one of the ingredients in a mixture of herbs and it is difficult to evaluate the effects of a single ingredient. The study design allowed the investigators to prescribe additional herbs as needed for each individual patient. This is a common practice in TCM because the treatment is personalized; however, it makes it difficult to interpret the efficacy results. In terms of the formulation or manufacturing-related matters, the identity and source of herbal ingredients, Latin binomials, methods of preparation of the capsule or liquid formulation were not sufficiently provided. Comparing the English and Chinese reports, the 2002 Beijing study written in English contained the most detailed information.

12.3.3.2 Reporting Quality Improvement

Over the last 5 years, progress has been made in improving the quality of reporting TCM RCTs. In 2007, Consolidated Standards for Reporting Trials of Traditional Chinese Medicine (CONSORT for TCM) was published in English and Chinese for public comment.[65] The Committee consisted of experts from China, Hong Kong, Canada and the USA. The 2011 Update recommended transparency in several aspects of TCM study design, including the rationale for study design, diagnostic criteria, key information about intervention, outcome definition and safety reporting.[66]

The Update emphasized the importance of clearly explaining the overall rationale of the study design so that readers can understand the validity of the study. Because TCM has a unique system of determining diagnosis based on identification of key *zhèng* (patterns) of disease manifestations, the diagnostic criteria for patient selection must be explained in terms of *zhèng* imbalance or, where possible, both TCM and Western terms. The rationale for selecting the herbal combinations for both therapeutic interventions and controls must be explained.

For TCM interventions with oral herbal prescriptions, detailed information about the herbal formula is crucial because ideally, the results must be reproducible. The authors consider the following to be the crucial information: the rationale for herbal combinations, botanical species identity, chemical identity and pharmacological evidence for each active ingredient, dosage, treatment regimen and, if a proprietary formula, the product name and the manufacturer's information. It is not uncommon for a TCM practitioner to modify herbal prescriptions according to an individual situation. In that case, the rationale must be clearly explained and the change recorded. If a positive control medicine was used, detailed information should be included. The medicinal herbs must be free of impurities and contamination. The methodology used to ensure the quality of herbs should be reported, including the source of the herbs, place of harvest, methods used for validating the active ingredients and the control used for standardization. When a placebo was manufactured, the inactive ingredients must be disclosed.

It is also emphasized that safety reporting of side effects and adverse events (AEs) is necessary. The known side effects of the herbs should be described and the assessment methods for each AE and rationale for treatment explained. If AEs occur, they must be disclosed and the interpretation should be made in terms of TCM theory and conventional medicine if applicable.

From the Chinese regulators, the initiatives to advance research in TCM were recently announced. On 25 April 2012, the Center for Drug Evaluation of the State Food and Drug Administration (SFDA) announced that a special project will be launched.[67] The purposes are to focus on the analysis of existing data from TCM-based diagnosis and therapy, to promote new clinical methods for evaluating the efficacies of new syndrome-based TCM, to evaluate standards for TCM diagnosis and therapy and to establish clinical trials guidelines unique to TCM.

As illustrated above, recently, international research experts from both TCM and Western medicine as well as the Chinese regulators have been working diligently towards promoting transparency in TCM. Probably the new standard will be a blend of the unique TCM syndrome-oriented system and the conventional disease-oriented approach.

12.4 Safety

Brown alga is known for its high content of iodine. The average for 10 *Laminaria* species obtained worldwide was 1,542 μg g^{-1} dry weight.[34] Even though dried seaweed is soaked and washed before cooking and the mineral contents absorbed may be reduced, people with pre-existing thyroid conditions are recommended to consult healthcare professionals before taking excessive amounts of dietary seaweed because iodine may cause negative effects on abnormal thyroid function.[68]

One recent prospective study in Japan showed that high consumption of dietary seaweed was associated with higher risks of thyroid papillary carcinoma in postmenopausal women, but not in premenopausal women.[69] The investigators speculated that the increased risk may have been due to an antiestrogenic activity of seaweed. As Chen *et al.* discussed, estrogens may promote thyroid cancer cell proliferation and growth *via* estrogen receptor alpha (ERα).[70] The expression of ERα on papillary thyroid carcinoma was higher in premenopausal than postmenopausal women. Also, *in vitro*, seaweed extract was able to block the binding of estradiol to ERα. Overall, the authors speculated that postmenopausal women having lower levels of ERα expression on the thyroid cancer cells received less protective advantage from dietary seaweed, which may have resulted in a higher risk of thyroid cancer.

Alternatively, it may have been possible that the higher risk of cancer was associated with a higher intake of iodine from dietary seaweed; however, this study was limited in several ways. Because older women in Japan tend to consume higher amounts of seaweed, in this study the higher seaweed consumers were also older and postmenopausal. The determination of seaweed consumption was performed solely using questionnaires and other possible dietary sources of iodine were not investigated, and no samples were collected for quantifying iodine in the body. The authors therefore concluded that a different prospective study is needed to investigate iodine intake and cancer risk.

Seaweed and seaweed-based dietary supplements have been shown to contain various bioactive substances. It is possible that any such substances may manifest physical effects and/or interact with pharmacological agents. It is therefore important to communicate to healthcare professionals or ask patients about consumption of seaweed or seaweed containing dietary supplements as recommended by Obasanjo.[71] For example, bladderwrack (*Fucus vesiculosus*) may influence gut motility due to alginates and may lower blood pressure because of ACE inhibitory activity. High regular consumption of brown algae may cause an increased intake of iodine, which may influence thyroid function tests. Also, glycemic response and blood glucose values could be modified. The anticoagulation effects of sulfated polysaccharides may influence blood coagulation parameters. Moreover, the presence of vitamins and micronutrients in seaweed may influence laboratory values.[72]

Even though seaweed is a good source of important minerals, it may accumulate undesirable heavy metals. Seaweed grown in seawater contaminated with heavy metals such as lead, mercury and cadmium may accumulate these pollutants. It has been shown that *hijiki* (*Sargassum fusiforme*) can accumulate inorganic arsenic harmful to human health. The Food Standards Agency in the U.K. advised consumers not to eat *hijiki* because of its content of inorganic arsenic.[73]

Seaweed for human consumption is harvested domestically or around the world. Contamination with pesticides and pollutants may occur during farming, harvesting or processing. Contamination can also occur during the manufacturing stage, including liquid extraction, formulation, encapsulation and packaging. Because assurance of product safety is the most important issue for consumers, a set of standards were established over the years through the consumer feedback system followed by industry improvement efforts and public health initiatives by regulators. As outlined by Lee *et al.*, a set of performance standards safeguards the quality of farmed food products, dietary supplements and pharmaceutical products such as Good Agricultural Practice, Good Manufacturing Practice, Good Supply Practice and, in the case of animal testing and human sample testing, Good Laboratory Practice and Good Clinical Laboratory Practice, respectively.[74]

12.5   Conclusion – Totality of Evidence

TCM uses seaweed as a medicine and as a health food. Modern biochemistry research has supplied abundant positive data supporting the health-promoting bioactivity of seaweed constituents (Section 12.3.1 and Table 12.4). Large epidemiological studies have shown that dietary seaweed appears to be associated with a lower risk of developing chronic diseases associated with diet. Recently, small controlled clinical trials were conducted to investigate dietary seaweed and health effects. Measurable pharmacodynamic effects were reported on glycemic response, type II diabetes, the metabolic syndrome, estrogen regulation and viral illness (Section 12.3.2 and Table 12.5). Over the past 20 years, increasing numbers of TCM hospitals have been reporting results from RCTs in Chinese and foreign journals; however, meta-analysis has shown that the results are difficult to interpret because of a paucity of key information relating to biochemistry and manufacturing of the herbal ingredients, and study design. Currently, therefore, the TCM uses of seaweed are validated mostly by translational biochemistry research, epidemiological observations and the data from a small number of non-TCM controlled human clinical studies (Table 12.6). While there is a strong historical and cultural basis for medicinal uses of seaweed, currently the positive results from TCM human clinical trials are difficult to interpret. That being said, it is a time of transformation. Over the past 5 years, dialogue between the experts from both TCM and biometric-oriented Western style clinical research has been actively taking place. A unique TCM-focused clinical trial design may be applied to seaweed TCM studies in the near future.

Acknowledgements

The author greatly appreciates several multilingual colleagues who corrected the pinyin and translations. The author is a medical and nutritional product development consultant and has no commercial interest in any of the products mentioned. The content of this chapter is for general information purposes only and gives no medical recommendations.

References

1. Y. V. Yuan, in *Marine Nutraceuticals and Functional Foods*, ed. C. Barrow and F. Shahidi, CRC Press, Boca Raton, FL, 2008, p. 259.
2. J. H. Fitton, M. R. Irhimeh and J. Teas, in *Marine Nutraceuticals and Functional Foods*, ed. C. Barrow and F. Shahidi, CRC Press, Boca Raton, FL, 2008, p. 345.
3. J. D. Adams Jr and C. Garcia, *eCAM*, 2005, **2** (2), 143.
4. D. J. McHugh and Australian Defence Force Academy, *A Guide to the Seaweed Industry*, Food and Agriculture Organization of the United Nations, Rome, 2003.
5. S. Skoler-Karpoff, G. Ramjee, K. Ahmed, L. Altini, M. G. Plagianos, B. Friedland, S. Govender, A. De Kock, N. Cassim, T. Palanee, G. Dozier, R. Maguire and P. Lahteenmaki, *Lancet*, 2008, **372** (9654), 1977.
6. B. Li, F. Lu, X. Wei and R. Zhao, *Molecules*, 2008, **13**, 1671.
7. S. La Barre, P. Potin, C. Leblanc and L. Delage, *Mar. Drugs*, 2010, **8**, 988.
8. S. Y. Hu, in *An Enumeration of Chinese Materia Medica*, eds. Y. C. Kong and P. P. H. But, Chinese University Press, Hong Kong, 1980, p. 191.
9. Complementary and Alternative Healing University, Chinese Medical Classics, http://alternativehealing.org/list_of_Chinese_medical_classics. htm#全國中草藥匯編 (last accessed 9 July 2012).
10. Eastland Press, *List of East Asian Medical Book Titles*, http://www. eastlandpress.com/resources/ (last accessed 9 July 2012).
11. Wikipedia, *Traditional Chinese Medicine*, http://en.wikipedia.org/wiki/ Traditional_Chinese_medicine (last accessed 9 July 2012).
12. Complementary and Alternative Healing University, *Traditional Chinese Medicine (TCM) Diagnostics: The Eight Outlines*, http://alternativehealing. org/tcm_diagnosis.htm (last accessed 9 July 2012).
13. Hal Yakkyoku Health 24, Hakkou Bensho, *Chinese Herbal Medicine*, http://www.hal.msn.to/kankaisetu/kanpokai1637.html (in Japanese) (last accessed 9 July 2012).
14. World Health Organization, WHO International Standard Terminologies on Traditional Medicine in the Western Pacific Region, 2007, http://www. wpro.who.int/publications/docs/WHOIST_26JUNE_FINAL.pdf (last accessed 9 July 2012).
15. L.-Y. Sheen and G. Fuentes, in *Functional Foods of the East*, eds J. Shi, C.-T. Ho and F. Shahidi, CRC Press, Boca Raton, FL, 2011, p. 1.
16. Y. Yang, *Chinese Herbal Medicines: Comparisons and Characteristics*, Churchill Livingstone, Edinburgh, 2002.
17. National Taiwan Museum, *Seaweeds Net of Taiwan*, http://web.ntm.gov. tw/seaweeds/home.asp (last accessed 9 July 2012)
18. X. Chen, *J. Zhejiang Coll. Tradit. Chin. Med.*, 2000, **24** (2), 65 (in Chinese).

19. C. Zeng and J. Zhang, *Hydrobiologia*, 1984, **116**, 1.
20. S. Clavey, *Fluid Physiology and Pathology in Traditional Chinese Medicine*, Elsevier Health Sciences, New York, 2003, p. 265.
21. Y. Li, G. Zhong, Q. Wang and H. Liu, *Nanjing Zhongyiyao Daxue Xuebao*, 2011, **27**, 4 (in Chinese; Abstract in English).
22. K. Nakachi, K. Imai, Y. Hoshiyama and T. Sasaba, *J. Epidemiol. Community Health*, 1988, **42** (4), 355.
23. Y. Hoshiyama and T. Sasaba, *Jpn. J. Cancer Res.*, 1992, **83** (9), 937.
24. Y. Hoshiyama, T. Sekine and T. Sasaba, *Tohoku J. Exp. Med.*, 1993 **171** (2), 153.
25. L. Brownlee, A. Fairclough, A. Hall and J. Paxman, Dietary seaweed and human health, in *Culinary Arts and Sciences VII: Global, National and Local Perspectives*, ed. H. H. Hartwell, P. Lugosi and J. S. A. Edwards, International Centre for Tourism and Hospitality Research, Bournemouth University, Poole, 2011, p. 82.
26. S. L. Holdt and S. Kraan, *J. Appl. Phycol.*, 2011, **23**, 543.
27. Food and Agriculture Organization, *Physiological Effects of Dietary Fibre in Carbohydrates in Human Nutrition*, Food and Nutrition Paper 66, 1998, http://www.fao.org/docrep/W8079E/W8079E00.htm (last accessed 9 July 2012).
28. J. H. Hehemann, G. Correc, T. Barbeyron, W. Helbert, M. Czjzek and G. Michel, *Nature*, 2010, **464**, 908.
29. J. H. Fitton, *Mar. Drugs*, 2011, **9**, 1731.
30. P. A. Hwang, S. Y. Chien, Y. L. Chan, M. K. Lu, C. H. Wu, Z. L. Kong and C. J. Wu, *J. Agric. Food Chem.*, 2011, **59** (5), 2062.
31. N. Araya. *Antiviral Ther.*, 2011, **16**, 89.
32. X. Wu, M. Yang and X. Huang, *Chin. J. Rad. Mediation. Protect.*, 2004, **01**, 1.
33. H.-I. Back, S.-Y. Kim, S.-H. Park, M.-R. Oh, M.-G. Kim, J.-Y. Jeon, S.-W. Chae, J. S. Bae, Y. S. Kim, Y. J. Hwang and S. T. Lee, *FASEB J.*, 2010, **24**, 1347.
34. T. T. Zava and D. T. Zava, *Thyroid Res.*, 2011, **4**, 14.
35. A. J. Smit, *J. Appl. Phycol.*, 2004, **16**, 245.
36. F. Shahidi, in *Marine Nutraceuticals and Functional Foods*, ed. C. Barrow and F. Shahidi, CRC Press, Boca Raton, FL, 2008, p. 23.
37. M.-S. Yoo, J.-S. Shin, H.-E. Choi, Y.-W. Cho, M.-H. Bang, N.-I. Baek and K.-T. Lee, *Food Chem.*, 2012, **135** (3), 967.
38. J. Peng, J. P. Yuan, C. F. Wu and J. H. Wang, *Mar. Drugs*, 2011, **9**, 1.
39. H. Maeda, M. Hosokawa, T. Sashima, K. Funayama and K. Miyashita, *Biochem. Biophys. Res. Commun.*, 2005, **332**, 392.
40. P. MacArtain, C. I. R. Gill, M. Brooks, R. Campbell and I. R. Rowland, *Nutr. Rev.*, 2007, **65**, 12.
41. Y. Li, Z.-J. Qian, M.-M. Kim and S.-K. Kim. *J. Food Biochem.*, 2011, **35**, 357.
42. W. A. J. P. Wijesinghe, S.-C. Ko and Y.-J. Jeon, *Nutr. Res. Pract.*, 2011, **5** (2), 93.

43. Y. Wei, C. Wang, J. Li, Q. Guo and H. Qi, *Chin. J. Oceanol. Limnol.*, 2009, **27** (3), 558.

44. C. Li, *Yaoxue Xuebao*, 1965, **12**, 10.

45. D. S. Bhakuni and D. S. Rawat, *Bioactive Marine Natural Products*. Springer, New York, 2005, p. 192.

46. T. Higa and M. Kuniyoshi, *J. Toxicol. Toxin Rev.*, 2000, **19** (2), 119.

47. S. M. M. Shanab, E. A. Shalaby and E. A. El-Fayoumy, *J. Biomed. Biotechnol.*, 2011, **2011**, 726405.

48. M. L. Cornish and D. J. Garbary, *Algae*, 2010, **25**, 155.

49. I. Jaswir and H. A. Monsur, *J. Med. Plants Res.*, 2011, **5**, 7146.

50. K. Wada, K. Nakamura, Y. Tamai, M. Tsuji, Y. Sahashi, K. Watanabe, S. Ohtsuchi, K. Yamamoto, K. Ando and C. Nagata, *Nutr. J.*, 2011, **10**, 83.

51. I. Goñi, L. Valdivieso and A. Garcia-Alonso, *Nutr. Res.*, 2000, **20** (10), 1367.

52. M. S. Kim, J. Y. Kim, W. H. Choi and S. S. Lee, *Nutr. Res. Pract.*, 2008, **2** (2), 62.

53. J. Teas, M. E. Baldeón, D. E. Chiriboga, J. R. Davis, A. J. Sarriés and L. E. Braverman, *Asia Pac. J. Clin. Nutr.*, 2009, **18** (2), 145.

54. C. F. Skibola, *BMC Comp. Alt. Med.*, 2004, 4, 10.

55. C. F. Skibola, J. D. Curry, C. VandeVoort, A. Conley and M. T. Smith, *J. Nutr.*, 2005, **135**, 296.

56. J. Teas, T.G. Hurley, J. R. Hebert, A. A. Franke, D. W. Sepkovic and M. S. Kurzer, *J. Nutr.*, 2009, **139**, 939.

57. R. Cooper, C. Dragar, K. Elliot, J. H. Fitton, J. Godwin and K. Thompson, *BMC Comp. Alt. Med.*, 2002, **2**, 11.

58. J. Teas and M. R. Irhimeh, *J. Appl. Phycol.*, 2012, **24**, 575.

59. J. He, L. Du, G. Liu, J. Fu, X. He, J. Yu and L. Shang, *Trials*, 2011, **12**, 122.

60. Y. Wang, F. Niu, M. Zhao, B. Zhong, J. Hao, W. Yan, X. Lu, W. Wang, Q. Li, Q. Lin, Y. Dai, Y. Zhao, M. Xiao, L. Tang, L. He, X. Jiang, W. Guo, S. Wang and W. Lu, *J. Trad. Chin. Med.*, 2002, **22**, 2.

61. J. Chen, J. Chen, X. Zhou, Z. Shen, J. Hu, *Zhejiang J. Int. Trad. Chin. Western Med.*, 2002, **12**, 3 (in Chinese; Abstract in English).

62. M. Peng and X. Yang, *Yí Chūn Xué Yuàn Xué Bào*, 2007, **6**, 122 (in Chinese).

63. W. Jia, *Xiandai Zhongyiyao, Mod. Chin. Med.*, 2006, **26**, 6 (in Chinese).

64. H. Zeng, S. Wang, W. Lin, H. Chen, Z. Yan, T. Liang, *Guangdong Med. J.*, 2010, **31**, 8 (in Chinese).

65. T. Wu, Y. Li, Z. Bian, T. Li, J. Li, S. Dagenais and D. Moher, *Chin. J. Evid. Based Med.*, 2007, **7**, 9.

66. Z. Bian, B. Liu, D. Moher, T. Wu, Y. Li, H. Shang and C. Cheng, *Front. Med.*, 2011, **5** (2), 171.

67. Center for Drug Evaluation, *News 20120425*, http://www.cde.org.cn/news.do?method=largeInfo&id=312664 (in Chinese) (last accessed 28 July 2012).

68. K. Welch, *Herbs for Potential Adjunct Treatment of Thyroid Disease*, www.encognitive.com/node/3668 (last accessed 21 June 2012).

69. T. Michikawa, M. Inoue, T. Shimazu, N. Sawada, M. Iwasaki, S. Sasazuki, T. Yamaji, S. Tsugane and Japan Public Health Center-Based Prospective Study Group, *Eur. J. Cancer Prev.*, 2012, **21** (3), 254.

70. G. G. Chen, A. C. Vlantis, Q. Zeng and C. A. van Hasselt, *Curr. Cancer Drug Targets*, 2008, **8** (5), 365.

71. O. Obasanjo, *Dietary Supplements and the Clinical Encounter*, American College of Preventive Medicine, www.medscape.org/viewarticle/752282 (last accessed 18 July 2012).

72. Majoria Drugs, *Wellness Library: Herbs and Supplements, Seaweed, Kelp, Bladderwrack (Fucus vesiculosus)*, www.majoria.com/ns/DisplayMonograph. asp?DocID=bottomline-bladderwrack&storeID=CLVHMWU1GD7C9 NC2U0ENDSKJCXMF7X28#SAFETY (last accessed 18 July 2012).

73. Food Standards Agency, *Consumers Advised Not to Eat Hijiki Seaweed*, http://www.food.gov.uk/news-updates/news/2010/aug/hijikiseaweed, 4 August 2010 (last accessed 28 July 2012).

74. F. S. C. Lee, X. Wang and P. P. Fu, in *Functional Foods of the East*, ed. J. Shi, C. T. Ho and F. Shahidi, CRC Press, Boca Raton, FL, 2011, p.431.

CHAPTER 13

The Preventive Effect of Traditional Chinese Medicinal Herbs on Type 2 Diabetes Mellitus

ZHIJUN WANG[a], PATRICK CHAN[b] AND
JEFFREY WANG*[a,c]

[a] Center for Advancement of Drug Research and Evaluation; [b] Department of
Pharmacy Practice and Administration; [c] Department of Pharmaceutical
Sciences, College of Pharmacy, Western University of Health Sciences,
Pomona, CA 91766, USA
*E-mail: jwang@westernu.edu

13.1 Introduction

13.1.1 Diabetes Statistics

Diabetes mellitus (DM) is a chronic disease resulting from endocrine dysfunction. According to the World Health Organization in 2011, ~ 364 million people globally suffer from DM, with projections that DM-related deaths will double from 2005 to 2030.[1] In 2004, 3.4 million people died directly from the consequences of high blood glucose. The growing concern is the epidemic growth of obesity and the increase in the elderly population, which will continue to increase the prevalence of DM.

RSC Drug Discovery Series No. 31
Traditional Chinese Medicine: Scientific Basis for Its Use
Edited by James David Adams, Jr. and Eric J. Lien
© The Royal Society of Chemistry 2013
Published by the Royal Society of Chemistry, www.rsc.org

The Centers for Disease Control estimate that in the USA alone, 25.8 million Americans, or 8.3% of the population, suffer from DM, with 7 million currently undiagnosed.[2] DM is higher, 26.9%, in the elderly (aged 65 years or older). However, it has also quickly become a disease observed in younger patients. Almost 2 million adults over the age of 20 years were newly diagnosed with DM in 2010. More alarmingly, 35% of adults over the age of 20 years and 50% of the elderly had pre-diabetes, which equates to 79 million people in the USA. DM is the primary cause of renal failure, non-traumatic lower-limb amputations and newly diagnosed retinopathy.

DM is the seventh leading cause of mortality of Americans. The risk of death is twice in those with DM than those without the disease of similar age. Both heart disease and stroke are 2–4 times higher in patients with DM compared with those without DM and about two-thirds of DM patients have hypertension. DM is related to 44% of all newly diagnosed cases of renal failure. Peripheral neuropathy is also prevalent in DM patients. Between 60 and 70% of DM patients have mild to severe nerve damage, resulting in reduced sensation in the feet or hands, decreased digestion of food, erectile dysfunction and carpal tunnel syndrome. Reduced blood flow and peripheral neuropathy resulting in nerve damage may result in infections that require amputation of lower limbs. Over 60% of non-traumatic lower-limb amputations are related to patients with DM. Cardiovascular-related diseases account for more than 70% of deaths in patients suffering from DM.[3]

13.1.2 Pathology and Therapeutics

DM is the result of a combination of genetic susceptibility, environmental influences and increased calorie consumption, of which fat is the most important element.

Type 1 DM (T1DM) is considered an autoimmune disorder which causes the destruction of pancreatic β-cells, although the exact pathogenesis is unknown. At present, there is no effective way to prevent T1DM.

Type 2 DM (T2DM), a rapidly growing health concern in both developed and developing nations, accounts for over 90% of diabetic cases globally.[4,5] T2DM is thought to start with a state of insulin resistance which will cause the normally insulin-responsive tissues to become relatively refractory to insulin's effect. Although insulin resistance can be overcome by increasing insulin secretion from pancreatic β-cells at early stages, β-cells eventually fail to compensate for the insulin resistance of the target tissues due to the loss of or decreased renewal of β-cells. The disorder typically develops gradually, without obvious symptoms at the onset, and is frequently diagnosed either by elevated blood glucose in routine screening tests or after the disease has become severe enough to cause polyuria and polydipsia. Obesity is the single most important factor for T2DM and 80% of patients are obese. In many cases, insulin resistance is the result of obesity and sedentary lifestyle, although the molecular predisposition is poorly understood. Many investigators have described insulin receptors and post-receptor signaling defects as other potential pathological mechanisms.

T2DM is associated with multiple comorbidities and complications. DM education, prevention and care are complex and should be designed to be patient specific. Physicians, nurses, pharmacists and dieticians are often recruited as a balanced healthcare team in managing a patient's DM. The American Diabetes Association promotes DM self-management education, a process in which the patient is equipped with the knowledge and skills to provide self-care, manage crisis and make lifestyle interventions.[4,5] Table 13.1 lists the most frequently prescribed drugs currently used in the treatment of T2DM.

13.2   Prevention of T2DM

The prevention of T2DM could be classified into four stages: primordial, primary, secondary and tertiary level strategies.[6] Primordial interventions include dietary and lifestyle modifications such as reduction of fat or salt intake, increase of physical activity and weight loss in the healthy population. Primary intervention is initiated at the stage of impaired fasting glucose/impaired glucose tolerance. At this juncture, it is still possible to prevent the progression to T2DM with the use of pharmacological agents in addition to the aforementioned lifestyle modifications. When patients require secondary prevention, it involves retarding the development of diabetic complications. Tertiary prevention encompasses the treatment of specific diabetic complications and to mitigate morbidity and mortality. Hence the prevention of diabetes can be applied in all stages of T2DM. Conceptually, there is no significant difference in the approaches between the treatment and prevention of T2DM. However, diabetic patients require more aggressive drug therapy and frequent monitoring.

The primordial stage is also termed pre-diabetes.[6-8] Patients with pre-diabetes have a fasting blood glucose (FBG) level of 100–125 mg dL^{-1} (5.6–6.9 mmol L^{-1}), a 2 h plasma glucose concentration following an oral glucose tolerance test (OGTT) of 140–199 mg dL^{-1} (7.8–11.0 mmol L^{-1}) or an HgbA$_{1C}$ of 5.7–6.4%. Treating pre-diabetes to prevent progression to overt diabetes could be significantly beneficial. Maintaining the blood glucose level in the normal range prevents diabetic complications.[8] Identification of individuals with pre-diabetes provides an opportunity to treat those who are at high risk for developing overt T2DM.

Lifestyle modifications are the most common and effective ways of T2DM prevention. However, certain pharmacological interventions have also demonstrated efficacy in lowering the incidence or delaying the onset of the diabetes.[9] In the Diabetes Prevention Program trial, the lifestyle modification was found to decrease the incidence by 58% (from 11.0 to 4.8% per year), while metformin (850 mg, twice daily) decreased the incidence of DM by 31% (from 11.0 to 7.8%) in 3234 non-diabetic persons during 2.8 years.[10] As such, metformin is the only

Table 13.1 Commonly prescribed anti-T2DM agents in the USA.

Class		Drug	Mechanism of action	Expected HgbA$_{A1C}$ decrease as monotherapy (%)[145,146]
Biguanides		Metformin	• Inhibits gluconeogenesis • Increases skeletal muscle uptake of glucose • Decreases intestinal absorption of glucose	1.0–2.0
Sulfonylureas	1st generation	Chlorpropamide, tolazamide, tolbutamide	• Stimulates insulin release from pancreatic β-cells • Decreases hepatic gluconeogenesis • Increases insulin sensitivity of peripheral tissues	1.0–2.0
	2nd generation	Glyburide, glimepiride, glipizide		
Thiazolidinediones		Pioglitazone, rosiglitazone	• Enhances peripheral insulin sensitivity by activating peroxisome proliferator-activated receptors (PPARs)	0.5–1.4
Meglitinides		Nateglinide, repaglinide	• Blocks ATP-dependent potassium channels in the membrane of β-cells • Stimulates rapid and short-lived release of insulin	0.5–1.5
α-Glucosidase inhibitors		Acarbose, miglitol	• Inhibits enzymatic breakdown of carbohydrates to monosaccharides • Delays intestinal glucose absorption	0.5–0.8

Table 13.1 (*Continued*)

Class	Drug	Mechanism of action	Expected HgbA$_{A1C}$ decrease as monotherapy (%)[145,146]
GLP-1 agonists	Exenatide, liraglutide	• Glucagon-like peptide (incretin) binds to GLP-1 receptors on β-cells • Increases glucose-mediated insulin secretion • Suppresses glucagon secretion • Slows gastric motility • Increases satiety • Reduces hepatic fat content	0.5–1.0
DPP-4 inhibitors	Sitagliptin Saxagliptin	• Blocks enzyme dipeptidyl peptidase-4 • Prevents degradation of GLP-1	0.5–0.8 0.4–0.5
Amylin analogs	Pramlinitide	• Synthetic analog of β-cell amylin • Slows gastric emptying • Inhibits glucagon production • Increases satiety	0.5–1.0

current medication that has been advocated to be used in the prevention of diabetes in high-risk populations such as those with a history of gestational diabetes or morbid obesity and those with progressive hyperglycemia. In addition, the α-glucosidase inhibitor acarbose reduced morbidity by 10% and delayed the development of T2DM in patients with impaired glucose tolerance in the STOP-Noninsulin-Dependent Diabetes Mellitus (STOP-NIDDM) trial.[11] The thiazolidinedione troglitazone reduced the risk by 56% in the Troglitazone in Prevention of Diabetes (TRIPOD) study.[12]

Of the anti-DM agents, metformin can mitigate the symptoms of T2DM. However, whether the incidence of T2DM can be reduced after discontinuation of the drug has not yet been determined. In addition, in the population with BMI <30 and older than 60 years, metformin was not as effective as diet or exercise.[13] Also, as with any prescription drug, it requires regular monitoring. It has undesirable side effects, such as diarrhea, nausea, chest discomfort, flushes, palpitations and hypoglycemia. Side effects must be considered especially when the drugs are being used to prevent or delay diabetes rather than to treat it.[14]

13.3   The Historical Use of TCM Herbs for DM Treatment

Five thousand years ago, the ancient Chinese invented a medical system based on the *Yin–Yang* Theory to diagnose diseases and used many herbs for treatment. This medical and pharmaceutical practice, dubbed traditional Chinese medicine (TCM), has continued without interruption until the present day. DM is identified in TCM due to lack of *qi*, toxin retention and blood stasis in the body. Thus, TCM herbs which replenish *qi*, remove toxins and promote blood circulation are frequently used to treat diabetic patients. TCM herbs provide a good alternative to Western medications for DM prevention due to their effectiveness, fewer side effects and lower cost.[15–19]

13.4   Frequently Used TCM Herbs

It has been reported that ∼800 plants are used in the control of DM as single herbal extracts or in complex formulas, and more than 400 have been demonstrated experimentally to be effective *in vitro* or *in vivo*.[19] Many of the pharmacological mechanisms can be classified as (1) reducing carbohydrate absorption, such as inhibition of α-glucosidase, α-amylase and aldose reductase, (2) stimulating insulin secretion, (3) increasing insulin sensitivity, (4) increasing glucose uptake, (5) potentiating endogenous incretins, (6) being antioxidant and decreasing β-cell apoptosis and (7) inhibiting hepatic glycogenolysis (see Figure 13.1).[19–21] Since multiple chemicals are present in each herbal preparation, some herbal medications may work *via* multiple mechanisms. In the following sections, we summarize the Chinese and English literature and list the most effective TCM herbs under these identifiable mechanisms. Those with less or uncertain evidence are listed in Table 13.2. Most of the studies have been conducted using *in vitro* systems and diabetic animals. However, there has been an increase in randomized placebo-controlled clinical trials. Table 13.3 lists recent and current DM clinical trials with TCM herbs.

13.4.1 Individual TCM Herbs

13.4.1.1 *Reducing Carbohydrate Absorption*

13.4.1.1.1 Barringtonia racemosa

Barringtonia racemosa is an evergreen mangrove tree that grows in Bangladesh, Sri Lanka and the west coast of India. Its bark and leaves are used for snake bites, rat poisoning, boils and gastric ulcers. Its seeds are aromatic and useful

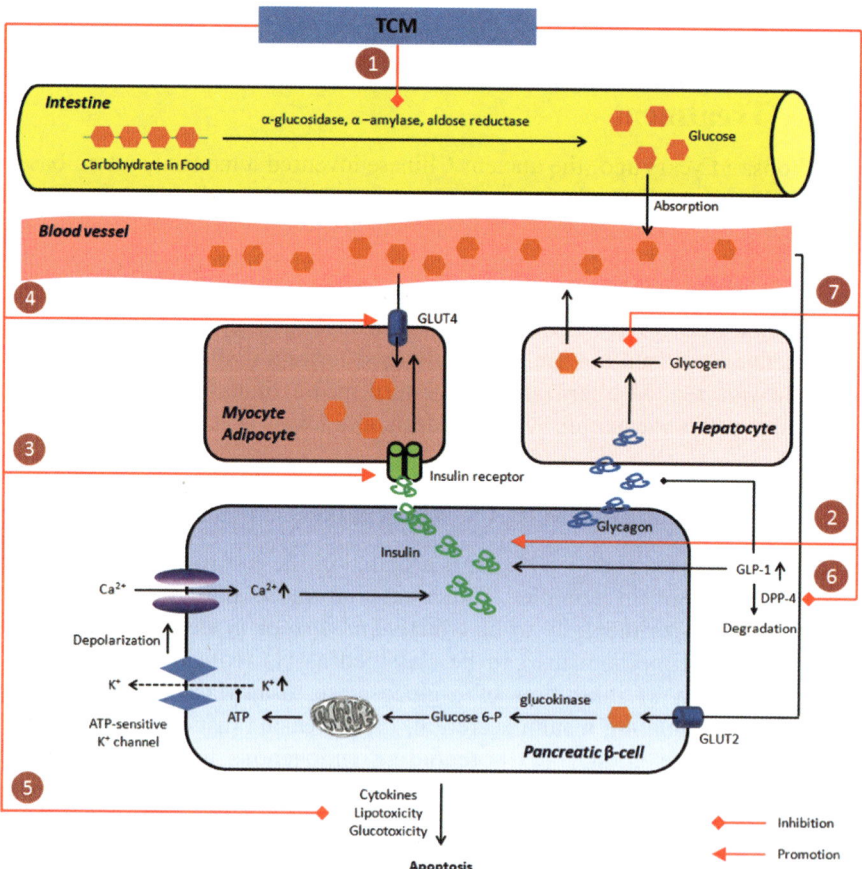

Figure 13.1   Mechanisms of anti-diabetic effects of TCM herbs.

Table 13.2 Less studied TCM herbs with potential antidiabetic effect.

Plant name	Major components	Therapeutic/preventive mechanism	Animal/clinical evidence	Ref.
Acosmium panamense	Quinolizidine alkaloids (acosmine, acosminine, hydroxysparteine), lupinane alkaloids, caffeic acid, desmethylyangonine	Inhibit α-glucosidase (IC$_{50}$ = 109 μg mL^{-1})	The aqueous and butanol extracts have significant hypoglycemic effect in STZ diabetic mice model	147
Barringtonia racemosa Roxb.	Bartogenic acid	Hexane, ethanol, methanol extract can inhibit α-glucosidase		22
Cleistocalyx operculatus Merr and Perry		Inhibit the rat-intestinal maltase and sucrase activities (IC50 = 0.70 and 0.47 mg mL^{-1})	In STZ-induced diabetic rats, aqueous extract can lower postprandial blood glucose at an oral dose of 500 mg kg^{-1}. The glucose level was decreases after 8 months of treatment (500 mg kg^{-1} per day)	148
Commelina communis L.	Isoquercitrin, isorhamnetin-3-O-rutinoside, isorhamnetic-3-O-β-D-glucoside, glucoluteolin, chrysoriol-7-O-β-D-glucoside, orientin, vitexin, isoorientin, isovitexin, swertisin, flavocommelin	α-Glucosidase inhibition	(1) α-Glucosidase activity reduced *in vitro*; decreased pre- and postprandial blood glucose in both diabetic and non-diabetic male ICR rats. (2) α-Glucosidase activity reduced in rat intestine; increased superoxide radical-scavenging activity	149,150
Geranium dielsiaum		Delay carbohydrate digestion		151
Glycine max (L.) Merr	Aqueous extract *via* sprouting or bioprocessing by diet fungus	Inhibit pancreatic α-amylase, slightly inhibit α-glucosidase		152

Table 13.2 (*Continued*)

Plant name	Major components	Therapeutic/preventive mechanism	Animal/clinical evidence	Ref.
Harungana madagascariensis	Prenylated 1,4-anthraquinone, prenylated anthranols (kengaquinone, kenganthranols), anthraquinones, anthrones and xanthones	Inhibit α-glucosidase		153
Hydnocarpus wightiana Blume	Hydnocarpin, luteolin, isohydnocarpin	Inhibit glucosidase and *N*-acetyl-β-D-glucosaminidase		154
Myrcia multiflora DC	Flavanone glucosides (myrciacitrin I and II), new acetophenone glucosides (myrciaphenone A and B)	Inhibit α-glucosidase and aldose reductase	Inhibit the increase of serum glucose level in sucrose-loaded rats and in alloxan-induced diabetic mice	155
Nerium indicum	3-*O*-Caffeoylquinic acid (chlorogenic acid), 5-*O*-caffeoylquinic acid, quercetin, catechins	Inhibit α-glucosideases in a non-competitive manner		156
Pelvetia babingtonii de Toni		Inhibit α-glucosidase	Suppress the elevation of blood glucose level after an oral dose of 1 g kg^{-1}	157
Pinus densiflora		α-Glucosidase inhibition	(1) α-Glucosidase inhibition demonstrated in porcine small intestine, yeast *Saccharaomyces cerevisiae* and bacterium *Bacillus stearothermophilus*	158
Rhizoma Pinelliae	Flavone *C*-glycoside	Inhibit aldose reductase at concentration of 100 μM		20
Ribes nigrum L.	Polyphenolic components	Inhibit α-amylase and α-glucosidase		159

Table 13.2 (*Continued*)

Plant name	Major components	Therapeutic/preventive mechanism	Animal/clinical evidence	Ref.
Salacia chinensis		Inhibit α-glucosidase, rat lens aldose reductase, formation of Amadori compounds and advanced glycation end products, nitric oxide production	Methanolic extract showed antihyperglycemic effects in oral sucrose- or maltose-loaded rate	160
Salacia reticulata	Thiocyclitol	Inhibit α-glucosidase	Lowered the postprandial glucose level in Wistar rats after maltose and sucrose loading	161
Syagrus romanzoffiana	13-Hydroxykompasinol A(1) and scirpusin C(4)	Inhibit α-glucosidase	Kompasinol A and 3,3',4,5',5'-pentahydroxy-*trans*-stilbene reduced the postprandial blood glucose level (10.2 and 12.1% at 10 mg kg^{-1} dose level)	162
Tournefortia hartwegiana	Methanol extract: β-sitosterol, stigmasterol, lupeol, ursolic acid, oleanolic acid, saccharose, *myo*-inositol	Inhibit α-glucosidase	Methanol extract reduced the plasma glucose level after taking glucose, sucrose and maltose (OGTT) following dosing of 310 mg kg^{-1} by gavage	163
Vaccinium corymbosum	Polyphenolic components	Inhibit α-amylase and α-glucosidase		164
Vigna angularis		Inhibit α-glucosidase and α-amylase, no effect on endogenous insulin level	Potential hypoglycemic activity in both normal mice and STZ diabetic rat	165

Table 13.2 (Continued)

Plant name	Major components	Therapeutic/preventive mechanism	Animal/clinical evidence	Ref.
Cordyceps	Parasitic fungus of Cordyceps sp. and larva host of Hepialus armoricanus	Increased insulin release, stimulation of cholinergic activation	(1) Blood glucose and body weight reduced in treated diabetic mice. (2) Western blot analysis showed increased insulin signaling (IRS-1 and GLUT-4) in treated male Wistar rats	166,167
Cortex Phellodendri	Berberine, palmatine, jatrorrhizine	Increased insulin release	Decreased blood glucose and increased insulin secretion in diabetic male Sprague–Dawley rats	168
Radix Stephaniae tetrandrae	Tetrandrine, fangchinoline	Potentiate the insulin release	Fangchinoline (0.3–3 mg kg^{-1}) decreased the blood glucose and increased the level of blood insulin in STZ diabetic mice. Formononetin (0.1 mg kg^{-1}) and calycosin (0.1 mg kg^{-1}) facilitated this effect	20,43
Radix Notoginseng	Ginsenoside Rg1, ginsenoside Re, notoginsenoside R1, ginsenoside Rb1, ginsenoside Rd	Increase insulin sensitivity	Lowered the plasma glucose level in alloxan diabetic mice, lowered FBGL, improved GT and smaller body weight incremental percentage after 30 day treatment in KK-Ay mice (200 mg kg^{-1})	20,169
Rhizoma Dioscoreae	Polysaccharides, diosgenin, linoleic acid, linolenic acid	Increase insulin sensitivity	Water decoction had an antihyperglycemic effect on diabetic mice	20,170
Rhizoma Fagopyri cymosi		Improve insulin sensitivity	Lowered blood glucose and lipids in diabetic animals and patients	171
Allium cepa Bulbus	Allyl propyl disulfide, S-methylcysteine sulfoxide	Lower blood cholesterol level and decrease lipid peroxidation	Onion diet improved the metabolic status in diabetic conditions	20

Table 13.2 (*Continued*)

Plant name	Major components	Therapeutic/preventive mechanism	Animal/clinical evidence	Ref.
Myrtus communis L.		Decreased glucoskinase enzyme activity	Decreased postprandial glucose, glucokinase enzyme activity and glycogen levels in diabetic New Zealand albino rabbits	172
Polygonum multiflorum (fleeceflower root)	2,3,5,4'-Tetrahydroxystilbene 2-*O*-β-D-glucoside	Inhibit the formation of advanced glycation end products in a dose-dependent manner by trapping reactive MGO under physiological conditions; meliorate the nephropathy by antioxidative effect		47
Radix Acanthopanacis senticosi	Saponins	Saponins possess antilipid peroxidation activity to protect the vascular endothelium and prevent diabetic complications	Doses of 10–200 mg kg^{-1} of senticoside A reduced the blood glucose level of diabetic mice and rat induced by glucose, alloxan and adrenalin	20
Radix Astragali seu Hedysari	Isoflavonoids and astragalosides	Reduce the apoptosis in combination with Radix Codonopsis and Cortex Lycii	Reduced blood glucose, triglyceride and myocardial calcium, improved the abnormalities of myocardial ultrastructure and the metabolism of diabetic rats and mice; delayed the onset of T1DM	48,173
Radix Clematidis		Protect RINm5F insulinoma cells	Prevented cytokine-induced β-cell damage in STZ mice	174

Table 13.2 (Continued)

Plant name	Major components	Therapeutic/preventive mechanism	Animal/clinical evidence	Ref.
Rhodiola sachalinensis A.	Polysaccharides, flavones, organic acids, saponins, hydroxybenzenes, terpenes and steroids, anthraquinones, alkaloids, lignins	Antioxidant effect and protection of β-islet cells	Acute decrease in blood glucose and chronic decrease in blood glucose; increase in insulin levels in diabetic male Wistar rats	175
Stigma Maydis	Tannins, proanthocyanidins, flavonoids	Antioxidant effect which may protect the β-cells	Significant decrease in blood glucose levels observed after a single administration of Stigma Maydis in STZ mice	20,176
Bombyx batryticatus	Conjugated linoleic acids, alkaloids	Stimulate liver lipid accumulation in the absence of insulin	CLA may reduce glucogenic activity and promote glycogenesis in diabetic mice	177,178
Radix Paeoniae rubra and Radix Paeoniae alba	Tannins, paeoniflorin, 1,2,3,4,6-penta-*O*-galloyl-β-D-glucose (PGG), paeonol	Inhibition of phosphoenolpyruvate carboxykinase (PEPCK) gene transcription	PEPCK gene transcription reduced in H4IIE cells	179
Radix Polygalae (*Polygala tenuifolia* Willd.)	Saponins	Antiglycation	Glycosylated hemoglobin was reduced in δ-Glu assay; advanced glycation end products (AGEs) evaluated by BSA glucose assay; increased inhibitory effect on AGEs assayed by *N*-acetylglycyllysine methyl ester (GK peptide)–ribose assay; blood samples obtained from male Sprague–Dawley rats	180
Rhizoma Polygoni cuspidate	Polydatin (resveratrol)	Dilate blood capillaries and improve micro-circulation		20

Table 13.2 (*Continued*)

Plant name	Major components	Therapeutic/preventive mechanism	Animal/clinical evidence	Ref.
Radix Ophiopogonis		Inhibit α-glucosidase, increased glucose uptake by tissue	α-Glucosidase activity reduced *in vitro*; glucose uptake reduced in intestinal border membrane vesicles (BBMV) and 3T3-L1 mouse adipocytes	181
Oriciopsis glaberrima Engl.	Acridone alkaloids (atalaphyllidine, oleanolic acid, butulinic acid), atalaphyllidine, oleanolic acid, butulinic acid, β-sitosterol, stigmasterol, stigmasterol glucoside, acridone	Inhibit α-glucosidase, moderate free radical scavenging activity against DPPH		182
Rhus verniciflua	Garbanzol, sulfuretin, fisetin, fustin, mollisacasidin, butein	Inhibit α-aldose reductase (butein); antioxidative effect (ethyl acetate extract and butein and sulfuretin)	Blood glucose level was decreased in STZ diabetic rats after treatment with an oral dose of 50 mg kg^{-1} for 4 weeks	50,178,183
Rhizoma Polygonati	Triterpenoid glycosides (senegins II, III and IV and desmethoxysenegin II)	Inhibit the activity of α-glucosidase in digestive canal and improve the metabolism of glucose and triglyceride	Decreased the glucose level of normal and KK-Ay mice but not STZ mice; RPO reduced fasting blood glucose, glycosylated hemoglobin (GHb) and improved the glucose tolerance in diabetic mice after 4 weeks of treatment	20,184

Table 13.2 (*Continued*)

Plant name	Major components	Therapeutic/preventive mechanism	Animal/clinical evidence	Ref.
Fructus Balsampear	Eleostearic acid, stearic acid	Modulation of pancreatic β-cells, increase insulin secretion; activation of the AMP-activated protein kinase system; Zn-free protein from MC has the insulinomimetic activities	Suppressed postprandial hyperglycemia in rats and decreased fasting and postprandial blood glucose in several clinical trials	41,42
Fructus Corni	Ursolic acid, oleanolic acid, moracin M, kaempferol-3-*O*-β glucopyranoside, quercetin-3-*O*-β-glucopyranoside	Increase the expression of GLUT-4 by promoting proliferation of pancreatic islets and increase postprandial secretion of insulin	AD effect in STZ-induced rats	20,97
Gynostemma Herba	Saponins, 1-ephedrine	Regenerate atrophied pancreatic islets and restore the secretion of insulin	Decreased the plasma glucose level and increased the plasma triglyceride level in diabetic rats at a dose of 1 mg kg^{-1}; indicated they have a hypoglycemic effect in high fat diet in C57BL/6J mice at a dose of 2 g kg^{-1}	20,185,186
Allii sativi Bulbus	Allicin, *S*-allylcysteine sulfoxide	Stimulate *in vitro* insulin secretion, improve glucose tolerance and increase liver glycogen synthesis	Lower blood sugar in normal and alloxan-induced diabetic rat and rabbit	20

Table 13.2 (Continued)

Plant name	Major components	Therapeutic/preventive mechanism	Animal/clinical evidence	Ref.
Radix Salvia Miltiorrhizae	Salvianolic acid B, tanshinone IIA	Increase insulin sensitivity, up-regulate GLUT-4 gene expression and attenuate hepatic PEPCK gene expression		122
Radix et Rhizoma Rhei	Rhein, emodin, rhubarb polysaccharides	Promote glucose uptake and usage in adipocytes and hepatocytes by increasing PPAR-γ, GLUT-2 transporter	Lowered the blood sugar levels of alloxan- and adrenaline-treated hyperglycemic mice. Treatment of diabetic nephropathy clinically	99
Cichorium intybus	Tannins, anthocyanins, phenols, sesquiterpenes (guaianolides, eudesmanolides and germacranolides)	Decreased glucose-6-phosphatase activity and regeneration of β-cells	(1) Blood glucose and glucose-6-phosphatase activity reduced; insulin levels unchanged in diabetic male Sprague–Dawley rats. (2) Increased glucose uptake into 3T3-L1 adipocytes as measured by radiolabeled glucose uptake assay; decreased protein tyrosine phosphatase 1B (PTP1B) signaling	187–189
Murraya koenigii	Chloroform, methanol, aqueous extract	Increase the activity of G-6-phosphate dehydrogenase activity leading to the glycogenesis; increase endogenous insulin secretion; inhibit α-amylase and α-glucosidase	MK reduced postprandial hyperglycemia and improved glucose metabolism in diabetic mice	21,190

Table 13.2 (Continued)

Plant name	Major components	Therapeutic/preventive mechanism	Animal/clinical evidence	Ref.
Prunella vulgaris L. (Labiatae)	Jiangtangsu	Repair β-cells of pancreatic islet to release insulin, prolong antihyperglycemic effect of exogenous insulin and induce glucokinase expression	Antihyperglycemic effect of ethanol extract to mice, decreased the blood glucose level by oral dose of 100 mg kg^{-1} in STZ mice	20,191,192
Cornus officinalis	Iridoid glycosides (morroniside, loganin, mevaloside, loganic acid, 5-hydroxymethyl-2-furfural, 7-O-galloyl-D-sedoheptulose)	Inhibit α-glucosidase; reduce gene expression for hepatic gluconeogenesis, protect β-cells against toxic challenge, enhance insulin secretion	Decreased blood sugar in STZ mice; reduced renal oxidative stress and glycation products in STZ-induced diabetic rats	193–195
Cyclocarya paliurus		Protect β-islet cells; promote insulin secretion; promote use of blood glucose as energy; inhibit α-glucosidase and glycogen phosphorylase ND[a]	Decreased blood glucose in KK-Aγ rats; decreased blood glucose in diabetic Kunming rats	196,197
Agrimonia pilosa Ledeb. (Rosaceae)	Quercitrin, quercetin, hyperoside, taxifoliol, luteolin-7-O-β-D-glucopyranoside, rutin	ND	Diabetes of some patients disappeared. The extract of the herb was proved experimentally to be effective in lowering blood glucose in normal and alloxan-induced diabetic mice	198,199
Amorphophallus spp. (Araceae)	Oligosaccharides, galactomannan	ND	Lowered blood glucose of animals and humans	20

Table 13.2 (*Continued*)

Plant name	Major components	Therapeutic/preventive mechanism	Animal/clinical evidence	Ref.
Aralia elata (Miq.) Seem.	Triterpenoids, glycosides (oleanolic acid, 16-β-hydroxy-18-β-D-oleanolic acid, oleanolic acid 28-*O*-β-D-glucopyranoside, 16-hydroxy-18-β-doleanolic acid 28-*O*-β-D-glucopyranoside, ursolic acids)	ND	Lowered the level of total serum cholesterol, glucose and triglyceride in normal and diabetic animals	20,200
Asparagus cochinchinensis	20-Hydroxyecdysone, ecdysone, ajugasterone C	ND	Can be used to treat diabetes	18,201
Cucurbita moschata (cushaw seed)	Tetrasaccharide glyceroglycolipids	ND	Exhibited glucose-lowering effects in streptozotocin- and high fat diet-induced diabetic mice	202
Ephedra sinica Stapf. and *Ephedra distachya* L.		ND	Fasting blood glucose and insulin levels reduced in treated human premenopausal females	203
Fructus Ligustri lucidi	Catechin, quercetin, verbascoside, aucubin, acteoside, betulinic acid, oleanolic acid, ursolic acid	ND	FLL showed antihyperglycemic effect in STZ-induced diabetic rats	20,50,204
Lonicera japonica Thunb. (honeysuckle flower)		ND	Fasting blood glucose, postprandial glucose, HgA1C, glucose disposal rate and insulin levels were reduced in treated humans	205

Table 13.2 (Continued)

Plant name	Major components	Therapeutic/preventive mechanism	Animal/clinical evidence	Ref.
Nelumbo nucifera Gaertn.	Sterols, alkaloids, glycosides, saponins, tannins, carbohydrates, flavonoids	ND	Postprandial blood glucose levels reduced in diabetic Wistar rats and fasting blood glucose reduced in male Wistar rats	206–208
Radix Asparagi Officinalis	Coumarins	ND	Can be helpful for diabetes care	20
Radix Panacis quinquefolii		ND	DM patients taking capsules showed reduction in postprandial glucose following glucose challenge	209
Radix Trichosanthis	Glycans (trichosans A, B, C, D and E)	ND	The aqueous extract of the roots of *Trichosanthes kirilowii* reduced plasma glucose levels in mice	46

*a*ND: not determined.

Table 13.3 Recent and current clinical trials of TCM herbs for diabetes and its complications.

Clinical trial identifier	Title	Sponsor	Intervention	Phase	Completion date
NCT01471275	The Study of TCM in Prevention and Promote of Community-based Diabetes	Guang'anmen Hospital of China Academy of Chinese Medical Sciences, China	Metformin (250 or 500 mg, t.i.d.). Jiang-Tang-Tiao-Zhi decoction (8 herbs, 15 g, b.i.d.)	–	December 2013
NCT01107171	Multi-Center Randomized Controlled and Double Blind Trial of Tang-Min-Ling Pills in the Treatment of Type 2 Diabetes	Guang'anmen Hospital of China Academy of Chinese Medical Sciences, China	Tang-Min-Ling pills, low dosage (6 g, t.i.d.). Tang-min-ling pills, high dosage (12 g, t.i.d.). Placebo	2	December 2008
NCT01087242	Randomized, Double-blind, Dose Parallel Controlled, Multicentre Clinical Trial of Tang-Min-Ling Pills and Placebo in Diabetes Mellitus with Liver–Stomach Heat Retention Syndrome	Guang'anmen Hospital of China Academy of Chinese Medical Sciences, China	Tang-Min-Ling pills, low dosage (6 g, t.i.d.). Tang-Min-Ling pills, high dosage (12 g, t.i.d.). Placebo	3	June 2010
NCT01296308	Effectiveness of a Complementary and Integrative Therapy in Diabetic Neuropathy Patients	Universität Duisburg-Essen, Germany	Inpatient integrative treatment	–	October 2012

Table 13.3 (Continued)

Clinical trial identifier	Title	Sponsor	Intervention	Phase	Completion date
NCT00567905	Effect of Green Tea Extract on Type 2 Diabetes: a Randomized, Double-Blind, Placebo-Controlled Clinical Trial	Taipei Hospital, Taiwan	Green tea extract (500 mg, t.i.d.). Placebo	2/3	August 2008
NCT01389362	The Healing Effect of a Two-Herb Recipe on Foot Ulcer in Chinese Patients with Type 2 Diabetes: a Randomized, Placebo-Controlled Study	Chinese University of Hong Kong, China	Radix Rehmanniae and Radix Astragali (2 sachets/day). Placebo	2/3	May 2011
NCT01219803	Dosage–Efficacy Relationship Clinical Trial of Ge Gen Qin Lian Decoction	Guang'anmen Hospital of China Academy of Chinese Medical Sciences, China	Ge-Gen-Qin-Lian decoction, 150 mL/bag, t.i.d. GGQL decoction, 3 × 150 mL/bag, t.i.d. Ge-Gen-Qin-Lian Decoction, 5 × 150 mL/bag, t.i.d. Placebo	—	August 2014
NCT00886665	The Evaluation of Safety and Efficacy of JWHGWT on Diabetic Neuropathy	Taichung Veterans General Hospital, Taiwan	JWHGWT powder (4 g, t.i.d.). Placebo	—	July 2009
NCT00904592	A Study on TCM Comprehensive Protocol of Prevention and Control in Diabetic Retinopathy	Chengdu University of Traditional Chinese Medicine, China	Qi-Ming granules (4.5 g, t.i.d.). Placebo	—	April 2011

Table 13.3 (Continued)

Clinical trial identifier	Title	Sponsor	Intervention	Phase	Completion date
NCT00393510	An Evidence Based Study on the Clinical Effects of Integrated Western Medicine and Traditional Chinese Medicine for Diabetic Foot Ulcer Treatment	Chinese University of Hong Kong	12-herb formulation including Radix Astragali (b.i.d.). Placebo	2	March 2006
NCT00131287	The Supplementary Effects of the Extract of *Agaricus blazei* Murrill on Type II Diabetes Mellitus (DM)	ECbiotech, Taiwan	*Agaricus blazei* Murill extract	3	December 2006
NCT01373476	Multicentre, Randomized, Double-Blind, Multiple-Dose, Placebo-Controlled, Parallel-Group Trial of Qideng Mingmu Capsule in the Treatment of Diabetic Retinopathy with Blood Stasis Syndrome and Deficiency of Qi-Yin Syndrome	Chengdu University of Traditional Chinese Medicine, China	Qi-Deng-Ming-Mu capsules (4, 2 or 1, t.i.d.). Placebo	2	December 2012
NCT00839865	Phase 1 Study of Herb Yuyang Ointment That Expressed Good Effect in Diabetic Foot Ulcer	Guiyang No. 1 People's Hospital, China	Yu-Yang ointment (every 2 days). Conventional procedure	1	March 2011

Table 13.3 (Continued)

Clinical trial identifier	Title	Sponsor	Intervention	Phase	Completion date
NCT00704236	Phase 4 Study on the Effect of Insulin Resistance with Traditional Chinese Treatment in Type 2 Diabetes	Shanghai Jiao Tong University School of Medicine, China	*Coptis chinensis* (50 mg), *Astragalus membranesceus* (30 mg) and *Lonicera japonica* (120 mg). Placebo	4	February 2008
NCT01248286	Effect of Whole Grain Diet on Insulin Sensitivity, Advanced Glycation End Products and Inflammatory Markers in Pre-Diabetes	Mount Sinai School of Medicine, USA	Whole grain rice in meals. Refined grain rice in meals	–	April 2011
NCT00951912	The Study of the Effects of Soy Isoflavones on the Metabolism of Glucose and Lipids in Postmenopausal Chinese Women with Impaired Glucose Regulation	Sun Yat-Sen University, China	Daidzein (50 mg). Genistein (50 mg). Placebo	–	November 2010
NCT00363233	The Effects of Flax Lignans on Lipid Profile and Glucose Management in Type 2 Diabetes: a Randomized Double-Blind Cross-Over Study	Chinese Academy of Sciences, China	Flax lignans	2/3	January 2007

Source: based on information accessed at http://clinicaltrials.gov/ on 23 June 2012.

in colic and ophthalmic disorders. One of the major components of *B. racemosa* is bartogenic acid. *In vitro* enzymatic studies showed that the hexane, ethanol and methanol extracts in addition to bartogenic acid could inhibit the intestinal α-glucosidase activity at concentrations ranging from 0.02 to 0.2 µg mL^{-1}. The methanol extract was most potent in suppressing the rise in the blood glucose level in rats receiving maltose.[22]

13.4.1.1.2 Cassia auriculata

Cassia auriculata L. (Cesalpinaceae; common name: Tanner's Cassia), a common plant in Asia, has been widely used in TCM as a cure for rheumatism, conjunctivitis and DM. The methanol extract of dried flowers of *C. auriculata* was shown to have α-glucosidase-inhibitory activity and lowered the blood glycemic response following maltose ingestion at a dose of 5 mg kg^{-1}.[23]

13.4.1.1.3 Syzygium cumini (L.) Skeels

Syzygium cumini (L.) Skeels, also known as *Eugenia jambolana*, Jamun, Jambu, Black Plum or Black Berry, has frequently been used to treat DM in India[17] and Brazil.[24] Blood glucose, postprandial glucose, cholesterol and free fatty acid reduction have been produced in experiments with diabetic rats.[25–27] Extracts of *S. cumini* have shown α-glucosidase inhibitory activities.[28,29] Hepatic enzymatic activities of G6P, glucokinase and phosphofructokinase were significantly reduced in diabetic animals.[29] Insulinase, which breaks down insulin, was inhibited in β-cells isolated from both non-diabetic and diabetic mice.[30] Blood glucose levels were reduced as rapidly as 30 min after treatment. Adenosine deaminase activity, which plays a role in modulating insulin, was reduced in human subjects.[31] However, two double-blind, randomized trials involving non-DM[32] and DM[33] patients did not reduce postprandial glucose levels.

13.4.1.2 Stimulating Insulin Secretion

13.4.1.2.1 Radix Rehmanniae Praeparata

Radix Rehmanniae, the root of *Rehmannia glutinosa* Libosch. (family Scrophulariaceae), has a wide spectrum of pharmacological activities on the blood, immune, endocrine, nervous and cardiovascular systems. It is commonly prescribed to relieve febrile diseases, DM, epistaxis and skin eruptions in China. The aqueous extract of Radix Rehmanniae was found to be effective in promoting diabetic foot ulcer healing in rats through the processes of tissue regeneration, angiogenesis and inflammation control and can potentially be used to aid in the healing of diabetic foot ulcers in DM patients.[34]

Radix Rehmanniae contains catalpol, rehmannioside A, B, C and D, phenethyl alcohol derivatives including leucosceptoside A and purpureaside C, monocyclic sesquiterpenes and their glycosides.[35]

Radix Rehmanniae showed hypoglycemic activity in normal and strepto-zotocin (STZ)-induced diabetic mice. The postulated mechanisms of action are stimulation of insulin secretion, regulation of glucose metabolism in DM rats and reduction of hepatic glycogen content of non-DM mice.[20,36]

13.4.1.2.2 Gymnema sylvestre (syn. Periploca sylvestris Retz.)

Gymnema sylvestre is a traditional medicinal plant that grows in southern China and India. It has been widely used as a remedy for DM and gastrointestinal and genitourinary ailments. The major chemical components include gymnemic acids I–VII, triterpenoid saponins (gymnemosides a–f and gymnemoside W1–2), conduritol A and dihydroxygymnemic triacetate.[37]

Body weight and also pancreas and liver weight were increased in alloxan-induced diabetic mice by oral administration of the leaf or callus extract of *G. sylvestre* at 200 mg kg^{-1}. The effect of the extracts was similar to 4 units kg^{-1} of insulin. The hepatic glycogen level was also increased (2.15–2.47 mg g^{-1} *versus* 1.35 mg g^{-1} for the extracts and control, respectively), which in turn could stimulate the secretion of insulin.[38] In STZ rats, the hexane, acetone and methanol extracts decreased plasma glucose levels. The acetone extract was found to be the most potent. Oral administration of 600 mg kg^{-1} of the acetone extract for 45 days decreased the glucose level from 443 to 114 mg L^{-1}. Dihydroxygymnemic triacetate was identified as the major active component; at 5–20 mg kg^{-1} it showed a significant effect on lowering the blood glucose level by increasing plasma insulin levels.[39]

In a clinical trial, *G. sylvestre* extract proved to be beneficial in the management of T2DM in patients. The subjects showed reduced polyphagia, fatigue, blood glucose (fasting and postprandial) and HgbA$_{1C}$ in comparison with the control group following an oral dose of 500 mg of herbal extract for a period of 3 months.[40]

In another study, *G. sylvestre* extracts were shown to be able to regenerate β-cells and increase the circulating insulin level by stimulating its secretion.[38]

13.4.1.2.3 Momordica charantia

Momordica charantia, also known as bitter melon, karela, balsam pear or bitter gourd, is a popular vegetable. It has been used for the treatment of DM-related conditions in southern China, India, the Caribbean and East Africa.[19,41]

The major components of *M. charantia* include eleostearic acid, stearic acid, charantin, vicine, polypeptide-p and alkaloids.[41]

The methanol extract of *M. charantia* exhibited hypoglycemic effects in diabetic male ddY mice at 400 mg kg^{-1}. It can also improve glucose tolerance by inhibiting the absorption of carbohydrates from the gastrointestinal tract. In several clinical trials, *M. charantia* decreased both fasting and postprandial blood glucose levels in DM patients.

M. charantia lowered blood glucose levels by increasing serum insulin concentrations and improving β-cell function by suppressing peroxidation and apoptosis. The ethyl acetate extract of *M. charantia* activated peroxisome proliferator-activated receptors (PPARα and -γ) and up-regulated the expression of the acyl-CoA oxidase gene in H4IIEC3 hepatoma cells. The extract could also enhance insulin sensitivity and lipolysis and suppress postprandial hyperglycemia in rats. In STZ rats, gluconeogenesis was inhibited by *M. charantia via* down-regulation of hepatic glucose-6-phosphatase and fructose-1,6-bisphosphatase activities and enhancement of glucose oxidation *via* up-regulation of hepatic glucose-6-phosphate dehydrogenase. In adipocytes, the momordicosides from *M. charantia* could stimulate glucose transporter-4 (GLUT4) translocation to the cell membrane. GLUT4 in turn increased the activity of adenosine monophosphate-activated protein kinase (AMPK), enhancing glucose uptake from the blood.[42]

13.4.1.2.4 Stephania tetrandra

The root of *Stephania tetrandra* S. Moore has been demonstrated to have anti-inflammatory, antiallergic and hypotensive effects in experimental animal studies.[43]

Its major components are alkaloids including tetrandrine, fanchinonline, bis-benzylisoquinoline, protoberberine, morphinane and phenanthrene. Fangchinoline at 0.3–3 mg kg^{-1} significantly decreased blood glucose and increased blood insulin in STZ-diabetic mice by potentiating insulin release.[20]

13.4.1.3 Increasing Insulin Sensitivity and Lowering Blood Glucose

13.4.1.3.1 Rhizoma Coptidis

The rhizome of *Coptis chinensis* Franch. of the Ranunculaceae family, recognized as Rhizoma Coptidis (RC) in the Chinese Pharmacopeia with the Chinese name Huang Lian, has been widely used in various formulas against intestinal infection, diarrhea and inflammation. It also possesses antihypertensive, antioxidant and hypoglycemic effects. Alkaloids including berberine, palmatine, jateorrhizine, epiberberine and coptisine have been identified as its major components. At an oral dose of 200 mg kg^{-1}, RC extract or berberine was found to decrease blood glucose and serum cholesterol levels significantly in high fat diet-fed mice. In alloxan-induced diabetic mice, berberine showed an antihypoglycemic effect. Berberine antagonized the hyperglycemic effect induced by intraperitoneal glucose or adrenaline in normal mice. The activity of berberine was similar to those of both sulfonylureas and biguanides.[20,44]

Berberine could activate AMPK and acetyl-CoA carboxylase and thus improve fatty acid oxidation. The antihyperglycemic effect might be related to improvement of insulin sensitivity *via* either activation of the AMPK pathway or induction of insulin receptor expression. Six quaternary protoberberine-type alkaloids were also found to be able to inhibit aldose reductase activity *in vitro* using rat lens or human recombinant aldose reductase with an IC_{50} <200 μM.[45,46] However, no *in vivo* studies have been performed to validate this mechanism.

13.4.1.3.2 Radix Astragali

Radix Astragali is the dried roots of perennial herbs *Astragalus membranaceus* (Fisch.) Bunge and *Astragalus mongholicus* Bunge (Fabaceae) from the Leguminosae family. The herb can be used to consolidate the exterior of the body and alleviate heat in the muscles by ascending positive *qi*. Other effects include promoting diuresis, relieving edema and promoting skin wound/ulcer healing and tissue/muscle regeneration.[34,47]

Radix Astragali contains many isoflavones and isoflavonoids (formononetin, calycosin and ononin), saponins (astragaloside IV, astragaloside II, astragaloside I and acetylastragaloside) and astragalus polysaccharides.[48]

Diabetic Sprague–Dawley rats showed an improvement in insulin sensitivity and attenuation of fatty liver after treatment with Radix Astragali decoction (500 mg kg^{-1} i.p. daily) for 2 months, although blood glucose levels, β-cell function and glucose tolerance were not improved. Astragalus polysaccharides could reduce hyperglycemia and lead to indirect preservation of β-cell function and mass via immunomodulatory effects on T1DM mice. In addition, astragalus polysaccharides restored glucose homeostasis in T2DM mice/rats by sensitizing the insulin effect. Formononetin, calycosin and ononin might exert a synergistic hypoglycemic effect with fangchinoline in STZ-diabetic mice, most likely by increasing insulin release.[34,47]

13.4.1.3.3 Eriobotrya japonica

The dried leaves of *Eriobotrya japonica* (Thunb.) Lindl., also called Folium Eriobotryae, have been utilized for treatment of various diseases including chronic bronchitis, cough and DM. The major components in Eriobotryae Folium are triterpenes, sesquiterpenes, flavonoids and megastigmane glycosides. These include ursolic acid, oleanolic acid, cinchonain Ib, procyanidin B-2, chlorogenic acid and epicatechin.[49–51] Co-fermentation of Folium Eriobotryae and green tea leaf led to a preparation which could reduce the blood glucose level by 23.8% at 30 min in maltose-loaded rats, but not in sucrose- and glucose-loaded rats. hence the possible mechanism is maltase inhibition.[52] The aqueous extract and also a pure component, cinchonain Ib, enhanced insulin secretion from INS-1 cells in a dose-dependent manner. An *in vivo* animal study showed that the aqueous extract could transiently reduce blood glucose levels at 15 and 30 min post-dose, although an insulin secretion-enhancing effect was not observed.[53] A 70% ethanol extract exerted a significant hypoglycemic effect on alloxan-induced diabetic mice at doses of 15, 30 and 60 g kg^{-1}.[54] In another study, Folium Eriobotryae extract with the major components of tormentic acid, maslinic acid, corosolic acid, oleanolic acid and ursolic acid was shown to ameliorate high fat-induced hyperglycemia, hyperleptinemia, hyperinsulinemia and hypertriglyceridemia. One of the possible mechanisms is the activation of AMPK leading to an increase in insulin sensitivity.[55]

13.4.1.4   Increasing Glucose Uptake

13.4.1.4.1   Ginkgo biloba

The *Ginkgo biloba* tree, called a "living fossil," can survive more than 1000 years.[56] EGb 761, an extract from its leaves, contains ginkgo flavonoid glycosides, terpene lactones and ginkgolic acids. It is one of the most frequently used over-the-counter (OTC) herbal supplements in the USA.[57] While ginkgo has been studied in many clinical trials, it has been mostly sought for its presumptive neuroprotective effects in the prevention and treatment of Alzheimer's disease, cerebrovascular disease and dementia[58–60] while having favorable side effect profiles.[61] Other uses include treating intermittent claudication, tinnitus and sexual dysfunction associated with antidepressant use and improving blood flow in response to cold.[62–65] EGb 761 also reduced atherogenesis in DM rats.[66] In DM rats, administration of EGb 761 increased glucose uptake into hepatic and muscle tissues.[67] *In vitro* assays with ginkgo showed a reduction in α-glucosidase and amylase activities.[68] Whereas *in vitro* and animal studies showed promise, varying responses have been observed in human subjects. Kudolo's group studying healthy human subjects found no reduction in blood glucose levels with an accompanying significant increase in plasma insulin levels.[69] However, in T2DM patients, ingestion of *Ginkgo biloba* extract might increase the hepatic metabolic clearance rate of insulin, which leads to reduced plasma insulin levels and elevated blood glucose.[70] A double-blind, randomized clinical study involving non-DM, pre-DM and DM patients showed that ginkgo did not increase insulin sensitivity or reduce blood glucose levels.[71] In another study, co-administration of ginkgo with metformin did not alter the pharmacokinetics of metformin and therefore it was considered safe for concomitant use.[72]

13.4.1.5   Potentiating Endogenous Incretins by Block DPP-4

The incretin degrading enzyme dipeptidyl peptidase-4 (DPP-4) is a new anti-DM drug target and DPP-4 inhibitors are one of the newest classes of anti-DM agents. Currently there are no reports showing inhibitory activity of TCM herbs towards DPP-4.

13.4.1.6   Antioxidant and Inhibiting β-Cell Apoptosis

13.4.1.6.1   Tinospora cordifolia

Tinospora cordifolia, also called Guduchi, is a herbaceous vine of the Menispermaceae family indigenous to the tropical areas of India, Myanmar and Sri Lanka. The bioactive ingredients are alkaloids (palmatine, jatrorrhizine and magnoflorine), diterpenoid lactones, glycosides, steroids, sesquiterpenoid, phenolics, aliphatic compounds and polysaccharides. It was reported to have antispasmodic, antipyretic, antiallergic, anti-inflammatory, immunomodulatory and antileprosy activities.[17] Its anti-DM potential has been investigated in diabetic mice. Treatment with a 70% ethanol extract (oral 100 or 200 mg kg^{-1} for 14 days) caused a significant blood glucose reduction and suppressed the increase in blood glucose levels after a 2 g kg^{-1} glucose loading in normal rats.[73] The underlying mechanism was amelioration of oxidative stress by reducing the production of thiobarbituric acid-reactive substances. At 200 mg kg^{-1}, the extract increased the expression of *TRx* and *GRx*. It can also adjust carbohydrate metabolism and reduce gluconeogenesis by inhibiting glucose 6-phosphatase and fructose 1,6-diphosphatase.[74] Other possible mechanisms include an increase of insulin release and inhibition of α-glucosidase.[75]

13.4.1.7   Inhibiting Hepatic Glycogenolysis

13.4.1.7.1   Trigonella foenum-graecum L.

Trigonella foenum-graecum L., better known as fenugreek seeds, is a traditional remedy used by ancient Egyptians.[76] Mishkinsky *et al.* first described the hypoglycemic effects in rats taking fenugreek seeds in 1974.[77] The amino acid (2*S*,3*R*,4*S*)-4-hydroxyisoleucine (ID-1101) was isolated from fenugreek seeds and demonstrated insulinotropic effects and increased peripheral glucose uptake *in vitro*.[78–80] A fraction containing soluble dietary fiber was shown to decrease serum fructosamine levels significantly without affecting insulin levels in murine DM models.[81] The activities of hepatic enzymes hexokinase, glucokinase, glucose 6-phosphatase and fructose-1,6-bisphosphatase were reduced in DM rats.[82,83] Trials involving non-DM patients and DM patients showed decreases in plasma glucose levels. In a group of obese patients, fenugreek fiber did not reduce postprandial blood glucose levels;[84,85] however, insulin levels were significantly higher in the fenugreek treatment arm than the placebo arm. Subjects in the treatment arm also stated significantly higher satiety and consumed less food.[86]

13.4.1.7.2   Catharanthus roseus

Catharanthus roseus is more commonly known as Cape periwinkle or rose periwinkle. *C. roseus* contains many bioactive compounds including the vinca alkaloids, which exert antimitotic and antimicrotubule actions that have been effective in the treatment of a number of cancers.[87–91] *In vivo* animal studies using diabetic mice, rats and rabbits showed reductions in fasting glucose levels.[93,94] Glycogen storage in the liver was increased in rats treated with an extract of *C. roseus*.[95] The activity of the hepatic enzyme succinate dehydrogenase was increased significantly whereas glucose 6-phosphatase and lactate dehydrogenase were not significantly affected.

13.4.1.8   Multiple Mechanisms

13.4.1.8.1   Radix Ginseng

Ginseng is found only in the northern hemisphere such as eastern Asia (northern China, Korea and eastern Siberia) and north America. Radix Ginseng is any one of 11 species of slow-growing perennial plants with fleshy roots. The bioactivities largely depend on the geographical location and processing method. The Chinese and Korean ginseng has higher hypoglycemic activities than the American ginseng.[19,96,97] At the 2003 American Diabetes Association Annual Meeting, the antidiabetic effects of Korean red ginseng and American ginseng were reported. American ginseng was able to lower glycosylated hemoglobin A1C, whereas Korean ginseng could enhance insulin sensitivity in tissues.[98]

More than 700 components have been identified in ginseng and, among them, the major active components include ginsenosides (Rb1, Re, Rd), polysaccharides, peptides and polyacetylenic alcohols.[97] Ginsenosides are the main active components in ginsengs, which comprise up to 4% of total material.[99]

Radix Ginseng has been widely investigated for its antidiabetic activity in both animal and human studies. The lipophilic components have better hypoglycemic activities than the aqueous extract. Korean red ginseng (0.1-1.0 g mL^{-1}) significantly stimulated insulin release from isolated rat pancreatic islets. American ginseng (100 mg kg^{-1}) decreased serum levels of glucose and HbA$_{1C}$ in diabetic rats.[100] In two clinical trials, ginseng reduced postprandial blood glucose, fasting blood glucose and HbA$_{1C}$ levels. Multiple mechanisms have been identified. Ginseng has been shown to have antioxidant properties. It also reduces β-cell apoptosis by up-regulating adipocytic peroxisome proliferator-activated receptor γ (PPAR-γ) protein expression. Ginseng impairs glucose absorption by decreasing glucosidase activity. It may also increase insulin sensitivity in peripheral tissues.[19] One of the active components, ginsenoside Rb1, could enhance glucose transport by inducing the differentiation of adipocytes *via* up-regulating the expression of PPAR-γ and C/EBP-α.[101] In addition, ginsenoside Rb1 could increase GLUT-4 activity leading to increased uptake of glucose from blood by adipocytes.[102]

13.4.1.8.2   Fructus Schisandrae

Fructus Schisandrae, with the common name of "five-flavor berry", is the fruit of a deciduous woody vine, *Schisandra chinensis*, native to forests of northern China and has been used in Chinese medicine to astringe the lungs and nourish the kidneys, and is also used as a tonic, sedative and anti-DM agent. The major components include lignans (schizandrin, gomisin A, angeloylgomisin H) and polysaccharides.[43]

FS-60, a sub-fraction from the 70% ethanol extract, was identified to be most potent as a PPAR-γ agonist in 3T3-L1 adipocytes. FS-60 contains schizandrin, gomisin A and angeloylgomisin H. In pancreatectomized diabetic rats, FS-60 lowered serum glucose levels during an OGTT which was close to the fasting stage. During hyperglycemic clamp, FS-60 increased the first-phase insulin secretion in diabetic animals, although it did not affect the phase II insulin secretion.[20] FS-60 increased first-phase insulin secretion (0–10 min) and increased insulin sensitivity by ameliorating insulin resistance *via* stimulating PPAR-γ activity. It could also improve glucose homeostasis in diabetic mice by inhibiting aldose reductase.[103]

13.4.1.8.3 Gegen

Gegen has been used for over 2000 years and is also known as Yegen, the kudzu root and kudzu vine root.[104] The original plant, *Pueraria lobata*, is abundant throughout the world and is considered an invasive species threatening the natural ecological system in various parts of the world.[105] The root contains isoflavone and puerarin and has a variety of pharmacological actions that may be effective in the treatment of hypertension, hyperlipidemia and DM.[106,107] Proposed mechanisms of action from *in vitro* and *in vivo* studies include α-glucosidase inhibition, increased expression and activity of PPAR-γ, up-regulation of GLUT-4 mRNA, increased plasma endorphins and preservation of pancreatic islets.[108–110]

13.4.1.8.4 Folium Mori

Folium Mori, the leaves of the mulberry tree (*Morus alba* L.), has been traditionally used to treat diabetic hyperglycemia. The main bioactive components are flavonoids, alkaloids and polysaccharides.[99] Research showed that Folium Mori exerts its anti-DM effect via multiple mechanisms. A major alkaloid, 1-deoxynorimycin, is a potent inhibitor of α-glucosidase.[111] In a study in human subjects, it was demonstrated that a food-grade mulberry powder enriched with this compound suppressed postprandial blood glucose surge.[112] A similar effect was observed in Goto-Kakizaki rats, a spontaneous non-obese animal model for T2DM.[113]

In an 8-week study with high sucrose-fed KK-Ay mice with spontaneous T2DM, mulberry leaf extract was shown to reduce insulin resistance in a dose-dependent manner. Fasting blood glucose levels and urinary glucose levels were significantly reduced by treatment with 3 and 6% extract in the diet. The plasma insulin levels were significantly lower in the 6% group in comparison with those of the control group and the 3% group.[114]

Mulberry leaf extract at 5-45 μg mL^{-1} was able to increase glucose uptake in a dose-dependent manner and enhanced the translocation of GLUT-4 in adipocytes.[115] It was also shown to ameliorate adipocytokines in white adipose tissue in *db/db* mice, possibly through the mechanism of inhibiting oxidative stress.[116]

13.4.2 TCM Herbal Formulas

13.4.2.1 Wu-Ling-San

Wu-Ling-San consists of five TCM herbs including hoelen (Sclerotium Poriae Cocos), alisma (Rhizoma Alismatis), polyporus sclerotium (Sclerotium Polypori Umbrellati), bighead atractylodes rhizome (Rhizoma Atractylodis Macrocephalae) and cinnamon twig (Ramulus Cinnamomi Cassiae). Wu-Ling-San decreased elevated plasma glucose levels in STZ diabetic mice at 2.5 g kg^{-1} per day. However the effect was not significant at a lower dose of 0.5 or 1.5 g kg^{-1} per day.[117] The mechanism of action of Wu-Ling-San has not been characterized.

13.4.2.2 Tong-Guan-Wan

Tong-Guan-Wan, consisting of cork, anemarrhena and cinnamon with the major constituents mangiferin, berberine, cinnamaldehyde, timosaponin BII and timosaponin AIII, significantly lowered blood glucose and HbA$_{1C}$ levels and improved glucose tolerance in C57BL/KsJ *db/db* mice at oral doses of 62, 125 and 250 mg kg^{-1}. Serum triglyceride levels in *db/db* mice were also significantly reduced, whereas high-density lipoprotein cholesterol levels were significantly increased, after treatment with Tong-Guan-Wan.[118]

13.4.2.3 Rehmannia-Six

Rehmannia-Six is a formula that is commonly used in TCM to treat patients with DM. The six herbs are *Rehmannia glutinosa*, Fructus Corni, *Dioscorea* sp., *Poria cocos*, *Alisma* sp. and *Paeonia suffruticosa*. Rehmannia-Six appeared to have beneficial effects on blood glucose, neuropathy and nephropathy. There is also evidence of anti-inflammatory and antioxidant effects.[119] Fructus Corni appeared to be the major active component contributing to lowering blood glucose levels. It could increase plasma insulin levels in a dose-dependent manner.[120]

13.4.2.4 Tang-Min-Ling

Tang-Min-Ling is an improved preparation of the Dachaihu decoction, which has a history of human use of more than 2000 years. Tang-Min-Ling's main constituents are *Coptis chinensis* Franch., *Scutellaria baicalensis* Georgi, *Rheum officinale* Baill. and *Bupleurum chinense* DC. Tang-Min-Ling has positive effects on regulating glucose metabolism in T2DM patients in multi-centered, randomized controlled clinical trials. Tang-Min-Ling also affects serum lipid, leptin and adiponcetin after oral administration for 12 weeks in Otsuka Long-Evans Tokushima fatty rats. The mechanism of action was improving insulin resistance by increasing the levels of skeletal muscle AMPK and GLUT-4.[121]

13.4.2.5 NHF

NHF, consisting of Rhizoma Polygonati, Radix Rehmanniae, Radix Salviae Miltiorrhizae, Radix Puerariae, Fructus Schizandrae and Radix Glycyrrhizae, was studied in neonatal STZ-induced T2DM rats. Administration of NHF significantly decreased blood glucose levels, food and water intake and body weight. NHF treatment also significantly increased plasma insulin levels and the number and size of insulin-immunoreactive cells in the pancreas. The possible mechanisms for the anti-DM effect of NHF are increased insulin sensitivity, GLUT-4 gene expression and an attenuation of hepatic PEPCK gene expression.[122]

13.4.2.6 ADHF

ADHF is a recent antidiabetes herbal formula, which consists of eight herbs (Radix Ginseng 17%, Radix Rehmannia 17%, Radix Astragali 10%, Radix Trichosanthis 10%, Radix Ophiopogonis 10%, Radix Puerariae 10%, Fructus Lycii 10% and Rhizoma Discoreae 10%).[123] In a C57BL/6J mouse model of diet-induced T2DM, intervention for 12 weeks using ADHF as a diet supplement (4 and 8%) resulted in a significant inhibition of DM-related changes in major organs and a significant reduction in circulating levels of glucose and insulin.[124] In this study, the 4 and 8% supplements showed similar efficacy. The index of insulin resistance was remarkably reduced in the treatment group (10-fold change). The blood insulin was increased in comparison with the regular diet group, although lower than the diabetes induction diet group. Blood glucose was reduced by 37% with the ADHF diet more than the diabetes induction diet group. A significant reduction of progressive damage to major target organs was also observed on addition of ADHF to the diet.[124]

13.4.2.7 Yu-San-Xiao

The ingredients of Yu-San-Xiao are winged euonymus twigs, Chinese clematis, burdock, lycium bark, lichi seed, ballon flower, turmeric and American ginseng. Yu-San-Xiao was used to treat 30 T2DM patients for 1 month and about 40% of patients showed significant improvements in blood glucose levels such that the mean fasting blood glucose and postprandial blood glucose were decreased by 32% and 41%, respectively. Other diabetic symptoms such as thirst and hunger disappeared and glucose tests in urine became negative. The preprandial insulin level was also decreased by oral Yu-San-Xiao, although the postprandial insulin level was not significantly affected.[18]

13.4.2.8 Yi-Jin

Yi-Jin, consisting of *Panax ginseng, Atractylodes macrocephala, Poria cocos* and *Opuntia dillenii,* showed similar efficacy in lowering blood glucose to phenformin, a biguanide, following multiple doses of 30 g kg^{-1} in diabetic mice. In a clinical trial completed in China, about 86% of patients showed clinical improvements after taking Yi-Jin. The hypoglycemic effect might be due to Yi-Jin's ability to restore the function of pancreatic β-cells.[18]

13.4.2.9 Jin-Qi-Jiang-Tang

Jin-Qi-Jiang-Tang is composed of three herbs: coptis, astragalus and honeysuckle. Jin-Qi-Jiang-Tang improved glucose tolerance in diabetic mice. Honeysuckle has the ability to decrease blood sugar. Berberine in coptis also possesses a glucose-lowering effect. Jin-Qi-Jiang-Tang improved insulin resistance caused by hydrocortisone and decreased serum insulin levels in mice. In a human study, Jin-Qi-Jiang-Tang improved insulin resistance. Jin-Qi-Jiang-Tang not only prevented the progression from pre-diabetes to diabetes but also delayed development of diabetic nephropathy. The combination of Jin-Qi-Jiang-Tang and gliclazide decreased fasting blood glucose levels in T2DM patients. Another randomized controlled trial is currently evaluating the effectiveness and safety of Jin-Qi-Jiang-Tang for the treatment of patients with pre-diabetes.[125]

13.4.3 Combination of TCM with Western Drugs

Xiao-Ke-Wan contains extracts from *Astragalus mongholicus*, *Ophiopogon japonicas* Ker-Gawl, *Pueraria lobata* and *Hirudo nipponia* and was combined with glibenclamide in DM treatment. In an animal study, the herbal extract improved microcirculation, which in turn might facilitate the effect of glibenclamide.[18]

The combinational effect of *Trigonella foenum-graecum* L. total saponins with sulfonylureas was explored in the treatment of T2DM patients who were not well controlled by the latter alone. In a human trial of 69 T2DM patients, 23 were assigned to receive the hypoglycemic drug only and the rest received the combination. The results showed that the combination had a better effect in ameliorating clinical symptoms and the therapy was relatively safe. The fasting blood glucose, 2 h postprandial blood glucose, HbA$_{1C}$ and clinical symptomatic quantitative scores were remarkably decreased in the combination group, whereas no significant differences were found in BMI and hepatic/renal functions.[126]

13.5 Concerns and Future Perspectives

The complexity of the chemistry and pharmacology of TCM herbs and TCM physicians' tendency to individualize formulas for different patients have significantly hindered the systematic scientific investigation according to Western medicine standards.[127] In comparison with their Western counterparts, the anti-DM efficacy of TCM herbs has not been extensively studied using randomized, double-blind clinical trials, although many animal studies have been carried out. Hence the efficacy of most TCM herbs has not been well validated.

TCM herbs are generally thought to be relatively safe and with mild side effects. On the other hand, their activities are usually not as potent as those of Western medications. Hence high doses (sometimes as high as 10 g per day) are necessary to achieve ideal therapeutic effects. In addition, the toxicity of herbal products cannot be ignored. In some cases, the side effects could be very serious. Most common complaints of herbal supplements ingestion are GI-related, including stomach upset, diarrhea, constipation, nausea and vomiting. Patients taking *Ginkgo biloba* were reported to be excessively energetic, resulting in insomnia, short-term memory loss, increased appetite and dermatological reactions.[69,128] Ginseng abuse syndrome is a result of chronic ingestion of excessive amounts of ginseng. This is characterized by hypertension and CNS stimulation, insomnia and nervousness.[129] There is an increased recognition that herbal medicines can alter the efficacy of co-administered conventional drugs by affecting drug absorption, metabolism and/or elimination *via* interacting with the metabolizing enzymes and/or drug transporters' activities, which may lead to variations in circulating drug concentrations. When used in combination with established anti-DM medications, herbal supplementation may predispose patients to an increased risk of hypoglycemia. *In vivo*, *in vitro* and clinical studies are demonstrating that TCM herbal medications may be involved in drug interactions. There are case reports of *Ginkgo biloba* increasing the risk of bleeding in patients taking anticoagulants[130–132] and non-steroidal anti-inflammatory drugs (NSAIDs).[133] *Ginkgo biloba* also interacts with selective serotonin reuptake inhibitors, resulting in the serotonin syndrome and thiazide diuretics resulting in decreased efficacy. Cytochrome P450 enzymes are induced by *Ginkgo biloba* in cell culture,[134] rats[135,136] and humans,[137] so there are possible interactions with other medications that are metabolized by the same enzymes.

There is evidence that fenugreek seeds also increase the risk of bleeding with anticoagulants.[138] Concurrent administration of ginseng and imatinib, an antineoplastic agent, may increase hepatotoxicity.[139] Ginseng has also been shown to increase resistance to loop diuretics[140] and produce additive effects with estrogens.[141]

Many TCM herbal supplements contain a variety of bioactive ingredients, providing additional therapeutic effects in addition to anti-DM activity. Owing to the difficulty of identifying the relevant bioactive markers, the development of methods for quality control remains a challenge.[142] Factors such as the source of the raw material and the subsequent extract preparation and delivery can affect constituents presented in the final herbal product. Without effective quality control methods, consistency of the herbal product may be compromised. Improved methods for quality control of herbal products, such as bioactivity-guided pharmacokinetic methods and genomic fingerprint techniques, have been explored. However, applying these techniques on a larger scale in the pharmaceutical industry remains a major challenge.

There are opportunities to advance TCM herbs for DM prevention and treatment. We would like to categorize the approaches into the following:

1. To advance TCM herbs in the Western way, we can follow the example of the biguanide metformin, the development of which into the most prescribed anti-DM drug, has its root in galegine, a guanine derivative present in the French lilac or goat's rue (*Galega officinalis*).[143] Most of the modern research published used this methodology in an effort to identify a single component for its anti-DM activity. However, this Western process has proven to be long, expensive and inefficient.

2. To advance TCM herbs in the Chinese way, we should focus our resources and efforts on a limited number of promising single herbs or formulas. A prerequisite of this approach is to develop a standardized product based on fingerprint or chemical/biological markers. Well-designed, placebo-controlled, randomized clinical trials should subsequently be conducted to confirm the efficacy.

3. An integrative way would combine a low dose of a proven Western drug such as metformin with a standardized TCM herbal extract to provide the benefit of proven efficacy with non- or low toxicity. This probably is a shortcut to modernizing TCM products. In principle, this method has long been adopted by clinicians in China to treat various diseases including diabetes,[144] albeit without standardization of the TCM preparations.

13.6   Conclusion

It is evident through this review of the literature that many TCM herbs possess anti-DM activities. In particular, because of their empirically known safety profiles, nutritional supplement status, multiple components for multiple drug targets, low cost and easy access, TCM herbs such as ginseng and mulberry are excellent candidates for long-term use for the prevention of DM. Consistency in product integrity and quality clinical trials are essential in translating their potential into reality.

References

1. World Health Organization, *Diabetes: Key Facts*, WHO, Geneva, 2011.
2. Centers for Disease Control and Prevention, *National Diabetes Fact Sheet: National Estimates and General Information on Diabetes and Prediabetes in the United States*, Centers for Disease Control and Prevention, Atlanta, GA, 2011.
3. G. Hu, T. A. Lakka, T. O. Kilpelainen and J. Tuomilehto, *Appl. Physiol. Nutr. Metab.*, 2007, **32** (3), 583.
4. J. Naowaboot, C. H. Chung, P. Pannangpetch, R. Choi, B. H. Kim, M. Y. Lee and U. Kukongviriyapan, *Nutr. Res.*, 2012, **32** (1), 39.

5. S. Inzucchi, R. Bergenstal, J. Buse, M. Diamant, E. Ferrannini, M. Nauck, A. Peters, A. Tsapas, R. Wender and D. Matthews, *Diabetologia*, 2012, **55** (6), 1577.

6. N. Choudhary, S. Kalra, A. Unnikrishnan and T. Ajish, *Indian J. Endocrinol. Metab.*, 2012, **16** (1), 33.

7. A. Tabák, C. Herder, W. Rathmann, E. Brunner and M. Kivimäki, *Lancet*, 2012, **379** (9833), 2279.

8. S. Grundy, *J. Am. Coll. Cardiol.*, 2012, **59** (7), 635.

9. L. Perreault, Q. Pan, K. Mather, K. Watson, R. Hamman and S. Kahn, *Lancet*, 2012, **379** (9833), 2243.

10. W. Knowler, E. Barrett-Connor, S. Fowler, R. Hamman, J. Lachin, E. Walker and D. Nathan, *N. Engl. J. Med.*, 2002, **346** (6), 393.

11. J. Chiasson, R. Josse, R. Gomis, M. Hanefeld, A. Karasik and M. Laakso, *Lancet*, 2002, **359** (9323), 2072.

12. T. Buchanan, A. Xiang, R. Peters, S. Kjos, A. Marroquin, J. Goico, C. Ochoa, S. Tan, K. Berkowitz and H. Hodis, *Diabetes*, 2002, **51** (9), 2796.

13. Position Statement, *Diabetes Care*, 2004, **27** (Suppl. 1), s47.

14. Lexicomp, *Database, Lexi-Drugs*, Lexicomp, Hudson, OH, 2012.

15. S. Ballali and F. Lanciai, *Int. J. Food Sci. Nutr.*, 2012, **63** (S1), 51.

16. A. Ceylan-Isik, R. Fliethman, L. Wold and J. Ren, *Curr. Diabetes Rev.*, 2008, **4** (4), 320.

17. J. Grover, S. Yadav and V. Vats, *J. Ethnopharmacol.*, 2002, **81** (1), 81.

18. W. Jia, W. Gao and L. Tang, *Phytother. Res.*, 2003, **17** (10), 1127.

19. P. Prabhakar and M. Doble, *Chin. J. Integr. Med.*, 2011, **17** (8), 563.

20. W. Li, H. Zheng, J. Bukuru and N. De Kimpe, *J. Ethnopharmacol.*, 2004, **92** (1), 1.

21. M. Bhat, S. Zinjarde, S. Bhargava, A. Kumar and B. Joshi, *Evid. Based Compl. Alt. Med.*, 2011, Article ID 810207.

22. P. Gowri, A. Tiwari, A. Ali and J. Rao, *Phytother. Res.*, 2007, **21** (8), 796.

23. K. Abesundara, T. Matsui and K. Matsumoto, *J. Agric. Food Chem.*, 2004, **52** (9), 2541.

24. A. Oliveira, D. Endringer, L. Amorim, L. das Gracas and M. Coelho, *J. Ethnopharmacol.*, 2005, **102** (3), 465.

25. P. Prince, N. Kamalakkannan and V. Menon, *J. Ethnopharmacol.*, 2004, **91** (2–3), 209.

26. P. Prince, V. Menon and L. Pari, *J. Ethnopharmacol.*, 1998, **61** (1), 1.

27. S. Prince, N. Kamalakkannan and V. Menon, *J. Ethnopharmacol.*, 2003, **84** (2–3), 205.

28. J. Shinde, T. Taldone, M. Barletta, N. Kunaparaju, B. Hu, S. Kumar, J. Placido and S. Zito, *Carbohydr. Res.*, 2008, **343** (7), 1278.

29. J. Grover, V. Vats and S. Rathi, *J. Ethnopharmacol.*, 2000, **73** (3), 461.

30. S. Achrekar, G. Kaklij, M. Pote and S. Kelkar, *In Vivo*, 1991, **5** (2), 143.

31. A. Bopp, K. S. De Bona, L. P. Belle, R. N. Moresco and M. B. Moretto, *Fundam. Clin. Pharmacol.*, 2009, **3** (4), 501.

32. C. C. Teixeira, C. A. Rava, P. Mallman da Silva, R. Melchior, R. Argenta, F. Anselmi, C. R. Almeida and F. D. Fuchs, *J. Ethnopharmacol.*, 2000, **71** (1–2), 343.

33. C. C. Teixeira, F. D. Fuchs, L. S. Weinert and J. Esteves, *J. Clin. Pharm. Ther.*, 2006, **31** (1), 1.

34. K. M. Lau, K. K. Lai, C. L. Liu, J. C. W. Tam, M. H. To, H. F. Kwok, C. P. Lau, C. H. Ko, P. C. Leung, K. P. Fung, S. K. S. Poon and C. B. S. Lau, *J. Ethnopharmacol.*, 2012, **141** (1), 250.

35. Z. J. Wang, S. K. Wo, L. Wang, C. B. S. Lau, V. H. L. Lee, M. S. S. Chow and Z. Zuo, *J. Pharm. Biomed. Anal.*, 2009, **50** (2), 232.

36. T. Kiho, T. Watanabe, K. Nagai and S. Ukai, *Yakugaku Zasshi*, 1992, **112** (6), 393.

37. X. M. Zhu, P. Xie, Y. T. Di, S. L. Peng, L. S. Ding and M. K. Wang, *J. Integr. Plant Biol.*, 2008, **50** (5), 589.

38. A. B. A. Ahmed, A. S. Rao and M. V. Rao, *Phytomedicine*, 2010, **17** (13), 1033.

39. P. Daisy, J. Eliza and K. A. M. Mohamed Farook, *J. Ethnopharmacol.*, 2009, **126** (2), 339.

40. S. N. Kumar, U. V. Mani and I. Mani, *J. Diet Suppl.*, 2010, **7** (3), 273.

41. L. Leung, R. Birtwhistle, J. Kotecha, S. Hannah and S. Cuthbertson, *Br. J. Nutr.*, 2009, **102** (12), 1703.

42. R. M. Hafizur, N. Kabir and S. Chishti, *Nat. Prod. Res.*, 2011, **25** (4), 353.

43. W. Ma, M. Nomura, T. Takahashi-Nishioka and S. Kobayashi, *Biol. Pharm. Bull.*, 2007, **30** (11), 2079.

44. X. Xia, J. Yan, Y. Shen, K. Tang, J. Yin, Y. Zhang, D. Yang, H. Liang, J. Ye and J. Weng, *PLoS One*, 2011, **6** (2), e16556.

45. H. A. Jung, N. Y. Yoon, H. J. Bae, B. S. Min and J. S. Choi, *Arch. Pharm. Res.*, 2008, **31** (11), 1405.

46. W. Xie, D. Gu, J. Li, K. Cui and Y. Zhang, *PLoS One*, 2011, **6** (9), e24520.

47. Z. M. Lu, Y. R. Yu, H. Tang and X. X. Zhang, *Sichuan Da Xue Xue Bao Yi Xue Ban*, 2005, **36** (4), 529.

48. J. Z. Song, H. H.W. Yiu, C. F. Qiao, Q. B. Han and H. X. Xu, *J. Pharm. Biomed. Anal.*, 2008, **47** (2), 399.

49. H. Ito, E. Kobayashi, S.-H. Li, T. Hatano, D. Sugita, N. Kubo, S. Shimura, Y. Itoh, H. Tokuda, H. Nishino and T. Yoshida, *J. Agric. Food Chem.* 2002, **50** (8), 2400.

50. E. H. Lee, D. G. Song, J. Y. Lee, C. H. Pan, B. H. Um and S. H. Jung, *Biol. Pharm. Bull.*, 2008, **31** (8), 1626.

51. E. N. Li, J. G. Luo and L. Y. Kong, *Phytochem. Anal.*, 2009, **20** (4), 338.

52. K. Tamaya, T. Matsui, A. Toshima, M. Noguchi, Q. Ju, Y. Miyata, T. Tanaka and K. Tanaka, *J. Sci. Food Agric.*, 2010, **90** (5), 779.

53. F. Qa'dan, E. J. Verspohl, A. Nahrstedt, F. Petereit and K. Z. Matalka, *J. Ethnopharmacol.*, 2009, **124** (2), 224.

54. W. L. Li, J. L. Wu, B. R. Ren, J. Chen and C. G. Lu, *Am. J. Chin. Med.*, 2007, **35** (4), 705.
55. C. C. Shih, C. H. Lin and J. B. Wu, *Phytother. Res.*, 2010, **24** (12), 1769.
56. B. P. Jacobs and W. S. Browner, *Am. J. Med.*, 2000, **108** (4), 341.
57. D. M. Eisenberg, R. B. Davis, S. L. Ettner, S. Appel, S. Wilkey, M. Van Rompay, R and C. Kessler, *JAMA*, 1998, **280** (18), 1569.
58. B. S. Oken, D. M. Storzbach and J. A. Kaye, Arch. Neurol., 1998, **55** (11), 1409.
59. P. L. Le Bars, M. M. Katz, N. Berman, T. M. Itil, A. M. Freedman, A. F. Schatzberg, *JAMA*, 1997, **278** (16), 1327.
60. S. L. Rogers, M. R. Farlow, R. S. Doody, R. Mohs and L. T. Friedhoff, *Neurology*, 1998, **50** (1), 136.
61. V. S. Sierpina, B. Wollschlaeger and M. Blumenthal, *Am. Fam. Physician*, 2003, **68** (5), 923.
62. A. J. Cohen and B. Bartlik, *J. Sex Marital Ther.*, 1998, **24** (2), 139.
63. S. Drew and E. Davies, *BMJ*, 2001, **322** (7278), 73.
64. M. H. Pittler and E. Ernst, *Am. J. Med.*, 2000, **108** (4), 276.
65. J. P. Roncin, F. Schwartz and P. D'Arbigny, *Aviat. Space Environ. Med.*, 1996, **67** (5), 445.
66. S. Lim, J. W. Yoon, S. M. Kang, S. H. Choi, B. J. Cho, M. Kim, H. S. Park, H. J. Cho, H. Shin, Y. B. Kim, H. S. Kim, H. C. Jang and K. S. Park, *PloS One*, 2011, **6** (6), e20301.
67. J. R. Rapin, R. G. Yoa, C. Bouvier and K. Drieu, *Drug Dev. Res.*, 1997, **40** (1), 68.
68. M. da S. Pinto, Y. I. Kwon, E. Apostolidis, F. M. Lajolo, M. I. Genovese and K. Shetty, *Bioresour. Technol.*, 2009, **100** (24), 6599.
69. G. B. Kudolo, *J. Clin. Pharmacol.*, 2000, **40** (6), 647.
70. G. B. Kudolo, *J. Clin. Pharmacol.*, 2001, **41** (6), 600.
71. G. B. Kudolo, W. Wang, R. Elrod, J. Barrientos, A. Haase and J. Blodgett, *Clin. Nutr.*, 2006, **25** (1), 123.
72. G. B. Kudolo, W. Wang, M. Javors and J. Blodgett, *Clin. Nutr.*, 2006, **25** (4), 606.
73. M. B. Patel and S. Mishra, *Phytomedicine*, 2011, **18** (12), 1045.
74. M. Sangeetha, H. Balaji Raghavendran, V. Gayathri and H. Vasanthi, *J. Nat. Med.*, 2011, **65** (3), 544.
75. A. D. Chougale, V. A. Ghadyale, S. N. Panaskar and A. U. Arvindekar, *J. Enzyme Inhib. Med. Chem.*, 2009, **24** (4), 998.
76. H. Bawadi, R. Tayyem, R. Tayyem, S. Maghaydah, *Pharmacog. Mag.*, 2009, **5** (18), 134.
77. J. S. Mishkinsky, A. Goldschmied, B. Joseph, Z. Ahronson and F. G. Sulman, *Arch. Int. Pharmacodyn. Ther.*, 1974, **210** (1), 27.
78. C. Broca, R. Gross, P. Petit, Y. Sauvaire, M. Manteghetti, M. Tournier, P. Masiello, R. Gomis and G. Ribes, *Am. J. Physiol.*, 1999, **277** (4 Pt. 1), E617.
79. M. V. Vijayakumar, S. Singh, R. R. Chhipa and M. K. Bhat, *Br. J. Pharmacol.*, 2005, **146** (1), 41.

80. C. Broca, V. Breil, C. Cruciani-Guglielmacci, M. Manteghetti, C. Rouault, M. Derouet, S. Rizkalla, B. Pau, P. Petit, G. Ribes, A. Ktorza, R. Gross, G. Reach and M. Taouis, *Am. J. Physiol. Endocrinol. Metab.*, 2004, **287** (3), E463.

81. J. M. Hannan, B. Rokeya, O. Faruque, N. Nahar, M. Mosihuzzaman, A. K. Azad Khan and L. Ali, *J. Ethnopharmacol.*, 2003, **88** (1), 73.

82. B. A. Devi, N. Kamalakkannan and P. S. Prince, *Phytother. Res.*, 2003, **17** (10), 1231.

83. M. V. Vijayakumar and M. K. Bhat, *Phytother. Res.*, 2008, **22** (4), 500.

84. Z. Madar, R. Abel, S. Samish and J. Arad, *Eur. J. Clin. Nutr.*, 1998, **42** (1), 51.

85. R. D. Sharma, T. C. Raghuram and N. S. Rao, *Eur. J. Clin. Nutr.*, 1990, **44** (4), 301.

86. J. R. Mathern, S. K. Raatz, W. Thomas and J. L. Slavin, *Phytother. Res.*, 2009, **23** (11), 1543.

87. F. Cardoso, P. L. Bedard, E. P. Winer, O. Pagani, E. Senkus-Konefka, L. J. Fallowfield, S. Kyriakides, A. Costa, T. Cufer and K. S. Albain, *J. Natl. Cancer Inst.*, 2009, **101** (17), 1174.

88. R. I. Fisher, E. R. Gaynor, S. Dahlberg, M. M. Oken, T. M. Grogan, E. M. Mize, J. H. Glick, C. A. Coltman and T. P. Miller, *N. Engl. J. Med.*, 1993, **328** (14), 1002.

89. G. D. Goss, D. M. Logan, T. E. Newman, W. K. Evans, *Cancer Prev. Control*, 1997, **1** (1), 28.

90. L. M. Jost and R. A. Stahel, *Ann. Oncol.*, 2005, **16** (Suppl. 1), i54.

91. R. A. Larson, R. K. Dodge, C. P. Burns, E. J. Lee, R. M. Stone, P. Schulman, D. Duggan, F. R. Davey, R. E. Sobol and S. R. Frankel, *Blood*, 1995, **85** (8), 2025.

92. S. Nammi, M. K. Boini, S. D. Lodagala and R. B. Behara, *BMC Complement. Altern. Med.*, 2003, **3**, 4.

93. S. C. Ohadoma and H. U. Michael, *Asian Pac. J. Trop. Med.*, 2011, **4** (6), 475.

94. K. Rasineni, R. Bellamkonda, S. R. Singareddy and S. Desireddy, *Pharmacog. Res.*, 2010, **2** (3), 195.

95. S. N. Singh, P. Vats, S. Suri, R. Shyam, M. M. Kumria, S. Ranganathan and K. Sridharan, *J. Ethnopharmacol.*, 2001, **76** (3), 269.

96. L. Jia, Y. Zhao and X. J. Liang, *Curr. Med. Chem.*, 2009, **16** (22), 2924.

97. L. W. Qi, C. Z. Wang and C. S. Yuan, *Nat. Prod. Rep.*, 2011, **28** (3).

98. J. Q. Zhang, *Zhong Xi Yi Jie He Xue Bao*, 2007, **5** (4), 373.

99. L. X. Yang, T. H. Liu, Z. T. Huang, J. E. Li and L. L. Wu, *Chin. J. Integr. Med.*, 2011, **17** (3), 235.

100. K. Kim and H. Y. Kim, *J. Ethnopharmacol.*, 2008, **120** (2), 190.

101. W. Shang, Y. Yang, B. Jiang, H. Jin, L. Zhou, S. Liu and M. Chen, *Life Sci.*, 2007, **80** (7), 618.

102. H. L. Kong, J. P. Wang, Z. Q. Li, S. M. Zhao, J. Dong and W. W. Zhang, *Acta Pharmacol. Sin.*, 2009, **30** (4), 396.

103. D. Y. Kwon, D. S. Kim, H. J. Yang and S. Park, *J. Ethnopharmacol.*, 2011, **135** (2), 455.

104. Q. Luo, J. D. Hao, Y. Yang and H. Yi, *Zhong Guo Zhong Yao Za Zhi*, 2007, **32** (12), 1141.

105. Global Invasive Species Database, *Puerarin montana var. lobata (vine, climber)*, http://www.issg.org/database/species/ecology.asp?si=81&fr=1&sts=&lang=EN (last accessed 23 September 2012).

106. F. L. Hsu, I. M. Liu, D. H. Kuo, W. C. Chen, H. C. Su and J. T. Cheng, *J. Nat. Prod.*, 2003, **66** (6), 788.

107. X. P. Song, P. P. Chen and X. S. Chai, *Acta Pharmacol. Sin.*, 1988, **9** (1), 55.

108. K. He, X. Li, X. Chen, X. Ye, J. Huang, Y. Jin, P. Li, Y. Deng, Q. Jin, Q. Shi and H. Shu, *J. Ethnopharmacol.*, 2011, **137** (3), 1135.

109. F. L. Xiong, X. H. Sun, L. Gan, X. L. Yang and H. B. Xu, *Eur. J. Pharmacol.*, 2006, **529** (1–3), 1.

110. J. Yeo, Y. M. Kang, S. I. Cho and M. H. Jung, *Chin. Med.*, 2011, **6**, 10.

111. T. Oku, M. Yamada, M. Nakamura, N. Sadamori and S. Nakamura, *Br. J. Nutr.*, 2006, **95** (5), 933.

112. T. Kimura, K. Nakagawa, H. Kubota, Y. Kojima, Y. Goto, K. Yamagishi, S. Oita, S. Oikawa and T. Miyazawa, *J. Agric. Food Chem.*, 2007, **55** (14), 5869.

113. J. M. Park, H. Y. Bong, H. I. Jeong, Y. K. Kim, J. Y. Kim and O. Kwon, *Nutr. Res. Pract.*, 2009, **3** (4), 272.

114. K. Tanabe, S. Nakamura, K. Omagari and T. Oku, *Nutr. Res.*, 2011, **31** (11), 848.

115. J. Naowaboot, P. Pannangpetch, V. Kukongviriyapan, A. Prawan, U. Kukongviriyapan and A. Itharat, *Am. J. Chin. Med.*, 2012, **40** (1), 163.

116. M. Sugimoto, H. Arai, Y. Tamura, T. Murayama, P. Khaengkhan, T. Nishio, K. Ono, H. Ariyasu, T. Akamizu, Y. Ueda, T. Kita, S. Harada, K. Kamei and M. Yokode, *Atherosclerosis*, 2009, **204** (2), 388.

117. I. M. Liu, T.-F. Tzeng, S.-S. Liou and C. J. Chang, *J. Ethnopharmacol.*, 2009, **124** (2), 211.

118. Y. H. Tang, Z. L. Sun, M. S. Fan, Z. X. Li and C. G. Huang, *Planta Med.*, 2012, **78** (1), 18.

119. T. Y. C. Poon, K. L. Ong and B. M. Y. Cheung, *J. Diabetes*, 2011, **3** (3), 184.

120. S. S. Liou, I. M. Liu, S. F. Hsu and J. T. Cheng, *J. Pharm. Pharmacol.*, 2004, **56** (11), 1443.

121. Z. Zhen, B. Chang, M. Li, F. M. Lian, L. Chen, L. Dong, J. Wang, B. Yu, W. K. Liu, X. Y. Li, P. J. Qin, J. H. Zhang and X. L. Tong, *Am. J. Chin. Med.*, 2011, **39** (1), 53.

122. J. O. Kim, G. D. Lee, J. H. Kwon and K. S. Kim, *Biol. Pharm. Bull.*, 2009, **32** (3), 421.

123. Y. S. Huo, S. J. Luo and W. D. Winters, *US Patent*, 6 093 403, 2000.

124. W. D. Winters, Y. S. Huo and D. L. Yao, *Phytother. Res.*, 2003, **17** (6), 591.

125. H. Cao, M. Ren, L. Guo, H. Shang, J. Zhang, Y. Song, H. Wang, B. Wang, X. Li, J. Hu, X. Wang, D. Wang, J. Chen, S. Li and L. Chen, Trials, 2010, **11**, 27.

126. F. R. Lu, L. Shen, Y. Qin, L. Gao, H. Li and Y. Dai, Chin, *J. Integr. Med.*, 2008, **14** (1), 56.

127. D. Pon, Z. Wang, K. N. Le and M.S. Chow, *Ther. Deliv.*, 2010, **1** (2), 335.

128. P. C. Chan, Q. Xia and P. P. Fu, *J. Environ. Sci. Health C, Environ. Carcinog. Ecotoxicol. Rev.*, 2007, **25** (3), 211.

129. R. K. Siegel, *JAMA*, 1979, **241** (15), 1614.

130. J. Rowin and S. L. Lewis, *Neurology*, 1996, **46** (6), 1775.

131. S. Vale, *Lancet*, 1998, **352** (9121), 36.

132. J. Benjamin, T. Muir, K. Briggs and B. Pentland, *Postgrad. Med. J.*, 2001, **77** (904), 112.

133. C. Meisel, A. Johne and I. Roots, *Atherosclerosis*, 2003, **167** (2), 367.

134. B. H. Hellum, Z. Hu and O. G. Nilsen, *Basic Clin. Pharmacol. Toxicol.*, 2009, **105** (1), 58.

135. T. Sugiyama, Y. Kubota, K. Shinozuka, S. Yamada, J. Wu and K. Umegaki, *Life Sci.*, 2004, **75** (9), 1113.

136. Y. Taki, Y. Yamazaki, F. Shimura, S. Yamada and K. Umegaki, J. *Pharmacol. Sci.*, 2009, **109** (3), 459.

137. O. Q. Yin, B. Tomlinson, M. M. Waye, A. H. Chow and M. S. Chow, *Pharmacogenetics*, 2004, **14** (12), 841.

138. J.-P. Lambert and J. Cormier, *Pharmacotherapy*, 2001, **21** (4), 509.

139. N. Bilgi, K. Bell, A. N. Ananthakrishnan and E. Atallah, *Ann. Pharmacother.*, 2010, **44** (5), 926.

140. B. N. Becker, J. Greene, J. Evanson, G. Chidsey and W. J. Stone, *JAMA*, 1996, **276** (8), 606.

141. M. N. Dukes, *Br. Med. J.*, 1978, **i** (6127), 1621.

142. L. Wang, Z. Wang, S. Wo, C. B. S. Lau, X. Chen, M. Huang, V. H. L. Lee, M. S. S. Chow and Z. Zuo, *Int. J. Pharm.*, **406** (1-2), 99.

143. L. A. Witters, *J. Clin. Invest.*, 2001, **108** (8), 1105.

144. S. S. Guo and L. Qu, Zhongguo, *Zhong Xi Yi Jie He Za Zhi*, 2005, **25** (6), 565.

145. S. Bolen, L. Feldman, J. Vassy, L. Wilson, H. C. Yeh, S. Marinopoulos, C. Wiley, E. Selvin, R. Wilson, E. B. Bass and F. L. Brancati, *Ann. Intern. Med.*, 2007, **147** (6), 386.

146. W. L. Bennett, N. M. Maruthur, S. Singh, J. B. Segal, L. M. Wilson, R. Chatterjee, S. S. Marinopoulos, M. A. Puhan, P. Ranasinghe, L. Block, W. K. Nicholson, S. Hutfless, E. B. Bass and S. Bolen, *Ann. Intern. Med.*, 2011, **154** (9), 602.

147. A. Andrade-Cetto, J. Becerra-Jiménez and R. Cárdenas-Vázquez, *J. Ethnopharmacol.*, 2008, **116** (1), 27.

148. T. T. Mai and N. V. Chuyen, *Biosci. Biotechnol. Biochem.*, 2007, **71** (1), 69.

149. J. Y. Youn, H. Y. Park and K. H. Cho, *Diabetes Res. Clin. Pract. Suppl.*, 2004, **66** (Suppl. 1):S149.

150. M. Shibano, K. Kakutani, M. Taniguchi, M. Yasuda and K. Baba, *J. Nat. Med.*, 2008, **62** (3), 349.

151. M. Karato, K. Yamaguchi, S. Takei, T. Kino and K. Yazawa, Biosci. Biotechnol. Biochem., 2006, **70** (6), 1482.

152. Y. Zang, H. Sato and K. Igarashi, *Biosci. Biotechnol. Biochem.*, 2011, **75** (9), 1677.

153. S. F. Kouam, S. N. Khan, K. Krohn, B. T. Ngadjui, D. G. W. F. Kapche, D. B. Yapna, S. Zareem, A. M. Y. Moustafa and M. I. Choudhary, *J. Nat. Prod.*, 2006, **69** (2), 229.

154. S. V. Reddy, A. K. Tiwari, U. S. Kumar, R. J. Rao and J. M. Rao, *Phytother. Res.*, 2005, **19** (4), 277.

155. F. Ye, Z. F. Shen, F. X. Qiao, D. Y. Zhao and M. Z. Xie, *Yao Xue Xue Bao*, 2002, **37** (2), 108.

156. A. Ishikawa, H. Yamashita, M. Hiemori, E. Inagaki, M. Kimoto, M. Okamoto, H. Tsuji, A.N. Memon, A. Mohammadi and Y. Natori, *J. Nutr. Sci. Vitaminol.*, 2007, **53** (2), 166.

157. T. Ohta, S. Sasaki, T. Oohori, S. Yoshikawa and H. Kurihara, *Biosci. Biotechnol. Biochem.*, 2002, **66** (7), 1552.

158. Y. M. Kim, M. H. Wang and H. I. Rhee, *Carbohydr. Res.*, 2004, **339** (3), 715.

159. G. J. McDougall, F. Shpiro, P. Dobson, P. Smith, A. Blake and D. Stewart, *J. Agric. Food Chem.*, 2005, **53** (7), 2760.

160. M. Yoshikawa, Y. Pongpiriyadacha, A. Kishi, T. Kageura, T. Wang, T. Morikawa and H. Matsud, *Yakugaku Zasshi*, 2003, **123** (10), 871.

161. H. Oe and S. Ozaki, *Biosci. Biotechnol. Biochem.*, 2008, **72** (7), 1962.

162. W. Benalla, S. Bellahcen and M. Bnouham, *Curr. Diabetes Rev.*, 2010, **6** (4), 247.

163. R. R. Ortiz-Andrade, S. García-Jiménez, P. Castillo-España, G. Ramírez-Ávila, R. Villalobos-Molina and S. Estrada-Soto, *J. Ethnopharmacol.*, 2007, **109** (1), 48.

164. M. H. Johnson, A. Lucius, T. Meyer and E. G. de Mejia, *J. Agric. Food Chem.*, 2011, **59** (16), 8923.

165. T. Itoh, N. Kita, Y. Kurokawa, M. Kobayashi, F. Horio and Y. Furuichi, *Biosci. Biotechnol. Biochem.*, 2004, **68** (12), 2421.

166. H. C. Lo, S. T. Tu, K. C. Lin and S. C. Lin, *Life Sci.*, 2004, **74** (23), 2897.

167. Y. W. Cheng, Y. I. Chen, C. Y. Tzeng, H. C. Chen, C. C. Tsai, Y. C. Lee, J. G. Lin, Y. K. Lai and S. L. Chang, *Phytother. Res.*, 2012, **26** (8), 1173.

168. H. J. Kim, M. K. Kong and Y. C. Kim, *BMB Rep.*, 2008, **41** (10), 710.

169. C. Y. Yang, J. Wang, Y. Zhao, L. Shen, X. Jiang, Z. G. Xie, N. Liang, L. Zhang and Z. H. Chen, *J. Ethnopharmacol.*, 2010, **130** (2), 231.

170. J. H. Hsu, Y. C. Wu, I. M. Liu and J. T. Cheng, *J. Ethnopharmacol.*, 2007, **112** (3), 577.

171. T. Gao and Q. You, *Zhong Yao Cai*, 2001, **24** (6), 424.

172. A. Sepici, I. Gurbuz, C. Cevik and E. Yesilada, *J. Ethnopharmacol.*, 2004, **93** (2–3), 311.

173. J. Y. W. Chan, P. C. Leung, C. T. Che and K. P. Fung, *Phytother. Res.*, 2008, **22** (2), 190.
174. E. K. Kim, M. Y. Song, T. O. Hwang, H. J. Kim, W. S. Moon, D. G. Ryu, H. S. So, R. Park, J. W. Park, K. B. Kwon and B. H. Park, *Int. J. Mol. Med.*, 2008, **22** (3), 349.
175. D. Gao, W. Li, Z. Liu, J. Feng, J. Li, Z. Han and Y. Duan, *Methods Find. Exp. Clin. Pharmacol.*, 2009, **31** (6), 375
176. Z. Maksimović, D. Malenčić and N. Kovačević, *Bioresour. Technol.*, 2005, **96** (8), 873.
177. T. Jourdan, L. Djaouti, L. Demizieux, J. Gresti, B. Vergès and P. Degrace, *J. Nutr.*, 2009, **139** (10), 1901.
178. K. Park, G. Jung, K. Lee, J. Choi, M. Choi, G. Kim, J. Hung and H. Park, *J. Ethnopharmacol.*, 2004, **90** (1), 73.
179. Y. Juan, C. Chang, W. Tsai, Y. Lin, Y. Hsu and H. Liu, *J. Ethnopharmacol.*, 2011, **137** (1), 592.
180. M. Xi, C. Hai, H. Tang, M. Chen, K. Fang and X. Liang, *Phytother. Res.*, 2008, **22** (2), 228.
181. L. Ding, P. Li, C. Lau, Y. Chan, D. Xu, K. Fung and W. Su, *Phytother. Res.*, 2012, **26** (1), 101.
182. J. Wansi, J. Wandji, L. Mbaze Meva'a, A. Kamdem Waffo, R. Ranjit, S. Khan, A. Asma and C. Iqbal, *Chem. Pharm. Bull.*, 2006, **54** (3), 292.
183. C. Jung, S. Zhou, G. Ding, J. Kim, M. Hong, Y. Shin, G. Kim and S. Ko, *Biosci. Biotechnol. Biochem.*, 2006, **70** (10), 2556.
184. H. Chen, R. Feng, Y. Guo, L. Sun and J. Jiang, *J. Ethnopharmacol.*, 2001, **74** (3), 225.
185. L. Xiu, A. Miura, K. Yamamoto, T. Kobayashi, Q. Song, H. Kitamura and J. Cyong, *Am. J. Chin. Med.*, 2001, **29** (3-4), 493.
186. N. Hamza, B. Berke, C. Cheze, A. Agli, P. Robinson, H. Gin and N. Moore, *J. Ethnopharmacol.*, 2010, **128** (2), 513.
187. P. Pushparaj, H. Low, J. Manikandan, B. Tan and C. Tan, *J. Ethnopharmacol.*, 2007, **111** (2), 430.
188. V. Muthusamy, S. Anand, K. Sangeetha, S. Sujatha, B. Arun and B. Lakshmi, *Chem. Biol. Interact.*, 2008, **174** (1), 69.
189. V. Muthusamy, C. Saravanababu, M. Ramanathan, R. Bharathi Raja, S. Sudhagal, S. Anand and B. Lakshmi, *Br. J. Nutr.*, 2010, **104** (6), 813.
190. M. Dusane and B. Joshi, *Can. J. Physiol. Pharmacol.*, 2012, **90** (3), 371.
191. K. Valentová, N. Truong, A. Moncion, I. de Waziers and J. Ulrichová, *J. Agric. Food. Chem.*, 2007, **55** (19), 7726.
192. J. Zheng, J. He, B. Ji, Y. Li and X. Zhang, *Asian Pac. J. Clin. Nutr.*, 2007, **16** Suppl. 1, 427.
193. C. Chen, C. Hsu, C. Chen and H. Liu, *J. Ethnopharmacol.*, 2008, **117** (3), 483.
194. L He, *Yakugaku Zasshi*, 2011, **131** (12), 1801.
195. N. Yamabe, K. Kang, C. Park, T. Tanaka and T. Yokozawa, *Biol. Pharm. Bull.*, 2009, **32** (4), 657.

196. H. Kurihara, H. Fukami, A. Kusumoto, H. Toyoda, H. Shibata, Y. Matsui, S. Asami and T. Tanaka, *Biosci. Biotechnol. Biochem.*, 2003, **67** (4), 877.

197. S. Li, J. Li, X. Guan, S. Deng, L. Li, M. Tang, J. Huang, Z. Chen and R. Yang, *Fitoterapia*, 2011, **82** (7), 1081.

198. X. Xu, X. Qi, W. Wang and G. Chen, *J. Sep. Sci.*, 2005, **28** (7), 647.

199. L. Zhu, J. Tan, B. Wang, R. He, Y. Liu and C.Zheng, *Chem. Biodivers.*, 2009, **6** (10), 1716.

200. H. Mi, W. Gao, *Zhongguo Zhong Yao Za Zhi*, 1993, **18** (1), 41–42, 63.

201. J. Wu, K. Cheng, H. Wang, W. Ye, E. Li and M. Wang, *Phytochem. Anal.*, 2009, **20** (1), 58.

202. Z. Jiang and Q. Du, *Bioorg. Med. Chem. Lett.*, 2011, **21** (3), 1001.

203. R. Hackman, P. Havel, H. Schwartz, J. Rutledge, M. Watnik, E. Noceti, S. Stohs, J. Stern and C. Keen, *Int. J. Obes.*, 2006, **30** (10), 1545.

204. H. Ju, S. Chen, K. Wu, H. Kuo, Y. Hseu, H. Ching and C. Wu, *Food Chem. Toxicol.*, 2012, **50** (3–4), 492.

205. M. Chao, D. Zou, Y. Zhang, Y. Chen, M. Wang, H. Wu, G. Ning and W. Wang, *Endocrine*, 2009, **36** (2), 268.

206. S. Dhanabal, M. Raja, M. Ramanathan and B. Suresh, *Fitoterapia*, 2007, **78** (4), 288.

207. K. Rajagopal and K. Sasikala, *Singapore Med. J.*, 2008, **49** (2), 137.

208. P. Subash-Babu, S. Ignacimuthu, P. Agastian and B. Varghese, *Bioorg. Med. Chem.*, 2009, **17** (7), 2864.

209. V. Vuksan V, J. Sievenpiper, V. Koo, T. Francis, U. Beljan-Zdravkovic, Z. Xu and E. Vidgen, *Arch. Int. Med.*, 2000, **160** (7), 1009.

CHAPTER 14

Translational Approach to Treating Diabetes Using Acupuncture or Electroacupuncture

PHILIP V. PEPLOW*[a] AND G. DAVID BAXTER[b]

[a] Department of Anatomy, University of Otago, Dunedin, New Zealand;
[b] Centre for Physiotherapy Research, School of Physiotherapy, University of Otago, Dunedin, New Zealand
*E-mail: phil.peplow@stonebow.otago.ac.nz

14.1 Introduction

Diabetes mellitus is an increasingly prevalent disease that can damage many structures in the body. It was estimated that more than 285 million people in the world had diabetes in 2009; this has risen to 347 million in 2011 and is expected to increase to 438 million by 2030.[1] A global health problem is predicted over the next 10 years from the increased numbers of patients requiring hospital treatment and/or nursing care at home. This dramatic and disturbing increase is likely due to the spread of a Western-style diet to developing nations which is causing rising levels of obesity, and also from an increase in life expectancy due to improvements in the health of individuals and populations and better access to health services. Complications of diabetes include cardiovascular disease, chronic renal failure, retinal damage, nervous system disease, amputation, fatigue, depression, and complications during

RSC Drug Discovery Series No. 31
Traditional Chinese Medicine: Scientific Basis for Its Use
Edited by James David Adams, Jr. and Eric J. Lien
© The Royal Society of Chemistry 2013
Published by the Royal Society of Chemistry, www.rsc.org

pregnancy. Many of the adverse effects of diabetes are caused by elevated blood levels of glucose and lipids, termed hyperglycemia and hyperlipidemia. There are three main types of diabetes. In type 1 diabetes, the β-cells of the pancreas produce insufficient or no insulin. In type 2 diabetes, the body is unable to use effectively the insulin produced by the pancreatic β-cells, and is termed insulin resistance. Gestational diabetes is a complication of pregnancy and develops when the pancreatic β-cell reserve is insufficient to compensate for decreased insulin sensitivity during pregnancy. Type 2 diabetes is the most common type (>90%). Traditional Chinese medicine (TCM) using acupuncture or herbs has been used to treat patients with diabetes for over 2000 years. Translational research aims at applying the findings of basic research studies to clinical practice to improve the health of patients through increased medical, physical and social outcomes. Such studies may involve acupuncture or electroacupuncture.

14.2   The Basis of Acupuncture and Electroacupuncture

Acupuncture dates back at least 2500 years and has been widely practiced in China and many other Asian countries.

The basic principles of acupuncture involve terms such as *yin* and *yang*, *qi*, meridians and organ systems. *Qi* acts as the life force that circulates through the body. It is accumulated, balanced, and enhanced by the dietary intake and air. Disorder and illness are considered to be caused by an unbalanced, obstructed and irregular flow of *qi*. Meridians are channels that carry *qi* throughout the body, there being 14 meridians in the body of which 12 are associated with organs and functions of the body. They are made up of acupuncture points that form a specific pathway, there being about 400 acupuncture points in the body.[2] The organ systems are energy systems in the body that receive, rebuild, and regulate energy or *qi*. The organ systems not only affect their associated organs, but also have an effect on other parts of the body. The organs that the meridians are named after do not indicate the physical area in which the meridian is located; rather, the organ name refers to the body function that the meridian affects.

Acupuncture involves inserting needles at specific acupuncture points along meridians on the body to correct imbalances of energy flow in the body.[3] Needle stimulation at acupoints located along meridians can have a therapeutic effect on the disorder or illness because it increases or decreases *qi* flow, rectifying the imbalance of energy. The meridians utilized in acupuncture correspond to the trigger point patterns associated with myofascial pain.[4]

Electroacupuncture stimulates the acupoints using acupuncture needles connected to a device that delivers a direct electrical current, and is effective in correcting energy imbalance in the body. This technique is identical in principle and practice with percutaneous electrical nerve stimulation.[5] When electro-acupuncture was used to stimulate the stomach 36 acupoint (ST36) in normal rats or rats with experimentally induced hypertension, an increase occurred in the production of nitric oxide (NO) and nitric oxide synthase (NOS) in the organs along the stomach meridian.[6,7] Hypertension in these animals was characterized by a deficiency in NO. The increased production of NO and NOS by electro-acupuncture balances the deficiency and appears to reduce the symptoms of hypertension. Furthermore, the depressor effect of electroacupuncture on experimental hypertension in dogs was prevented by pretreatment with naloxone, indicating that the inhibition was related to endogenous opioids released in the central nervous system during electroacupuncture.[8] In contrast, acupuncture decreased NOS expression in the periaqueductal gray area of rats with induced diabetes,[9] a disease which was characterized by high NO production.[10] Hence acupuncture at specific acupoints rather than at random points on the skin can create chemical changes within the body to correct an imbalance.

Several studies have found that when acupuncture stimulates the sensory nerves in the muscle, the spinal cord, midbrain, and pituitary gland are activated and release neurochemicals such as endorphins, serotonin and noradrenaline to block pain messages from reaching the brain.[11] Through the secretion of various chemicals from the brain that affect other organs in the body, acupuncture exerts a strong homeostatic effect in normalizing the vital signs of body temperature, pulse rate, respiration rate and blood pressure, and also the rate of formation of urine, metabolic rate, sweating, blood electrolyte balance, and many other parameters.

14.3 Diabetes Causation

In type 1 diabetes, the insulin-producing β-cells of the pancreas are attacked and destroyed by the immune system. Insufficient or no insulin is then produced by the pancreas. Type 1 diabetes develops most often in children and young adults, but can appear at any age. Type 2 diabetes is the most common form of diabetes, with 90–95% of people with diabetes having type 2. This form of diabetes is associated with older age, obesity, family history of diabetes, previous history of gestational diabetes, physical inactivity, and ethnicity. About 80% of people with type 2 diabetes are overweight. Type 2 diabetes is increasingly being diagnosed in children and adolescents. The pancreas is usually producing sufficient insulin in type 2 diabetes, but the body is not able to use the insulin effectively (insulin resistance). Gestational diabetes develops in pregnant women who have never had diabetes before; it may precede the development of type 2 diabetes.

As to the causes of diabetes, 70% has been attributed to aging and 30% to the increased prevalence of other factors, with high blood pressure, obesity and body mass the most important. The increasing prevalence of obesity is caused by excessive calorie intake, suboptimum dietary quality, and sedentary lifestyle.

14.4 Diabetes Prevention and Treatment

Adequate treatment of diabetes is extremely important, in addition to blood pressure control and lifestyle factors such as smoking cessation and maintaining a recommended body weight.

Patients with high body mass index (BMI) are at risk of developing diabetes and this disease has become increasingly more common as a greater proportion of the population in many developed countries has less physical exercise and increased dietary intake of saturated fats, leading to obesity and overweight. It is essential for therapies to be established that lead to adequate control of hyperglycemia and hyperlipidemia in patients with diabetes.

All forms of diabetes have been treatable with insulin, and type 2 diabetes may be controlled with medications, diet modification, and exercise. Pancreas transplants have had limited success in type 1 patients. Gastric bypass surgery has been successful in many grossly obese patients with type 2 diabetes. Recently, it has been shown that type 2 diabetes can be reversed in as little as 7 days by patients on a very low calorie diet of 600 calories per day. This was associated with a lowering of plasma glucose and insulin and decreased pancreatic and liver triacylglycerol stores.[12] A lowering of blood glucose and lipid levels may be attributable to normalization of pancreatic β-cell function and insulin sensitivity. Increased physical activity, including brisk walking, also significantly improves glycemic control among patients with diabetes.[13] Gestational diabetes is a form of diabetes that occurs during the second half of pregnancy and usually resolves after birth of the baby. A woman who develops gestational diabetes is more likely than other women to develop type 2 diabetes later in life.

14.5 The Traditional Approach to the Treatment of Diabetes

In TCM, diabetes is caused by an imbalance in the flow of *qi* in the meridians and organ systems. This particular imbalance produces heat that depletes the body's fluids and causes symptoms such as fatigue, unexplained weight loss, excessive thirst (polydipsia), excessive urination (polyuria), excessive eating (polyphagia), impaired wound healing, susceptibility to infections, altered mental status and blurry vision. A variety of techniques may be used during treatment, including acupuncture, Chinese herbal medicine, lifestyle/dietary changes and physical exercise. It focuses on regulating the circulation of blood and *qi* and balancing the organ systems to improve pancreatic function and to regulate internal heat and the depletion of fluids. The acupoints used to treat diabetes are all over the body and lie on several meridians.

Electroacupuncture has been reported to improve abnormal lipid metabolism and reduce fat accumulation effectively in obese rats, possibly closely associated with its effect in regulating the balance between leptin and insulin concentrations.[14] Muscle contractions occurring during electroacupuncture function like a type of exercise and promote the release of endogenous opioid peptides in rats.[15,16] Exercise enhanced the secretion of β-endorphins and insulin sensitivity in obese Zucker rats.[17]

14.6 Clinical Trials and Experimental Animal Studies

The results of recently published studies to assess the effectiveness of acupuncture and electroacupuncture to treat diabetes are summarized in Tables 14.1–14.6. The studies are of clinical trials and experimental animal models with type 1 or type 2 diabetes published up to December 2010.[18-36]

14.6.1 Acupuncture

Six clinical trials and one experimental animal study in rats were identified, which together with the acupoints used in the clinical trials are summarized in Tables 14.1–14.3.

14.6.1.1 Clinical Trials (Tables 14.1 and 14.2)

14.6.1.1.1 Blood glucose

Of the six human studies, three reported on blood glucose in patients with type 1 or type 2 diabetes and in two of these blood glucose levels were lowered in fasting patients by acupuncture (2/2 fasting, 100% successful in having hypoglycemic effect), whereas in the third study involving patients not reported as being fasted there was no change in blood glucose. The other studies were performed in patients with cancer or cardiovascular disease, senile patients with impaired glucose tolerance and healthy subjects. For patients with cancer or cardiovascular disease, blood glucose was lowered in patients with hyperglycemia by acupuncture; for senile patients with impaired glucose tolerance and measurements performed 2 h after breakfast there was a decrease in blood glucose in patients receiving acupuncture in addition to controlling diet, whereas for healthy subjects who were not reported as being fasted there was no change in blood glucose compared with normal values. Thus, in both of the two studies with patients having hyperglycemia, acupuncture lowered blood glucose (2/2, 100% successful in having hypoglycemic effect).

14.6.1.1.2 Blood insulin

Two studies reported on blood insulin in patients with type 1 or type 2 diabetes receiving acupuncture. In one of these studies, for patients with high insulin secretion, plasma insulin levels were reduced at all test intervals except for fasting plasma insulin; for the inadequate insulin group, plasma insulin increased at all test intervals except at 3 h after load; for the deficient insulin secretion group, plasma insulin was not changed at any test period of the oral glucose tolerance test (OGTT). In the other study in which patients with diabetes were not reported as being fasted, there was no notable change in blood insulin. In a third study involving healthy subjects not reported as being fasted, plasma insulin was decreased by acupuncture.

Table 14.1 Clinical trials using acupuncture.

Study	Acupuncture treatment	Subjects Normal	Diabetic	Measured outcomes of acupuncture treatment Blood glucose
Omura et al., 2007[18]	Needle inserted into Omura's ST36 (true ST36) acupoint	Cancer, cardiovascular disease		Decrease in high blood glucose in patients with hyperglycemia
Ahn et al., 2007[19]	Patients randomized to two styles of acupuncture: 4 patients received Japanese style (Kiiko-Matsumoto's form) and 3 received TCM style. Treatments delivered weekly for a total of 10 treatments		Type 1 and type 2, not reported as being fasted	No notable change in blood glucose
Wu et al., 2006[20]	64 patients randomly divided into two groups. 32 patients in the control group treated with interference therapy of controlling diet. 32 patients in the observation group given acupuncture at Yishu, Pishu, Sanyingjiao, Zusanli (ST36) and other points in addition to interference therapy of controlling diet. Acupuncture performed 10 times as a course of treatment	Senile with impaired glucose tolerance; patients not fasted with measurements performed 2 h after breakfast		Decrease in postprandial blood glucose in observation group of patients receiving acupuncture in addition to controlling diet and also in control group of patients treated by controlling diet, with a lower level in observation group than in control group
Jiang et al., 2006[21]	All patients given hypoglycemic agents. When blood glucose was stable for more than 3 months and medication for diabetic peripheral neuropathy was withdrawn for more than 2 weeks, patients were divided randomly into 3 groups. 30 patients received wrist-ankle acupuncture (Group 1), 30 body acupuncture (Group 2) and 30 Western medical treatment (Group 3). Treatment was given once daily, 7 treatments constituting a course, with a 2 day interval between courses. In total 3 courses of treatment were given		Type 2 with peripheral neuropathy with patients fasted for blood glucose measurements	Decrease in blood glucose in all 3 groups
Chen, 1985[22]	3 patients had juvenile-onset, 18 adult-onset and 5 old-age-onset diabetes. Acupoints were Pishu (UB20), Geshu (UB17) and Zusanli (ST36) with additional points selected according to symptoms and signs		Type 1 and type 2, fasting and non-fasting patients	In the effective cases (21 out of 26), fasting and 2 h postprandial blood sugar lowered by acupuncture treatment. For high insulin secretion group (3 cases), blood glucose values at all test intervals of oral glucose tolerance test (OGTT) were decreased; for inadequate insulin secretion group (12 cases), blood glucose values at all test intervals of OGTT decreased; for deficient insulin secretion group (3 cases), blood glucose showed a tendency to increase
Szczudlik and Lypka, 1984[23]	10 patients received needles inserted s.c. to Du20, Extra 1, Extra 2 - symmetr., GB20-symmetr., LI4 - symmetr., followed by injection of 0.1 mL 10% NaCl. Needles kept in position for 15 min	Adults, not reported as being fasted		Serum glucose not changed from normal values in investigated subjects

Table 14.1   Extended.

Blood insulin	Blood lipids	Blood flow	Pain score	Nerve conduction velocity
	Decrease in high triglycerides in patients with hypertriglyceridemia	Improved circulatory disturbances in patients with hypertension		
No notable change in blood insulin			Decreased neuropathy-associated pain in patients receiving Japanese acupuncture, whereas the group receiving TCM acupuncture reported minimal effects; both acupuncture styles lowered pain according to McGill Short Form Pain Score	
	Decrease in blood lipids in acupuncture groups but not in group receiving Western medicine	Decrease in all pathological changes relating to blood flow in acupuncture groups, but in group receiving Western medicine only blood viscosity reduced		NCV increased in acupuncture groups but not in group receiving Western medicine
For high insulin secretion group, plasma insulin levels reduced at all test intervals except fasting plasma insulin; for inadequate insulin group, plasma insulin increased at all test intervals except at 3 h after load; for deficient insulin secretion group, plasma insulin was not changed at any test period of OGTT				
Plasma insulin decreased at 5 min after end of acupuncture and increased after next 30 min				

Table 14.2 Acupoints in clinical trials using acupuncture.

Study	Subjects	Acupuncture treatment	Acupoints	Blood glucose	Blood insulin	Blood lipids
Ahn et al., 2007[19]	Type 1 and type 2 diabetes and patients, not reported as being fasted	In Japanese style, points all over the body		No change	No change	
		In Traditional Chinese style, mostly leg points	ST36, ST40, SP4, SP9, LR3, KI3, BL22–25	No change	No change	
Jiang et al., 2006[21]	Type 2 diabetes with peripheral neuropathy and patients fasted for blood glucose measurements	Group 1: wrist–ankle acupuncture	Bilateral Upper 2 and Lower 2, added with symptomatological points for the upper limb, for the head part, for medial aspect of the lower limb, for lateral aspect of the lower limb, for the knee	Decreased		Decreased
		Group 2: body acupuncture	Local points combined with symptomatological points were selected, including SP6, SP10, KI3, LI11, GB34; for upper limb LI15, SJ14, LI11, TE5, LI4; for lower limb GB30, ST36, GB34, ST41, ST44	Decreased		Decreased

Table 14.2 (*Continued*)

Study	Subjects	Acupuncture treatment	Acupoints	Blood glucose	Blood insulin	Blood lipids
Chen, 1985[22]	Type 1 and type 2 diabetes and patients fasted	Body and leg acupuncture	UB20, UB17, ST36	Decreased	Decreased for high insulin secretion group at all test intervals except for fasting plasma insulin; increased for inadequate insulin group at all test intervals except at 3 h after load; no change for deficient insulin secretion group at any test period of OGTT	
Omura et al., 2007[18]	Patients with cancer or cardiovascular disease	Leg acupuncture	Omura's ST36 (true ST36)	Decreased in patients with hyperglycemia		
Wu et al., 2006[20]	Senile patients with impaired glucose tolerance, patients not fasted	Body and leg acupuncture	EX-B3, BL20, SP6, ST36 as main points and BL13, BL17, BL21 and BL23 as adjunct points	Decreased in group receiving acupuncture in addition to controlling diet		
Szczudlik and Lypka, 1984[23]	Normal subjects	Head and hand acupuncture	Du20, Extra 1, Extra 2, GB20, LI4	No change	Decreased	

14.6.1.1.3 Blood lipids

One study reported on blood lipids in patients having type 2 diabetes with peripheral neuropathy and there was a decrease in blood lipids in the two groups of patients receiving acupuncture (wrist–ankle acupuncture, body acupuncture) but not in the group receiving Western medicine. Another study was performed in patients with cancer or cardiovascular disease and there was a decrease in triglyceride levels in those patients with hypertriglyceridemia by acupuncture. Thus, acupuncture decreased blood lipids or triglyceride levels in these two studies (2/2, 100% successful in lowering blood lipid/triglycerides).

14.6.1.1.4 Blood flow

One study reported on blood flow in fasted patients having type 2 diabetes with peripheral neuropathy and there was a decrease in all pathological changes relating to blood flow in groups receiving acupuncture (wrist–ankle acupuncture, body acupuncture), but in the group receiving Western medicine only blood viscosity was lowered. Another study reported on blood flow in patients with cancer or cardiovascular disease and there was an improvement in circulatory disturbances in patients with hypertension receiving acupuncture. Hence acupuncture brought about improvements relating to blood flow in both studies (2/2, 100% successful in improvements to blood flow).

14.6.1.1.5 Pain score

One study reported on neuropathy-associated pain in patients with type 1 and type 2 diabetes who were not reported as being fasted, with the pain being lowered by acupuncture according to the McGill Short Form Pain Score.

14.6.1.1.6 Nerve conduction velocity

One study reported on nerve conduction velocity (NCV) in patients with type 2 diabetes with peripheral neuropathy and there was an increase in NCV in the groups receiving acupuncture but not in the group receiving Western medicine.

14.6.1.2   Experimental Animal Studies (Table 14.3)

Following acupuncture, blood glucose was not lowered in adult rats with streptozotocin (STZ)-induced type 1 diabetes and not reported as being fasted. Blood flow in the footpad was lowered in weeks 1 and 2 but was not changed in weeks 3 and 4 and NCV was increased at the third or fourth weekend of acupuncture.

Table 14.3 Experimental animal studies using acupuncture.

	Experimental animals		Measured outcomes of acupuncture treatment						
Study	Species, acupuncture treatment	Normal	Diabetic	Blood glucose	Blood insulin	Blood lipids	Blood flow	Pain score	Nerve conduction velocity
Lin et al., 2005[24]	Rat: 15 STZ animals were randomly selected and received needle insertion to Zusanli (ST36), Yinlingquan (SP9), Sanyinjiao (SP6) and Taixi (K3) in the right lower limb		STZ adults, type 1, not reported as being fasted	No change in blood glucose for diabetic rats following acupuncture			Blood flow in footpad reduced for diabetic rats receiving acupuncture compared with normal rats in weeks 1 and 2 of treatment, but not changed in weeks 3 and 4		NCV increased for diabetic rats at 3rd or 4th weekend of acupuncture compared with diabetic control rats

14.6.2   Electroacupuncture

Two clinical trials and 14 experimental animal studies ($n = 11$, rat, of which two included mouse; $n = 1$, mouse; $n = 1$, rabbit; $n = 1$, sand rat) were identified and together with the electroacupuncture parameters and acupoints in the rat studies are summarized in Tables 14.4–14.6.

14.6.2.1   Clinical Trials (Table 14.4)

16.6.2.1.1   Blood glucose and blood insulin

In one of the studies involving obese women receiving a 1450 kcal diet restriction for 20 days, blood glucose was decreased and serum insulin increased in the electroacupuncture group compared with the placebo electroacupuncture group. In the other study, serum glucose in healthy subjects was not changed from normal values and plasma insulin was decreased.

14.6.2.2   Experimental Animal Studies (Tables 14.5 and 14.6)

14.6.2.2.1   Blood glucose

For the studies in rats, blood glucose was decreased in type 2 diabetic animals by electroacupuncture (3/3, 100% successful in having hypoglycemic effect); blood glucose was mostly unchanged in type 1 diabetic animals by electroacupuncture (5/6, 83% having no significant effect); blood glucose was lowered in normal adult rats by electroacupuncture (6/8, 75% successful in having hypoglycemic effect). These findings were for both non-fasted and fasted rats.

For the study with obese diabetic mice, blood glucose was decreased. In the two studies using non-fasting and fasting normal mice, blood glucose was lowered by electroacupuncture (2/2, 100% successful in having hypoglycemic effect).

14.6.2.2.2   Blood insulin

Blood insulin was increased in type 2 diabetic and normal rats by electroacupuncture (3/3, 100% and 2/3, 67% successful in increasing blood insulin, respectively).

14.6.2.2.3   Blood lipids

In the study with obese diabetic mice, free fatty acid levels in blood were decreased by electroacupuncture, but with no significant effect on triglycerides or total cholesterol.

In non-diabetic mice, electroacupuncture had no effect on free fatty acids, triglycerides or total cholesterol in blood.

Table 14.4 Clinical trials using electroacupuncture (EA).

Study	EA treatment	Subjects		Measured outcomes of electroacupuncture treatment					
		Normal	Diabetic	Blood glucose	Blood insulin	Blood lipids	Blood β-endorphin	Blood flow	Nerve conduction velocity (NCV)
Cabioglu and Ergene, 2006[25]	52 healthy women allocated to 3 groups: placebo EA group (n = 15), EA group (n = 20), diet restriction group (n = 20). 2 Hz, 30 min/day, 20 days; ear points Hunger and Shen Men and body points LI4, LI11, ST36, ST44	Obese women; a 1450 kcal diet for women in EA group, placebo EA group and diet restriction group for 20 days		Blood glucose decreased	Serum insulin increased				
Szczudlik and Lypka, 1984[23]	10 patients. 1 Hz, 15 min; needle insertion to Du20, Extra 1, Extra 2 - symmetr., GB20- symmetr., LI4 - symmetr. and twirled to obtain "degi" (Tei Chi) effect; current increased gradually to obtain painless tingling sensation in stimulated area, avoiding muscle contraction	Adults		Serum glucose not changed from normal values in investigated subjects	Serum insulin decreased at 5 min after end of acupuncture and increased after next 30 min				

Reproduced by permission of *Journal of Acupuncture and Meridian Studies* (*JAMS*, 2012, **5**, 1–10).

Table 14.5 Experimental animal studies using electroacupuncture.

Study	Species, EA treatment	Experimental animals Normal	Experimental animals Diabetic
Liang *et al.*, 2011[26]	Mouse: males, 16 db/db divided into two groups, *n* = 8/group; 14 db/m divided into two groups. 3 Hz/0.5–0.8 mA, 10 min/day, 5 treatments/week, 8 weeks. Acupuncture needles inserted at Zusanli (ST36) and Guanyuan (CV4) points; needles at these two points were linked to Zusanli on the other side on the following day	Non-diabetic heterozygote litter mates	Genetic obese diabetic
Pai *et al.*, 2009[27]	Rat: 36 males randomly divided into groups, *n* = 9/group. 15 Hz/ 10 mA, 30 min. Two acupuncture needles inserted into anterior tibial muscles at both Zusanli (ST36) points	Adults, fasted for 12 h	STZ neonatal type 2, fasted for 12 h
Ishizaki *et al.*, 2009[28]	Rat: 27 males divided into EA group (*n* = 13) and control group (*n* = 14). 15 Hz/10 mA, 90 min. Acupuncture needles inserted into muscle layer at two points between top of xiphoid process and upper border of pubic symphysis		Genetic type 2, fasted for 12–20 h
Higashimura *et al.*, 2009[29]	Rat: 25 males; 2 weeks after STZ, randomly divided into groups. 20 Hz/10 mA, 10 min. Two acupuncture needles inserted into right tibialis anterior muscle		STZ adults, type 1, not reported as being fasted (*n* = 7)
Chang *et al.*, 2006[30]	Rat: males, normal and STZ diabetic rats divided randomly into experimental and control groups, *n* = 8/group. 15 Hz, 30 and 60 min. Acupuncture needles inserted into anterior tibia muscles at both Zusanli (ST36) points	Adults	STZ adults, type 1, fasted for 12 h
Tseng *et al.*, 2005[31]	Rat: 7 animals. 2 Hz, 30 min, then after 90 min 2 Hz, 30 min (*n* = 2), 100 Hz, 30 min (*n* = 2). Acupuncture needle inserted 5 mm deep at bilateral Zusanli (ST36) acupoints; 2 Hz (*n* = 3) at non-acupoints	Adults, fasted for 12 h	
Lin *et al.*, 2005[24]	Rat: males divided randomly into groups, including normal (*n* = 12) and EA groups (*n* = 11/group). 4 or 20 Hz, 30 min/50 μA/ day, 4 weeks. Acupoints ST36, SP9 (positive charge), SP6, K3 (negative charge) in right lower limb	Adults	STZ adults, type 1
Chang *et al.*, 2006[32]	Rat: males. 2 Hz, 30 min. Mouse: 2 Hz, 30 min. Acupuncture needles inserted into anterior tibia muscles at both Zusanli (ST36) points	Adults Adults	STZ adults, type 1
Lin *et al.*, 2004[33]	Rat: males, randomly divided into adrenalectomized (ADX) and sham groups (*n* = 6/group). Mouse: males 15 Hz, 30 min. Acupuncture needles inserted into muscle layer at depth of 2 mm at Zhongwan and Gwanyuan acupoints on abdomen	Sham and ADX adults	

Reproduced by permission of *Journal of Acupuncture and Meridian Studies* (*JAMS*, 2012, **5**, 1–10).

Table 14.5 *(Extended)*

Measured outcomes of acupuncture treatment

Blood glucose	Blood insulin	Blood lipids	Blood β-endorphin	Blood flow	Nerve conduction velocity
Blood glucose decreased in diabetic mice	Insulin levels maintained and increased insulin sensitivity in diabetic mice	Free fatty acid levels decreased in blood of diabetic mice with no significant effect on triglyceride or total cholesterol; no effect on free fatty acid, triglyceride or total cholesterol levels in blood of non-diabetic mice			
Blood glucose decreased in normal rats and type 2 diabetic rats	Insulin secretion increased in normal rats and type 2 diabetic rats				
Fasting blood glucose decreased	Plasma insulin increased during fasting period				
No change in plasma glucose; response of plasma glucose to insulin increased					
Blood glucose decreased in normal rats and type 1 diabetic rat; response of plasma glucose to insulin increased in insulin challenge test in normal rats and type 1 diabetic rats	No change in normal rats and no increase in normal rats undergoing i.v. GTT				
After 2 Hz EA blood glucose decreased but 60 min after end of EA blood glucose was raised; blood glucose not changed after a second 2 or 100 Hz stimulus at 90 min					
No change in blood glucose for normal and diabetic rats following electrical stimulation of acupoints				No change of blood flow in footpad for diabetic rats following electrical stimulation of acupoints compared with normal rats	For diabetic mice, at 3rd or 4th weekend of electrical stimulation NCV was increased
Plasma glucose decreased in non-fasting ($n = 8$) and fasting normal rats ($n = 7$) but not in fasting type 1 diabetic rats ($n = 7$); a small decrease in plasma glucose occurred with EA at non-acupoint in normal rats ($n = 8$); plasma glucose decreased in non-fasting ($n = 8$) and fasting normal mice ($n = 7$)	Plasma insulin increased in fasting normal rats		Plasma β-endorphin increased in fasting normal rats		
Plasma glucose decreased in sham rats ($n = 6$) and mice ($n = 9$); adrenalectomy partially reduced plasma glucose response to EA in rats ($n = 6$) and mice ($n = 8$)	Plasma insulin not changed in ADX rats		Plasma β-endorphin not changed in ADX rats		

Table 14.5 *(Extended)*

Study	Species, EA treatment	Experimental animals Normal	Experimental animals Diabetic
Zeng and Li, 2002[34]	Rabbit: 14 males, 10 females. 25 min. Acupuncture needle inserted into Weiwanxiashu and/or Zusanli acupoint		Alloxan adults, type 1
Lin *et al.*, 2002[16]	Rat: males, divided into ADX (n = 9) and sham groups (n = 8). Mouse: males ADX (n = 8) and sham (n = 7). 2 Hz, 30 min. Acupuncture needles inserted into muscle layer at Zhongwan and Gwanyuan acupoints	Sham and ADX adults, fasted	
Shapira *et al.*, 2000[35]	Sand rat: 29 diabetic males randomly assigned to 3 groups: abdominal EA, back EA and control. 15 Hz/80 mA, 30 min. Acupuncture needles inserted 2 mm into muscle layer at Zhongwan and Gwanyuan acupoints		Type 2
Chang *et al.*, 1999[15]	Rat: males, divided into experimental groups and control group (EA at non-acupoint). 15 Hz/ 10 mA, 30 min.). Acupuncture needles inserted into muscle layer at Zhongwan (positive charge) and Gwanyuan acupoints (negative charge	Adults	Adults, genetic type 1 (BB/W), STZ adults, type 1, and STZ neonatal, type 2
Ikeda *et al.*, 1991[36]	Rat: 56 males. 50 Hz/5 V, 10 min. Two acupuncture needles inserted to deep muscle layer at Chuin acupoint; other acupoints in abdomen also chosen and to front paw and to leg and thigh	Adults	

14.6.2.2.4   Blood β-endorphin

Blood β-endorphin was increased in normal rats by electroacupuncture (2/2, 100% successful in increasing blood β-endorphin).

14.6.2.2.5   Blood flow

There was no change in blood flow in footpad for type 1 diabetic rats following electroacupuncture.

14.6.2.2.6   Nerve conduction velocity

NCV was increased in type 1 diabetic rats at the third or fourth weekend of electroacupuncture.

Table 14.5 *(Extended)*

Measured outcomes of acupuncture treatment

Blood glucose	Blood insulin	Blood lipids	Blood β-endorphin	Blood flow	Nerve conduction velocity
Plasma glucose and glucagon decreased following acupuncture at Weiwanxiashu point (*n* = 8) but not when acupuncture given at Zusanli point alone (*n* = 8); decrease in plasma glucose and glucagon more pronounced when acupuncture given at both Weiwanxiashu and Zusanli (*n* = 8)					
Plasma glucose decreased in sham rats and mice; no effect in ADX rats and mice	Plasma insulin increased in sham rats; no effect in ADX rats		Plasma β-endorphin increased in sham rats; no effect in ADX rats		
Decrease in blood glucose between real (abdominal EA) and placebo groups (back EA) throughout the 3 weeks	Serum insulin not changed between real and placebo groups	Plasma triglycerides and cholesterol not changed between real and placebo groups			
Plasma glucose decreased in normal rats (*n* = 8) and type 2 diabetic rats (*n* = 8); no effect in type 1 diabetic rats (BB/W, *n* = 8; STZ induced, *n* = 7)	Plasma insulin increased in normal rats and type 2 diabetic rats		Plasma β-endorphin increased in normal rats and type 2 diabetic rats; also increased in STZ type 1 diabetic rats		
Blood glucose levels increased following EA of middle upper abdominal area; stimulation of other areas of abdomen or front paw, leg or thigh did not alter blood sugar level					

14.7 Adipokines and Toxic Lipids

Recent studies have shown that alteration in the production of cytokines from adipose tissue, also known as adipokines, may be responsible for initiating a proinflammatory state in obese individuals and type 2 diabetic patients that is responsible for the development of insulin resistance and endothelial dysfunction, which is the initial stage in the development of atherosclerosis. Adipokines include adiponectin, leptin, tumor necrosis factor alpha (TNF-α), interleukin-6 (IL-6) and plasminogen activator inhibitor-1 (PAI-1). Increased adipose tissue mass was associated with overexpression of TNF-α, IL-6 and PAI-1 and underexpression of adiponectin in adipose tissue. Reduction of adipose tissue mass through weight reduction in association with exercise lowered TNF-α, IL-6 and PAI-1, increased adiponectin and was associated with improved insulin sensitivity and endothelial function.[37] In most patients with gestational diabetes, abnormally high levels of leptin are found and might increase the inflammatory process. Decreased levels of adiponectin in gestational diabetes might further exacerbate insulin resistance.[38] Evidence

Table 14.6 Acupoints in rat studies using electroacupuncture.

Study	Experimental animals	Electroacupuncture parameters	Acupoints	Measured outcomes of treatment		
				Blood glucose	Blood insulin	Blood β-endorphin
Type 1 diabetes						
Higashimura et al., 2009[29]	STZ adults, type 1, not reported as being fasted	20 Hz/10 mA, 10 min, leg acupoints	Needles inserted into right tibialis anterior muscle	No change; response of plasma glucose to insulin increased		
Chang et al., 2006[30]	STZ adults, type 1, fasted for 12 h	15 Hz, 30 and 60 min, leg acupoints	ST36	Decreased; response of plasma glucose to insulin increased in insulin challenge test		
Lin et al., 2005[24]	STZ adults, type 1, fasted	4 or 20 Hz/50 μA/day, 30 min, 4 weeks, right leg acupoints	ST36, SP9	No change		
Chang et al., 2005[32]	STZ adults, type 1, fasted	2 Hz, 30 min, leg acupoints	ST36	No change		
Chang et al., 1999[15]	Adults, genetic, type 1, not reported as being fasted	15 Hz/10 mA, 30 min, body acupoint (abdomen)	Zhongwan (CV12)	No change		
Chang et al., 1999[15]	STZ adults, type 1, not reported as being fasted	15 Hz, 10 mA, 30 min, body acupoint (abdomen)	Zhongwan (CV12)	No change		

Table 14.6 (*Continued*)

Study	Experimental animals	Electroacupuncture parameters	Acupoints	Measured outcomes of treatment		
				Blood glucose	Blood insulin	Blood β-endorphin
Type 2 diabetes						
Pai et al., 2009[27]	STZ neonatal, type 2, fasted for 12 h	15 Hz/10 mA, 30 min, leg acupoints	ST36	Decreased	Increased	
Ishizaki et al., 2009[28]	Adults, genetic, type 2, fasted for 12–20 h	15 Hz/10 mA, 90 min, body acupoint (abdomen)	Zhongwan (CV12)	Decreased	Increased	
Chang et al., 1999[15]	STZ neonatal, type 2, not reported as being fasted	15 Hz, 10 mA, 30 min, body acupoint (abdomen)	Zhongwan (CV12)	Decreased	Increased	
Normal (non-diabetic)						
Pai et al., 2009[27]	Adults, normal, fasted for 12 h	15 Hz/10 mA, 30 min, leg acupoints	ST36	Decreased	Increased	
Chang et al., 2006[30]	Adults, normal, fasted for 12 h	15 Hz, 30 and 60 min, leg acupoints	ST36	Decreased; plasma glucose decreased from 15 to 90 min in animals undergoing i.v. GTT	No change; no increase in animals undergoing i.v. GTT	
Tseng et al., 2005[31]	Adults, normal, fasted for 12 h	2 Hz, 30 min, then after 90 min 2 Hz, 30 min (n = 2), 100 Hz, 30 min (n = 2), leg acupoints	ST36	Decreased		

Table 14.6 (Continued)

Study	Experimental animals	Electroacupuncture parameters	Acupoints	Measured outcomes of treatment		
				Blood glucose	Blood insulin	Blood β-endorphin
Lin et al., 2005[24]	Adults, normal, not reported as being fasted	4 or 20 Hz/50 µA/day, 30 min, 4 weeks, right leg acupoints	ST36, SP9	No change		
Chang et al., 2005[32]	Adults, normal, non-fasted and fasted	2 Hz, 30 min, leg acupoints	ST36	Decreased in non-fasting and fasting states		Increased in fasting state
Lin et al., 2004[33]	Adults, normal (sham), fasted for 12 h	15 Hz, 30 min, body acupoint (abdomen)	Zhongwan (CV12)	Decreased		
Lin et al., 2002[16]	Adults, normal (sham), fasted	2 Hz, 30 min, body acupoint (abdomen)	Zhongwan (CV12)	Decreased	Increased	Increased
Ikeda et al., 1991[36]	Adults, normal, not reported as being fasted	50 Hz/5 V, 10 min, body acupoint (abdomen)	Chuin Right sides of middle abdomen 6 other abdominal acupoints	Increased Increased No change		

Reproduced by permission of *Journal of Acupuncture and Meridian Studies* (*JAMS*, 2012, **5**, 1–10).

from human patients and animal models with type 2 diabetes indicates that elevated glucose levels cause oxidative stress on pancreatic β-cells[39,40] and that the β-cells are very vulnerable during times of oxidative stress, having relatively low activities of the major antioxidant enzymes superoxide dismutase, catalase and glutathione peroxidase.[41]

Type 2 diabetes in human patients is frequently associated with obesity, hyperlipidemia and hyperglycemia. Fatty acids are detrimental to β-cells when present at elevated concentrations for prolonged periods of time, with adverse effects such as decreased glucose-induced insulin secretion,[42] impaired insulin gene expression[43] and increased cell death.[44] Lipotoxicity occurs only in the context of hyperglycemia, whereas glucotoxicity can occur in the absence of hyperlipidemia.[45]

14.8 Conclusion

Central to the management of diabetes are treatment strategies targeting blood glucose, blood pressure, triglycerides and low-density lipoprotein cholesterol levels,[46,47] together with lifestyle modifications. Such strategies lower the risk of cardiovascular complications and also reduce the risk or slow the progression of microvascular complications such as retinopathy and nephropathy.[48] The use of pharmacological agents and lifestyle modifications including exercise, weight loss, reduced caloric and dietary fat intake and cessation of smoking are strategies currently in use. Acupuncture has long been used in TCM to treat hyperglycemia in human subjects.[22] Several early studies in both normal and diabetic rodents showed that electroacupuncture applied on the abdomen or the legs had marked hypoglycemic effects by increasing insulin secretion. Electrical stimulation on the abdomen accelerated glucose consumption during an intravenous glucose tolerance test (GTT) in non-diabetic rats and increases in both insulin secretion and sensitivity were observed.[49] Furthermore, electroacupuncture applied to the hind limb of STZ-induced type 1 diabetic rats for 10 min, although not causing any significant change in plasma glucose concentration, increased the response of plasma glucose to insulin *via* excitation of somatic afferent nerves.[29] It has been suggested that glucose uptake in response to insulin can be augmented *via* two mechanisms, muscle contraction itself and an afferent nerve-mediated mechanism evoked by the muscle contraction (and/or muscle stimulation).[29] In anesthetized rats, high-frequency (20 Hz) electroacupuncture applied to hind limb reflexively increased hepatic glucose output.[50] Therapies involving acupuncture alone or together with electrical stimulation could form an important treatment strategy for the management of diabetic patients.

Of the six clinical trials using acupuncture, three were performed with diabetic patients, one with patients having cancer or cardiovascular disease, one with senile patients having impaired glucose tolerance and one with normal healthy subjects. Acupuncture was successful in reducing blood glucose in fasting diabetic patients, in hyperglycemic patients with cancer or cardiovascular disease and in senile patients with impaired glucose tolerance given in combination with a controlled diet. For fasting diabetic patients, blood insulin after acupuncture was only slightly elevated at fasting and at 30 min after glucose load in OGTT–IRT (oral glucose tolerance test and insulin release test). Most of the diabetic patients treated had type 2 diabetes. Interestingly, blood glucose was not changed in normal patients receiving acupuncture but there was a decrease in blood insulin. In type 2 diabetic patients with peripheral neuropathy, blood lipids were decreased by both wrist–ankle acupuncture and body acupuncture. In the study with STZ-induced type 1 diabetic rats there was no change in blood glucose following acupuncture. These studies indicate that acupuncture is an effective treatment for patients with type 2 diabetes, hyperglycemic patients with cancer or cardiovascular disease and senile patients with impaired glucose tolerance.

Electroacupuncture treatment of obese women with caloric restriction diet resulted in a decrease in blood glucose and an increase in blood insulin. Although low-energy electroacupuncture did not change blood glucose in normal healthy subjects, there was a decrease in blood insulin. The studies performed in the rat indicated that electroacupuncture had a hypoglycemic effect in type 2 diabetic and normal rats, with an increase in blood insulin. However, it was only successful in reducing blood glucose in a small percentage of type 1 diabetic rats. These findings suggest that electroacupuncture could be an effective treatment for obese patients with hyperglycemia and also for patients with type 2 diabetes.

In acupuncture, point selection is the key to treatment. Specific points such as Yishu (EX-B3), Pishu (BL20), Zusanli (ST36) can explicitly improve glucose metabolism and enhance glucose tolerance, and Sanyinjiao (SP6) and Zusanli (ST36) are selected to nourish the kidney and strengthen the spleen.[51] Yishu acupoint (EX-B3) is on the back, 1.5 cm lateral to the lower border of the spinous process of the eighth thoracic vertebra. Pishu (BL20) is also on the back, below the spinous process of the eleventh thoracic vertebra, 1.5 cm lateral to the posterior midline. Zusanli acupoint (ST36) is located on the front of the leg just below the knee and one finger width lateral from the anterior border of the tibia. Sanjinjiao acupoint (SP6) is located four finger widths above the tip of the medial malleolus of the tibia.

In TCM, the Zusanli (ST36) acupoint is considered a primer to control blood glucose,[52] whereas the Sanyinjiao (SP6) acupoint regulates not only blood glucose but also gynecological conditions.[53] In the human studies where a decrease in blood glucose was caused by acupuncture treatment, the Zusanli acupoint (ST36) was one of the acupoints used. Moreover, electroacupuncture 2 Hz stimulation at the Zusanli acupoint (ST36) in rats produced a greater hypoglycemic response than stimulation at the Zhongwan acupoint which is located on the upper abdomen, on the anterior midline 4 cm above the center of the umbilicus.[54] There was no change in blood glucose in fasting type 1 diabetic rats receiving 2 Hz stimulation for 30 min at acupuncture needles inserted at both Zusanli (ST36) acupoints. However a decrease in blood glucose occurred using 15 Hz for 30 min at acupuncture needles inserted at both Zusanli (ST36) acupoints. Thus frequency of stimulation is an important factor in electroacupuncture response and 15 Hz stimulation provided an effective treatment for lowering blood glucose in fasting type 1 diabetic rats which was not found using 2 Hz stimulation or acupuncture alone.

Acupuncture and electroacupuncture can produce a prolonged and sustained hypoglycemic effect by increasing insulin levels and/or elevating insulin sensitivity. The insulin-dependent hypoglycemic effect produced by electroacupuncture is possibly mediated by endogenous opioid peptides such as β-endorphin and activation of serotonin.[32] Acupuncture and electroacupuncture can also lower free fatty acids and lipid levels of blood plasma by altering citrate or glucose metabolism in the liver.[52,55] However, very few of the studies examined reported on free fatty acids and lipid levels. Of those that did, acupuncture decreased blood lipids in type 2 diabetic patients with peripheral neuropathy and also lowered triglyceride levels in patients having cancer or cardiovascular disease with hypertriglyceridemia. Future studies to examine the influence of acupuncture and electroacupuncture on the expression in adipose tissue of adiponectin, leptin, TNF-α, IL-6 and PAI-1 may provide further understanding of the hypoglycemic effects of these two treatments.

Recently, preoperative application of transcutaneous electrical nerve stimulation (TENS) at Zusanli (ST36) and Sanyinjiao (SP6) acupoints has been shown to prevent a hyperglycemic response in human patients during anesthesia.[56] Electroacupuncture had a greater hypoglycemic effect than TENS in STZ diabetic rats.[57] During hospitalization, ~30% of patients will experience hyperglycemia, with many not having diabetes or being undiagnosed. When a patient is critically ill, for example after heart surgery, blood glucose levels tend to increase – a condition termed stress hyperglycemia. This worsens outcomes in seriously ill patients with higher medical costs, greater incidence of infection and readmission to the hospital and increased mortality rates. Furthermore, a recent study reported that 40% of elderly patients admitted to hospital had hyperglycemia; excluding those patients already diagnosed with diabetes, 25% of these patients were hyperglycemic.[58] Electrical stimulation at selected acupoints could have important clinical application in reducing hyperglycemia in patients during anesthesia or surgery. Furthermore, acupuncture and electroacupuncture have analgesic effects through the release of endorphins and enkephalins, thereby increasing anesthesia in patients undergoing surgery. The current chapter is an expansion and reevaluation of previous work.[59]

References

1. F. B. Hu, *Lancet*, 2011, **378** (9786), 101.
2. W. Cai, *Am. J. Chin. Med.*, 1992, **20** (3), 331.
3. W. Zhou and J. C. Longhurst, *Future Cardiol.*, 2010, **2** (3), 287.
4. P. T. Dorsher, *J. Pain*, 2009, **10** (7), 723.
5. M. Cummings, *Acupunct. Med.*, 2001, **19** (1), 32.

6. S.-X. Ma, J. Ma, G. Moise and X.-Y. Li, *Brain Res.*, 2005, **1037** (1–2), 70.
7. H. S. Hwang, Y. S. Kim and Y. H. Ryu, *Evid.-Based Comp. Alt. Med.*, 2011, Article ID 130529, doi: 10.1093/ecam/nen064.
8. S. X. Lin and P. Li, *Acta Physiol. Sin.*, 1981, **33**, 335.
9. M.-H. Jang, M. C. Shin, G. S. Koo, C. Y. Lee, E. H. Kim and C. J. Kim, *Neurosci. Lett.*, 2003, **337** (3), 155.
10. A. Maree, G. Peer and A. Iaina, *Clin. Sci. (Lond.)*, 1996, **90** (5), 379.
11. L. Wensel, *Acupuncture for Americans*. Reston Publishing, Reston, VA, 1980, p. 18.
12. E. L. Lim, K. G. Hollingsworth, B. S. Aribisala, M. J. Chen, J. C. Mathers and R. Taylor, *Diabetologia*, 2011, **54** (10), 2506.
13. N. J. Snowling and W. G. Hopkins, *Diabetes Care*, 2006, **29** (11), 2518.
14. S. J. Wang, H. Z. Xu and H. L. Xiao, *Zhen Ci Yan Jiu*, 2008, **33** (3), 154.
15. S. L. Chang, J. G. Lin, T. C. Chi, I. M. Liu and J. T. Chen, *Diabetologia*, 1999, **42** (2), 250.
16. J. G. Lin, S. L. Chang and J. T. Cheng, *Neurosci. Lett.*, 2002, **326** (1), 17.
17. C. F. Su, Y. Y. Chang, H. H. Pai, I. M. Liu, C. Y. Lo and J. T. Cheng, *Diabetes Metab. Res. Rev.*, 2005, **21** (2), 175.
18. Y. Omura, Y. Chen, D. P. Lu, Y. Shimotsura, M. Ohki and H. Duvvi, *Acupunct. Electrother. Res.*, 2007, **32** (1–2), 31.
19. A. C. Ahn, T. Bennani, R. Freeman, O. Hamdy and T. J. Kaptchuk, *Acupunct. Med.*, 2007, **25** (1–2), 11.
20. Y. Wu, M. Fei, Y. He, C. Zhang, W. Zheng, Y. Wu and W. Li, *J. Trad. Chin. Med.*, 2006, **26** (2), 110.
21. H. Jiang, K. Shi, X. Li, W. Zhou and Y. Cao, *J. Trad. Chi. Med.*, 2006, **26** (1), 8.
22. J. Chen, *J. Trad. Chin. Med.*, 1985, **5**, 79.
23. A. Szczudlik and A. Lypka. *Acupunct. Electrother. Res.*, 1984, **9** (1), 1.
24. C. C. Lin, M. C. Chen, S. N. Yu and M. S. Ju, *Conf. Proc. IEEE, Eng. Med. Biol. Soc.*, 2005, **4**, 4271.
25. M. T. Cabioglu and N. Ergene, *Am. J. Chin. Med.*, 2006, **34** (3), 367.
26. F. Liang, R. T. Chen and A. Nakagawa, *Evid.-Based Complement. Alt. Med.*, 2011, doi: 10.1155/2011/735297.
27. H. C. Pai, C. Y. Tzeng, Y. C. Lee, C. H. Chang, J. G. Lin, J. T. Cheng and S. L. Chang, *J. Acupunct. Meridian Stud.*, 2009, **2** (2), 147.
28. N. Ishizaki, N. Okushi, T. Yano and Y. Yamamura, *Metab. Clin. Exp.*, 2009, **58** (10), 1372.
29. Y. Higashimura, R. Shimoju, H. Maruyama and M. Kurosawa, *Autonom. Neurosci. Basic Clin.*, 2009, **150** (1–2), 100.
30. S. L. Chang, K. J. Lin, R. T. Lin, P. H. Hung, J. G. Lin and J. T. Cheng, *Life Sci.*, 2006, **79** (10), 967.
31. C. S. Tseng, W. C. Shen, F. C. Cheng, G. W. Chen, T. S. Li and C. L. Hsieh, *Am. J. Chin. Med.*, 2005, **33**, 767.
32. S. L. Chang, C. C. Tsai, J. G. Lin, C. L. Hsieh, R. T. Lin and J. T. Cheng, *Neurosci. Lett.*, 2005, **379** (1), 69.

33. J. G. Lin, W. C. Chen, C. L. Hsieh, C. C. Tsai, Y. W. Cheng, J. T. Cheng and S. L. Chang, *Neurosci. Lett.*, 2004, **366** (1), 39.
34. Z. Zeng and Y. Li, *J. Trad. Chin. Med.*, 2002, **22**, 134.
35. M. Y. Shapira, E. Y. Appelbaum, B. Hirshberg, Y. Mizrahi, H. Bar-On and E. Ziv, *Diabetologia*, 2000, **43** (6), 809.
36. H. Ikeda, M. Kawatani and C. Takeshige, *Acupunct. Electrother. Res.*, 1991, **16** (3–4), 127.
37. W. Aldhahi and O. Hamdy, *Curr. Diabetes Rep.*, 2003, **3** (4), 293.
38. K. Miehle, H. Stepan and M. Fasshauer, *Clin. Endocrinol.*, 2012, **76** (1), 2.
39. C. S. Shin, B. S. Moon, K. S. Park, S. Y. Kim, S. J. Park, M. H. Chung and H. K. Lee, *Diabetes Care*, 2001, **24** (4), 733.
40. H. Sakuaba, H. Mizukami, N. Yagihashi, R. Wada, C. Hanyu and S. Yagihashi, *Diabetologia*, 2002, **45** (1), 85.
41. Y. Ihara, S. Toyokuni, K. Uchida, H. Odaka, T. Tanaka, H. Ikeda, H. Hiai, Y. Seino and Y. Yamada, *Diabetes*, 1999, **48** (4), 927.
42. T. M. Mason, T. Goh, V. Tchipashvili, H. Sandhu, N. Gupta, G. F. Lewis and A. Giacca, *Diabetes*, 1999, **48** (3), 524.
43. S. Jacqueminet, I. Briaud, C. Rouault, G. Reach and V. Poitout, *Metabolism*, 2000, **49** (4), 532.
44. R. Lupi, F. Dotta, L. Marselli, S. Del Guerra, M. Masini, C. Santangelo, G. Patané, U. Boggi, S. Piro, M. Anello, E. Bergamini, F. Mosca, U. Di Mario, S. Del Prato and P. Marchetti, *Diabetes*, 2002, **51** (5), 1437.
45. R. P. Robertson, J. Harmon, P. O. Tran and V. Poitout, *Diabetes*, 2004, **53** (Suppl. 1), S119.
46. S. Bansal, J. E. Buring, N. Rifai, S. Mora, F. M. Sacks and P. M. Ridker, *JAMA*, 2007, **298** (3), 309.
47. B. G. Nordestgaard, M. Benn, P. Schnohr and A. Tybjærg-Hansen, *JAMA*, 2007, **298** (3), 299.
48. American Diabetic Association, *Diabetes Care*, 2008, **31** (Suppl. 1), S12.
49. N. Ishizaki, T. Yano and Y. Yamamura, *J. Jpn. Soc. Balneol. Climatol. Phys. Med.*, 2006, **69** (2), 109.
50. R. Shimoju-Kobayashi, H. Maruyama, M. Yoneda and M. Kurosawa, *Auton. Neurosci.*, 2004, **115** (1–2), 7.
51. Y. Wu, M. Fei, Y. He, C. Zhang, W. Zheng, Y. Wu and W. Li, *J. Trad. Chin. Med.*, 2006, **26** (2), 110.
52. R. T. Lin, C. Y. Tzeng, Y. C. Lee, W. J. Ho, J. T. Cheng, J. G. Lin and S. L. Chang, *BMC Complement. Altern. Med.*, 2009, **9**, 26.
53. C. L. Wong, K. Y. Lai and H. M. Tse, *Complement. Ther. Clin. Pract.*, 2010, **16** (2), 64.
54. S. L. Chang, J. G. Lin, C. L. Hsieh and J. T. Cheng, *J. Chin. Med.*, 2002, **13**, 111.
55. Y. Y. Liao, K. Seto, H. Saito and M. Kawakami, *Am. J. Chin. Med.*, 1980, **8** (4), 354.
56. K. M. Man, S. S. Man, J. L. Shen, K. S. Law, S. L. Chen, W. J. Liaw and C. T. Lee, *Eur. J. Anaesthesiol.*, 2011, **28** (6), 420.

57. X. Mo, D. Chen, C. Ji, J. Zhang, C. Liu and L. Zhu, *Chen Tzu Yen Chiu Acupunct. Res.*, 1996, **21** (3), 55.
58. P. Iglesias, A. Polini, A. Munoz, A. Dardano, F. Prado, M. Castiglioni and M. T. Guerrero, *Int. J. Clin. Pract.*, 2011, **65** (3), 308.
59. P. Peplow and G. Baxter, *Jams J. Acupuncture Meridian Stud.*, 2012, **5** (1), 1.

Subject Index

References to figures are given in *italic* type. Reference to tables are given in **bold** type.